S0-CDT-683

By GEORGE R. KELLER

Published by the Editors of Hydraulics & Pneumatics Magazine

Hydraulic System Analysis

PREFACE

HYDRAULIC SYSTEM ANALYSIS fills an educational gap between practical texts that emphasize hardware and applications at the expense of theory, and theoretical fluid dynamics texts that have little relationship to the real world of functioning systems.

The material was developed from a course that is part of the continuing education program carried out at Boeing Aircraft Co. The original course was intended for use by graduate engineers working with hydraulic systems. The material has been modified and broadened to become a one-year course for third-year technical students.

Because this is a text about system analysis, an extensive amount of time may be allotted to a study of the techniques of actuator and system design, and servo analysis, as outlined in Chapters 9, 11, and 12. Shorter periods may be spent on the descriptive material such as in Chapters 2 and 10.

Where appropriate, problems are given at the ends of the chapters. The problems are straight-forward, rather than subject to individual interpretation, and are representative of the situations faced by design engineers working with advanced systems.

The emphasis in the problems is on analytical technique, and in most cases a knowledge of the calculus is not necessary. Complete solutions are given in the Appendix.

The graduate engineer will update or expand his knowledge of this field by applying himself to those individual chapters which may be pertinent, or to the text as a whole. The problem solutions can be of significant help because they are as self-explanatory as possible.

George R. Keller
Seattle, Washington
February, 1970

TABLE OF CONTENTS

CHAPTER 1

Pressure and Fluid Flow

CHAPTER 2

Hydraulic System Materials

CHAPTER 3

Power Generators

TABLE OF CONTENTS

CHAPTER 6

Transmission Lines – Components

TABLE OF CONTENTS

CHAPTER 7

Contamination and Filtration

CHAPTER 8

Control Valves

TABLE OF CONTENTS

CHAPTER 9

Power Output Transducers

CHAPTER 10

Seals and Sealing Devices

CHAPTER 11

Hydraulic Servomechanisms

CHAPTER 12

System Concepts

APPENDICES

CHAPTER ONE

Pressure and Fluid Flow

1.1 ADVANTAGES OF HYDRAULIC SYSTEMS

Paralleling the improvement in hydraulic fluids and other materials has been an increase in the complexity and sophistication of hydraulic systems. Most early hydraulic systems were no more complex than the system in an automobile hydraulic jack. The jack contains a hand operated pump which draws oil from a reservoir. Oil is pumped past a check or non-return valve into a linear actuator. The oil flowing into the actuator forces the piston to move. The moving piston can do work on an outside load. A valve controls flow from the piston back to the reservoir.

Through the use of power driven pumps and directional control valves these simple systems have grown so that a number of loads can be handled independently or sequentially from the same power source. Power hydraulics is a power amplifier in which a small effort applied to a control device results in a large amount of work performed on a load. With boost servos, hydraulic power aids the operator's effort. Further refinement has led to the irreversible servo-mechanism in which the operator's effort is separated from the load by the rigidity of the oil in the actuating cylinder or motor.

What are the reasons for using hydraulic power in heavy duty industrial, mobile and aircraft designs? Even though pneumatic control systems preceded hydraulic servos, and electrical power is in more widespread use than hydraulic, systems which require high power amplification or which must pack great energy into a small space are almost invariably hydraulic.

If hydraulic power transmission and amplification were to be compared with pneumatic, electrical and mechanical systems, these are some of the reasons for specifying hydraulic power:

Hydraulic motors possess an extremely high ratio of torque to inertia. Because of this, very little of the available torque is dissipated in accelerating the motor; most of the torque can be applied directly to the load. Acceleration rates for unloaded hydraulic motors[1], for example, vary from about 0.3 x 10^5 radians per second squared for larger sizes at 3000 psi to 3.4 x 10^5 radians per second squared for smaller motors. No other transmission system produces motors having acceleration rates anywhere near these values. Higher supply pressures develop accordingly higher acceleration rates.

Hydraulic motors develop no counter electro-motive force as do electrical motors. Therefore, the acceleration rate is essentially constant over their design speed ranges.

Hydraulic devices can generate much more power in the same volume than can electrical devices. The reason for this lies in the characteristics of the materials used in electrical equipment. The very best armature steels can develop only the equivalent of about 200 psi in tractive power before they become saturated. By comparison, 3000 psi hydraulic equipment is very common and pressures up to 5000 psi are not rare.

Because of the greater pressure density, hydraulic equipment in power sizes greater than 1 horsepower tend to be much smaller than electrical motors and somewhat smaller than pneumatic motors.

Hydraulic motors support heavy or springy loads much better than do electrical or pneumatic motors. Looking from the load back into the motors, electrical fields or compressed gasses are quite soft. The relatively incompressible oil is, however, stiff.

Hydraulic transmission systems efficiently transmit power farther than do mechanical systems, but not as far as electrical systems.

Hydraulic transmission systems may be less hazardous than electrical systems and are far safer than mechanical power transmission systems.

Hydraulic oil is a good heat conductor so that the heavily loaded working area stays relatively cool. Heat generated at the working interface is conducted away rapidly. The working areas can therefore be made small because heat absorbing mass is not needed. Because of this and because of the high pressure density, hydraulic motors weigh less than one pound per horsepower. Horsepower to weight ratios of up to 2.5 are not uncommon.

Hydraulic power systems also have some drawbacks. Hydraulic oils are messy. Leaks are inevitable. The great lubricity of most hydraulic oils makes leaks and spillage very hazardous. Certain oils are excellent removers of paint and electrical insulation.

Hydraulic systems, particularly the high performance servo systems which use hydraulic power so effiiciently, are extremely vulnerable to dirt and contamination. They must constantly be watched to prevent the addition or generation of contamination and hydraulic systems need expensive filtration to keep them clean.

A rupture or burst of an element of a high pressure hydraulic system can be very dangerous. The high velocity stream of oil which results can damage eyes, pierce skin, and harm other equipment. The author has seen a piece of tubing three feet long and 1½ inches in diameter thrown whirling for more than fifty feet when a rupture occurred in a 3000 psi line.

Unless fire-resistant fluids are used, hydraulic leaks create very real fire hazards. Most petroleum based fluids are highly flammable. High temperatures can be generated where a high velocity stream of oil from a rupture impacts on a flat surface.

Finally, hydraulic power is so compact that developing and controlling fractional horsepowers is difficult unless the pressure density is reduced to levels where electrical power is extremely competitive.

1.2. POWER SYSTEM ANALOGIES

Hydraulic systems—within the laminar flow regime at least—function in ways which are quite analogous to mechanical and electrical systems. Later in these discussions the equations of analogy will be developed. At this point, illustrations show the analogies between systems. Fig. 1.1 shows the analogies relating hydraulic parameters to those of electrical systems and translatory and rotary mechanical system. Similar analogies could be developed for pneumatic, thermal or other energy systems. These elements can be combined in ways which—at least in electrical systems—are termed "passive" networks. They are passive because no energy is added to the system in the network though energy is lost through conversion to heat in the resistance elements. These networks affect the stability of transmission systems at various frequencies when subjected to periodic oscillation such as sinusoidal cycling.

Because of the increasing importance of the dynamics of the hydraulic system in designing for high performance, this book will relate—in a practical way—those factors which affect the dynamical operation of the system to the design of components and to the disposition of the several components within the system. No attempt will be made to develop design techniques other than to show previous solutions to typical problems. Instead, this book will relate the discussion of the several parts of the hydraulic system to basic engineering principles. A sincere attempt will be made to show how these basic principles can be brought to bear on the solving of typical component and system design problems.

Where relevant, recourse will be made to the calculus to support the development of equations but, in most cases, a knowledge of the calculus will not be necessary in order to solve problems. Certain books [2, 3, 4, 5, 6] may be of value in gaining additional detailed understanding of specific subjects. This book, however, is intended to stand on its own and all reasonable efforts have been made to obviate the need for additional resources.

Some of the practices which have developed in the hydraulics industry make problem-solving difficult. Among these are the use of feet head of water as a measure of pressure and the use of Saybolt Universal seconds or metric units for viscosity. In this book

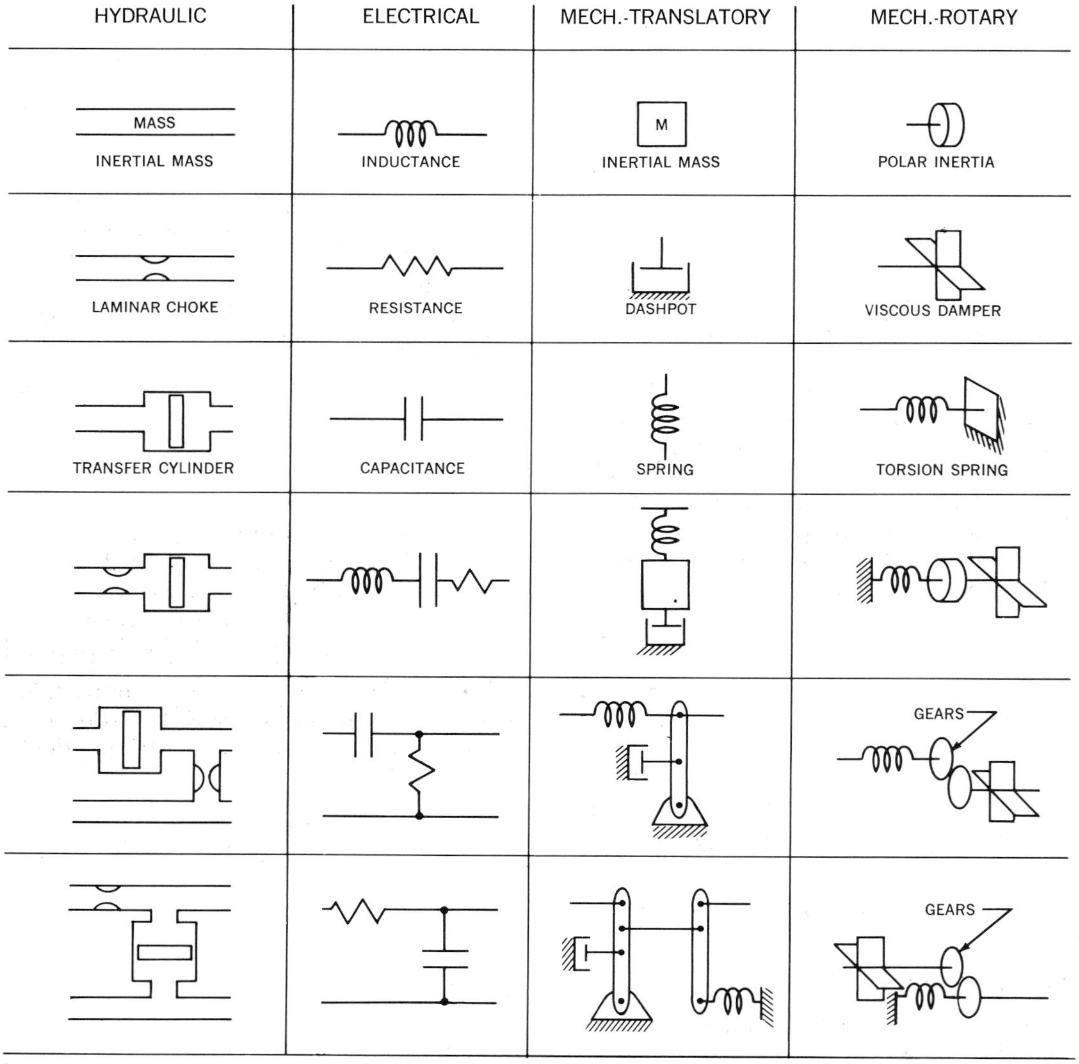

HYDRAULIC	ELECTRICAL	MECH.-TRANSLATORY	MECH.-ROTARY
MASS INERTIAL MASS	INDUCTANCE	M INERTIAL MASS	POLAR INERTIA
LAMINAR CHOKE	RESISTANCE	DASHPOT	VISCOUS DAMPER
TRANSFER CYLINDER	CAPACITANCE	SPRING	TORSION SPRING
			GEARS
			GEARS

Fig. 1.1. Hydraulic parameters compared with electrical and mechanical systems.

most such units have been converted into their English equivalents. Through personal preference, the author avoids confusion in conversion by putting everything into units of pound-force, inches and seconds. Wherever this might conflict with established practice—such as using units of Newts for kinematic viscosity-conversion formulae are provided.

Where the text material is appropriate, problems are given to allow the student to practice the achieving of solutions. These problems nearly always represent forms of difficulty which have faced the author at various times. They are practical problems in advanced hydraulics. In many cases, a solution is possible only because simplifying assumptions were made. Because of these assumptions, the hydraulics engineer should beware of blind application of the formulae. Careful consideration of the effect of practical deviations from the assumed conditions is necessary if success in overcoming real difficulties is to be achieved.

1.3. FLUID POWER FUNDAMENTALS

For many years a hydraulic system was considered the result of plumbing together a number of components. Also, a hydraulic system has been considered merely the source of power for control servos. Each of these concepts can lead to basic errors. Following either of these guide lines can result in systems which are inadequate, unstable, or a combination of the two.

From an operational standpoint, any hydraulic system can be divided into four logical segments: The power input segment, the power transmission system, the control devices and the power output portion. The power input segment usually consists of pumps and accumulators. The transmission system includes tubing, tubing connectors, swivel joints and flexible hose. Flow control valves, pressure control valves and relief valves make up the control elements. The power output portion usually consists of rotary and oscillatory motors or linear actuators. Because almost all practical hydraulic systems can be divided into these same four basic segments, they will be considered in detail later.

1.4. HYDROSTATICS

Nearly all power hydraulic systems, as distinct, for instance, from automotive torque converters, operate as practical applications of Pascal's law which states:

The pressure at any point in a fluid continuum is the same in any direction.

The important corollaries to Pascal's law are:

1. In any homogeneous fluid system at rest, the pressure increases with the depth of the fluid.
2. Pressure at any point in a homogeneous fluid system at rest acts perpendicularly to surfaces in contact with the fluid. For curved surfaces, the pressure at any point acts in the direction of the normal to the surface at that point.

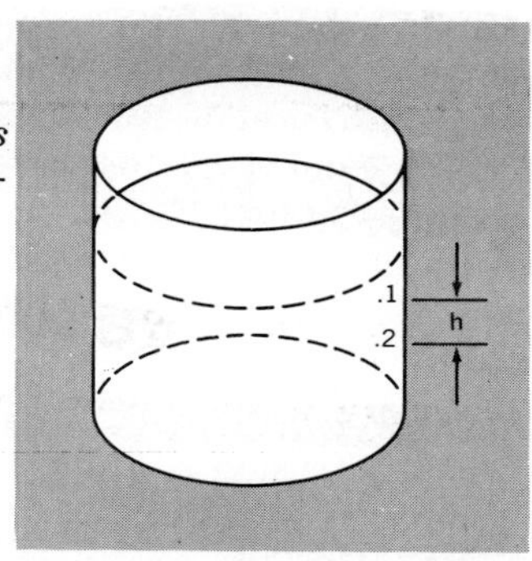

Fig. 1.2. Pressure increases with depth to give the relationship of formula 1.1.

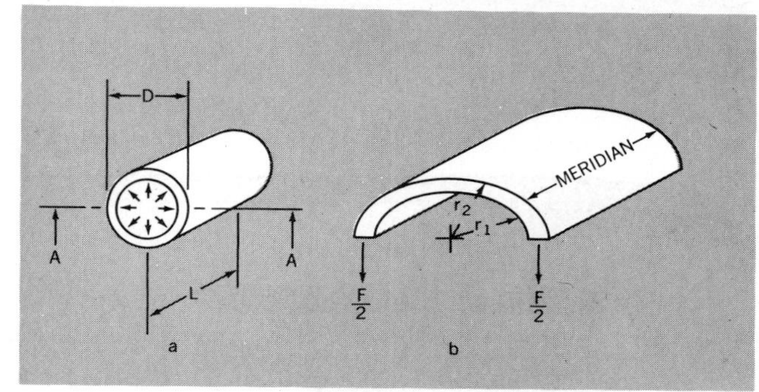

Fig. 1.3. Effect of pressure on curved surface: tubing, actuator, accumulator, reservoir.

The first corollary can be simply restated that pressure increases with depth. This is not news to swimmers or divers who have felt the increasing pressure on their eardrums as they swam to greater depths. It does establish the basis for the formal relationships shown in Fig. 1.2.

$$P_2 = P_1 + \rho g h \qquad 1.1$$

Where: P_1 = pressure at point 1
P_2 = pressure at point 2
ρ = fluid mass density
g = acceleration of gravity
h = vertical distance between 1 and 2

The second corollary is very important in understanding the effect of fluid pressure on reservoirs, actuator pistons and other working devices having curved surfaces. The useful pressure is exerted on the projection of the curved surfaces in the plane of action. Consider two examples:

Fig. 1.3a shows a cylindrical member of inside diameter D of unit length. It can be a portion of a hydraulic line, a segment of an actuator, accumulator, or reservoir. If forces in an axial direction are ignored, corollary 2 states that the forces tending to burst this pressure-containing member can be obtained by summing the forces in a direction perpendicular to line A-A. The total force is

$$F = PD \qquad 1.2$$

If the wall of the cylindrical member were thin enough ($t < 0.1\ r_1$) so that a stress gradient did not exist radially across the wall, then the stress per unit length which tends to rupture the walls through circumferential tension is

$$S_1 = \frac{Pr_1}{t} \qquad 1.3$$

S_t is called the hoop tension. As shown in Fig. 1.3b the hoop tension is the unit force acting tangentially on any free-body segment of the tubing. The pressure also develops a force in a meridional direction. In the thin walled tube the tension caused by this force is

$$S_2 = \frac{Pr_1}{2t} \qquad 1.4$$

The radial displacement (swelling) of the wall under these conditions is

$$\textit{Radial Disp.} = \frac{r_1}{E} \ (S_1 - \mu\ S_2) \qquad 1.5$$

Where: E = modulus of Elasticity
μ = Poisson's Ratio

As a second example, look at the curved head of a piston in a hydraulic actuator, Fig. 1.4. Despite the increase in working surface of the piston provided by the curved surface, the actuator force output is exactly the same, $F = PA$, as if the face of the piston were perfectly flat.

1.5. HYDRODYNAMICS

A series of relationships among velocity, flow and pressure can be developed from the basic equations of motion for a falling body.

$$h = \tfrac{1}{2}\ gt^2 \qquad 1.6$$

$$v = gt \qquad 1.7$$

By combining these, Torricelli's equation of frictionless flow from an orifice can be derived.

$$v^2 = 2gh \qquad 1.8$$

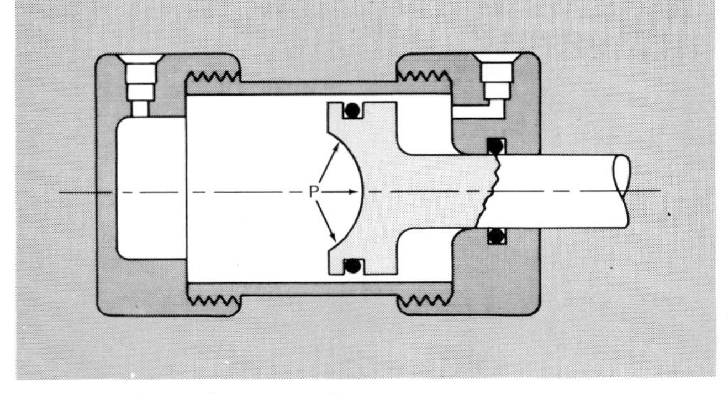

Fig. 1.4. Curved surface does not increase actuator force.

This equation states that the velocity of a fluid medium flowing under the influence of gravity is proportional to the square root of the height of the fluid column above the point where measurements are being taken.

Among other uses of 1.8 is the analysis of the time required to empty a reservoir. The flow rate of liquid out of a reservoir is

$$Q = A_o v = \frac{dV}{dt} \qquad 1.9$$

where: Q = flow rate
V = volume in reservoir
A_o = area of opening
v = flow velocity

Because

$$V = A_t h,\ dV = A_t dh \text{ and } v = \sqrt{2gh} \qquad 1.10$$

where A_t = Area of tank

$$A_t dh = A_o \sqrt{2gh}\ dt \qquad 1.11$$

$$dt = \frac{A_t \quad dh}{A_o \sqrt{2gh}} \qquad 1.12$$

$$t = \frac{A_t}{A_o \sqrt{2g}} \int_o^H \frac{dh}{\sqrt{h}} \qquad 1.13$$

$$t = \frac{\sqrt{2}\ A_t \sqrt{H}}{A_o \sqrt{g}} \qquad 1.14$$

Equation 1.8 can also be expressed in pressure units.

$$P = \frac{\rho v^2}{2} \qquad 1.15$$

This relationship gives the fluid velocity which derives from any pressure differential or it gives the pressure which will develop from a change in velocity. This pressure is called the "velocity head" of a system.

By substituting $v = Q/A$ into 1.8 and introducing an orifice coefficient c, the basic flow equation for circular orifices is obtained.

$$Q = cA \sqrt{\frac{2\ \Delta\ P}{\rho}} = K\sqrt{\Delta P} \qquad 1.16$$

The orifice coefficient is necessary because the actual flow path past the orifice is smaller in cross section than the orifice itself. Also due to friction, the average velocity is slightly less than given by equation 1.8. The orifice coefficient is the product of these two effects. It varies between about 0.6 to 1.0.

Equations (1.1) and (1.15) can be added to the basic system pressurization to produce Bernoulli's equation of flow for incompressible, frictionless fluids. This equation, in a form applicable to power hydraulic systems is

$$P = P_s + \rho gh + \frac{\rho v^2}{2} \qquad 1.17$$

Bernoulli's equation states that, in a system in which no work is being added or subtracted, the pressure at any point is the sum of (1) the system pressurization, (2) the difference in elevation between the point of interest and the point where system pressurization is measured and, (3) the velocity head at the point of interest.

Equation 1.17, in the form given is valid only for the point of interest. It can be made to apply to any pair of points in a liquid continuum and, by corollary, to all points in the liquid continuum by adding the pressure loss due to flow friction between the two points. The pressure total is then a constant for all points of the system.

$$P = constant = P_s + \rho g h + \rho \frac{v^2}{2} + P_f \qquad 1.18$$

Where: P_f is the pressure loss due to friction

1.6. FLOW RELATIONSHIPS

All practical power hydraulic systems are conservative. That is, all aspects of energy, mass, and work are identifiable and retained throughout any operation. For an understanding of the conservative nature of hydraulic systems, one can consider flow through portions of a typical hydraulic system.

1.7. CONSERVATION OF MASS

Fig. 1.5 shows a portion of a hydraulic transmission line containing an accumulator. The accumulator, for this discussion, is merely an element of the fluid transmission system in which the oil volume is variable. For any such system the mass rate into the system equals the mass rate out of the system plus the rate of accumulation of mass within the system.

$$\rho Q_i = \rho Q_o + \rho \frac{dV}{dt} \qquad 1.19$$

This relationship is basic to the design of any hydraulic system. For instance, the flow from a pump must go somewhere. If the flow isn't used in doing work on the outside world then it must go into an accumulator or into a reservoir or other variable volume element in the system. If such provisions are not made for accommodating excess flow, the system pressure will rise until either the system relief valve opens to allow for flow to the reservoir or tank or else the system fails by rupturing.

Another example of the conservation of mass is to assume that the tank of Fig. 1.2 has an inlet line and an outline line. The flow rate into the tank ρQ_i is unsteady, that is, it varies with time. If the level of the liquid rises, the inlet flow must be greater than the outlet flow by $\rho A_t \ (dh/dt)$. The total relationship is then

$$\rho Q_i - \rho Q_o = \rho A_t \frac{dh}{dt}$$

This is the same as equation 1.19.

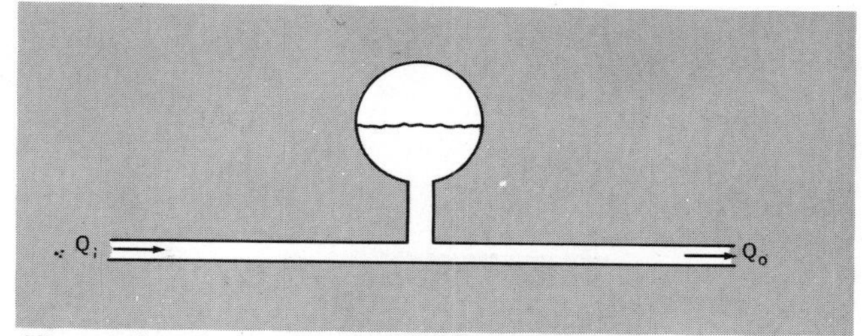

Fig. 1.5. Elemental transmission line.

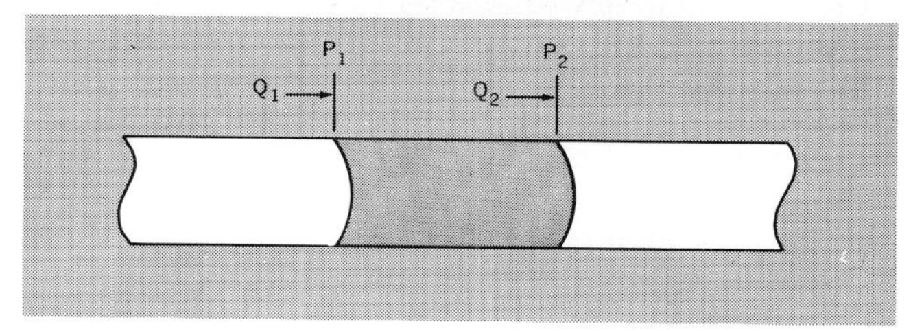

Fig. 1.6. Fluid in a cylindrical passage.

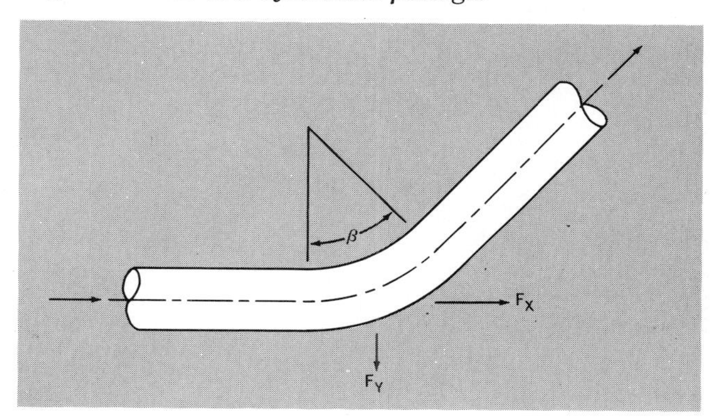

Fig. 1.7. Forces in a curved passage.

1.8. CONSERVATION OF MOMENTUM

In any system, of which Fig. 1.5 is but one example, the rate of change of momentum in the system equals the net applied external force.

$$F = M \frac{dv}{dt} \qquad 1.20$$

This equation looks—and is—the same as the relationship

$$F = Ma \qquad 1.21$$

known to all engineering students. Perhaps the most familiar example of conservation of momentum is the toy balloon which is blown up and released. The change in velocity of the air when multiplied by the mass rate of escapement must equal the mass of the balloon times its change in velocity. Result? The balloon flies away!

Assume a frictionless, incompressible liquid in a cylindrical passage such as is shown in Fig. 1.6. The flow rate is being accelerated uniformly with time

$$Q_2 = Q_1 + \int_o^t \frac{dQ}{dt} \qquad 1.22$$

The force balance is, from equation 1.20

$$P_1A - P_2A = \frac{d}{dt}(\rho\ ALv) \qquad 1.23$$

Because $Q = Av$

$$P_1 - P_2 = \frac{\rho L}{A}\frac{dQ}{dt} \qquad 1.24$$

Equation 1.24 says that a fluid in a passage cannot be accelerated (undergo a change in momentum) without having a pressure differential exist across the section of the fluid. This will be important later when studying the stability of certain forms of valves.

A similar reasoning applies to the flow of fluid through curved passages. Because velocity is a vector quantity it contains elements of both speed and direction. Therefore a change in the direction of flow, as occurs when hydraulic oil flows through a bent tubing, Fig. 1.7, is a change in momentum as defined in equation 1.20. The forces can be resolved into a component F_x which is axial to the inlet direction and a component F_y which is normal to the inlet direction.

$$F_x = \rho Qv(1\text{-}\cos\ \beta) \qquad 1.25$$

$$F_y = \rho Qv\ \sin\ \beta \qquad 1.26$$

For a 90° bend $F_x = F_y$.

As an example consider a line 1″ in diameter having .049 walls. Oil is flowing around a 90° bend at a rate of 20 gpm. The oil weighs 7.4 pounds per gallon.

The mass density

$$\rho = \frac{w}{g} = \frac{7.4}{(231)\ (12)\ (32)} = 8.35 \times 10^{-5}\ \text{lb-sec}^2/\text{in.}^4$$

The flow rate is

$$\frac{20 \times 231}{60} = 77 \text{ cubic inches per sec. (cis)}$$

The inside diameter of the tube is

$$1.00 - 2(0.049) = 0.902 \text{ inches}$$

This corresponds to an area of 0.642 sq. inches. The flow speed is

$$\frac{77}{0.642} = 120 \text{ inches per second (ips)}$$

For a 90° bend $\sin\beta = 1$, then

$$F_x = (8.35 \times 10^{-5})\ (77)\ (120)\ (1) = 0.77 \text{ lbs.}$$

1.9. CONSERVATION OF ENERGY

In many hydraulic systems, a pump accepts oil at a low pressure and injects it into a system which is held at high pressure. The internal energy level of the system has been increased by the work done on the system by the pump. Compression of the oil and friction losses in the pump increase the temperature of the oil. At the higher temperature, the oil contains more energy. As the oil moves out the transmission line, it may lose heat to the outside world. This loss of heat is a loss of internal energy. If the pressurized oil is used to perform work by moving an actuator or by rotating a motor against a load, its energy level will be reduced and the oil will be discharged to a return line at a relatively low pressure. These several changes in energy level illustrate the principle of conservation of energy. Any change in internal energy of a conservative system must equal the work done on or by the system plus the net change in heat.

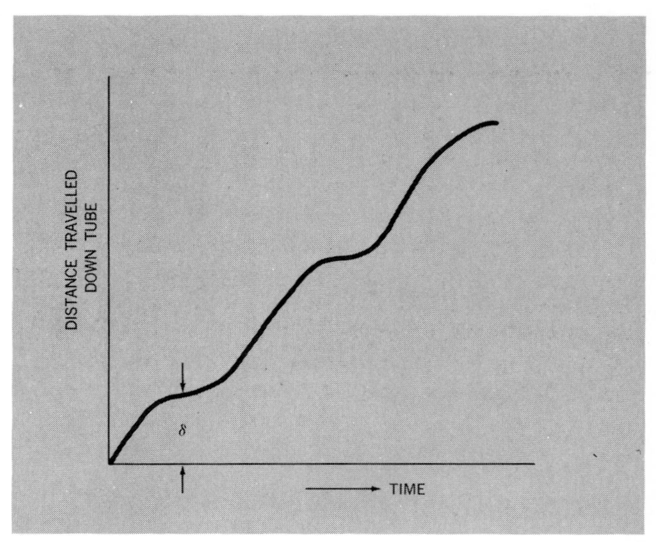

Fig. 1.8a. Oil particle's path due to servo cycling.

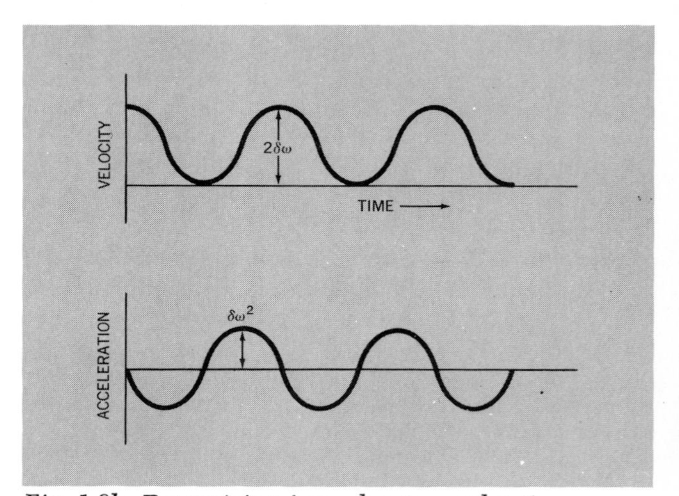

Fig. 1.8b. Determining force due to acceleration.

$$E_i = E_o + \Delta W + \Delta H \qquad 1.27$$

This concept is extremely important to the system designer. For example, any closed segment of a hydraulic system which is subjected to temperature rises must have, as a part of its design, some means for relieving excess energy.

Space—usually in the reservoir—must be provided

in a closed system to allow volumetric expansion of a system subjected to temperature rises due to outside heat or internal energy losses. The size of accumulators and reservoirs and the setting of relief valves must be carefully balanced against external heat, pump inefficiency and pressure losses in lines and orifices.

This also means that, if all energy changes are properly accounted for, the hydraulic system must always have an energy balance. Therefore, if a pump is supplying power into a system at a known rate, energy must be taken out of the system at about the same rate either as work or through a heat exchanger. Conversely, if the system exhibits an increase in fluid temperature, the engineer should look for internal leaks or inefficient pumps or valves.

Sometimes the input energy shows up in ways other than just heat. The author has seen 1.5″ OD x 0.079″ wall stainless steel tubing oscillate back and forth over an amplitude of ±3 inches during a time when the hydraulic system was supplying oil to a large servosystem. Such oscillation can cause damage to tubing and adjacent parts through banging. It can also overstress connections and line hardware. In this instance the oscillation was controlled by tieing down the tubing firmly every 20 inches. A study was then initiated to determine the cause of the violent motion.

This problem occurred when operating a large servo at 3000 psi. The servo was being cycled sinusoidally at 2 Hz (cycles per second). Flow measurements showed that a minimum flow of about 2 gpm due to leakage rose during cycling to a maximum of 43 gpm. The mass density of the fluid under the operating conditions was 8.35×10^{-5} lb-sec^2/in^4. Concern was expressed that the oscillation was caused by the acceleration of the mass of the fluid in the lines.

For the condition encountered, the maximum flow was 43 gpm. This corresponds to 167 cubic inches per second. The inside diameter of the tubing is 1.500–(2)(.079) = 1.342 inches. This corresponds to an area of 1.41 square inches. The maximum flow speed is then 167/1.41 = 118 inches per second.

The force due to the acceleration of the fluid in the lines can be determined by assuming that, due to servo cycling, a particle of oil would progress down the tube as shown in Fig. 1.8a. Though this path does not appear sinusoidal, the plot of the time derivative of this path is a cosine wave in which all values are positive. The time derivative of the velocity curve is a sinusoidal acceleration curve. Fig. 1.8b shows these relationships.

The amplitude of sinusoidal type of movement of a fluid particle can be established from knowledge of the servo actuator effective area and stroke as modified by the line inside area. In the absence of servo actuator data, the amplitude of movement can be derived from the equations of sinusoidal movement.

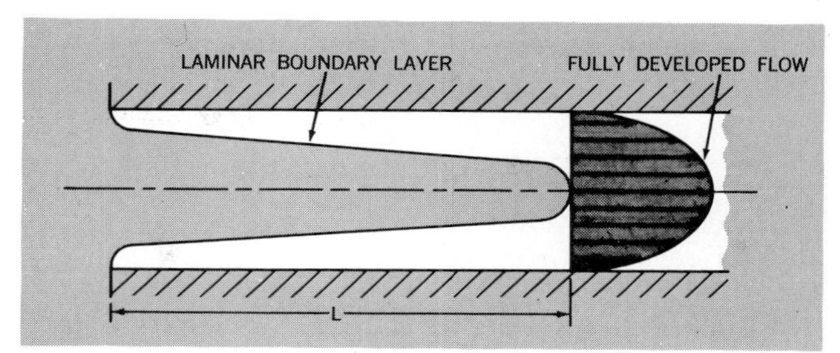

Fig. 1.9. Development of laminar flow.

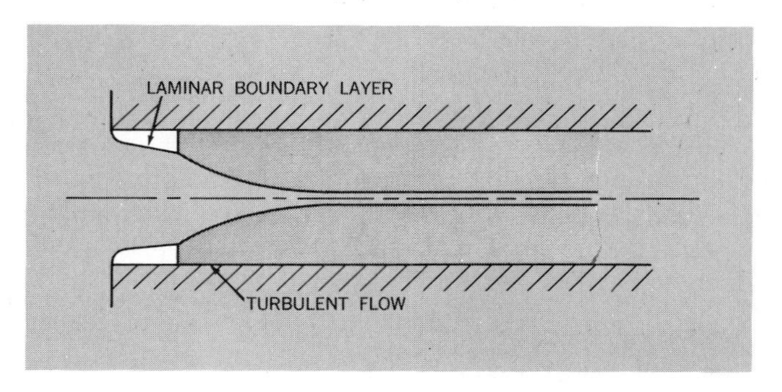

Fig. 1.10. Development of turbulent flow.

$$x = \delta \sin \omega t \quad (1.28)$$
$$v = \delta\omega \cos \omega t \quad (1.29)$$
$$a = -\delta\omega^2 \sin \omega t \quad (1.30)$$

where δ = maximum displacement in inches
ω = frequency in radians/sec.
x = momentary displacement
v = momentary velocity
a = momentary acceleration

Equation 1.29 can be used to establish the magnitude of δ because the velocity and frequency are known. The velocity due to cycling derives from the difference between the maximum and minimum flow rates. 43 gpm–2 gpm = 41 gpm = 158 cubic inches per second. For the internal area of 1.41 square inches, this flow differential yields a maximum velocity due to servo cycling of 158/1.41 = 112 inches per second. The maximum variation of velocity from the mean value is 112/2 = 56 ips. The cycling rate of 2 Hz = 12.56 radians per second. When velocity is at a maximum, $\cos \omega t = 1$. Then from (1.29)

$$\delta = \frac{v}{\omega} = \frac{56}{12.56} = 4.46 \text{ inches}$$

The maximum acceleration a can then be determined from (1.30) under the condition where $\sin \omega t = 1$

$$a = -4.46\,(12.56)^2$$
$$= -705 \text{ in/sec}^2$$

The negative sign merely relates to the vector direction of the acceleration and, for the purposes of this analysis, can be ignored. The mass of the oil can be computed if the length of the line is known. In the case being studied, the transmission line was 50 feet long.

$$Mass = (50)\ (12)\ (1.41)\ (8.35 \times 10^{-5})$$
$$= 0.071 \text{ lb} - \text{sec}^2/\text{inch}$$

The force along the tube is

$$F = Ma = (0.071)\ (705) = 50 \text{ lbs.}$$

Therefore, the forces due to fluid acceleration can become significantly large.

1.10. FLUID FLOW

If a small amount of colored dye were to be introduced in a stream of oil or water which is flowing slowly through a transparent tube, the filament of color formed by the dye would flow smoothly with the fluid. However, if the flow rate were to be increased, a flow rate would be reached at which the filament colored by the dye would break up and be lost in swirls. The initial smooth, streamline flow is called laminar. The swirling flow is turbulent. Laminar flow is characterized by viscous flow forces. Inertially dominated flow is usually turbulent, though in some instances, such as flow through an orifice, the flow is streamline and non-turbulent.

Flow conditions are often defined by means of a dimensionless ratio called Reynolds Number N_R. Reynolds Number relates the inertial flow forces to the viscous flow forces.

$$f_i = \phi\ (D^2\ \rho\ v^2) \qquad 1.31$$

$$f_{vis} = \phi\ (\mu\ vD) \qquad 1.32$$

$$N_R = \frac{D^2\ \rho\ v^2}{\mu\ vD} = \frac{\rho\ vD}{\mu} = \frac{vD}{\nu} \quad where\ \nu = \frac{\mu}{\rho} \qquad 1.33$$

At small values of Reynolds Number, the viscous forces predominate and the flow is laminar. As Reynolds Number is increased, the effect of the inertial forces becomes sufficiently great to break up streamline flow. In general, flow is laminar when the Reynolds Number is less than 1400 and turbulent when N_R is greater than 3000. Flow conditions in the $1400 < N_R < 3000$ range depend on local conditions and previous flow history. The two flow conditions are very dissimilar. With laminar flow, the pressure loss due to friction has a first-order relationship with flow. It is analogous, in this way, with DC electricity, in which resistance has a first-order relationship with voltage and current. When the flow is turbulent, the pressure loss becomes proportional to the square of the flow. Obviously, calculations of laminar flow are far simpler than those of turbulent flow.

If oil flows into a tube or pipe from a large chamber the initial velocity will be quite uniform over the cross section as shown in Fig. 1.9. A shearing action soon takes place between the oil layer—or lamina—in contact with the tubing wall and the next lamina away from the wall. The oil in contact with the wall will resist movement, while the oil farther away will try to move with the main stream. This region of rapid velocity change is called the "boundary layer." As the oil flows farther down the tube the boundary layer thickens and eventually a stable velocity profile develops. When the flow is laminar, the velocity profile in a circular tube is a paraboloid of revolution. The development of this parabolic velocity profile occurs at a distance down the tube which is dependent upon the tube inside diameter and on Reynolds Number.

$$l = 0.058\ d\ N_R$$

When Reynolds Number is high enough so that turbulent flow develops, the transition from laminar flow to turbulent flow occurs at some point before the boundary layer fills the tube as shown in Fig. 1.10. For most engineering calculations, however, turbulent flow may be considered to occur throughout the whole length of the transmission line.

REFERENCES

1. Data Summary SE-18D "Basic Characteristics of Vickers Aircraft—Type Hydraulic Pumps and Motors", Vickers, Inc. 2-10-65
2. Keller, G. R. "Aircraft Hydraulic Design" Industrial Publishing Corp., 1957
3. Blackburn, J. F.; Reethof, G.; Shearer, J. L.; "Fluid Power Control," Wiley 1960
4. Lewis, Ernest E.; Stern, Hansjoerg "Design of Hydraulic Control Systems" McGraw-Hill 1962
5. Merritt, H. E., "Hydraulic Control System" Wiley, 1967
6. Walters, R., "Hydraulic and Electro-Hydraulic Servo Systems", CRC Press, 1967

CHAPTER 1

PROBLEMS

1. What is the pressure on the ears of a scuba diver when he is 60 feet below the the surface of the ocean? Salt water weighs 64 lb/cu. ft.

2. Fresh water (weight 62.4 lb/cu. ft.) is poured into a pipe having an ID of 1 inch at a rate of 4 gpm. The pipe outlet is constricted so that when the pipe is full the outflow exactly matches the inflow. What is the pressure on the pipe 40 feet below the inlet?

3. In an aircraft hydraulic system, the quiescent supply pressure is 3000 psi at the pump. MIL-H-5606 (weight 0.03 lb/ cu. in.) is used. What is the quiescent pressure at the wheel well 10 feet below the pump?

4. A section of 1/2-inch OD x 0.020-inch wall tubing contains oil at 3000 psi. What is the hoop tension in the tubing?

5. A hydraulic oil having a mass density of 1×10^{-4} lb -sec^2 /in^4 is flowing through an orifice 0.025 inch in diameter at a rate of 130 cubic inches per minute. The orifice coefficient c is 0.80. What is the pressure differential across the orifice?

6. A hydraulic system reservoir contains 45 gallons of oil. The initial oil level is 3 feet above a 1-inch diameter drain hole. If the drain were opened, and if the oil were assumed frictionless, how quickly would the tank drain?

7. A reservoir is 2 feet in diameter. If 3.5 gpm of oil is flowing into the reservoir and 2 gpm is being drawn out, how rapidly is the oil level rising?

8. Oil having a mass density of 9×10^{-5} lb-sec^2/in^4 is flowing around a 45° bend in a 3/4-inch tube having 0.035-inch walls at a rate of 15 gpm. What are the forces on the tube?

9. A balanced hydraulic cylinder is being driven sinusoidally at 6 Hz at a half amplitude of 2 inches. The cylinder piston area is 3 sq. in. If the oil weighs 7.5 lbs per gallon and the supply line is 20 feet of tubing with a 3/4-inch OD and a 0.049-inch wall, what is the reaction force due to the acceleration of the oil?

CHAPTER TWO

Hydraulic System Materials

2.1. INTRODUCTION

The materials from which a hydraulic system is fabricated are intimately related to the use of the system. A small, high-performance missile might well (as an example,) require a fluid which provides far greater stiffness to the control system than would be needed for a die casting machine. The thermal and chemical nature of materials in systems operating at high temperatures must be far more stable than those of materials in low temperature systems. The interaction of fluids with metals and elastometers determines the temperature band within which the system can work.

2.2. HYDRAULIC FLUIDS

Clearly, the single most important material in the hydraulic system is the working fluid. Modern hydraulic fluids are often complex formulations carefully prepared to meet their exacting tasks. In addition to a base fluid they may contain additives to improve viscosity characteristics, to minimize system corrosion, to make the fluids more stable to their environments and to reduce foaming and flammability.

Dr. Roger Hatton[1], in his excellent book "Introduction to Hydraulic Fluids", gives the following as characteristics of importance to a hydraulic fluid:

Good lubricity
Chemically and environmentally stable
Possesses favorable viscosity
Compatible with other system materials
Heat transfer capability
High bulk modulus
Low volatility
Low foaming tendencies
Low weight
Fire resistant
Good dielectric properties
Nontoxic or allergenic
Nonmalodorous
Low in cost
Easily available

This is a formidable list. No working fluid possesses all of these good characteristics. The task of the system designer is to choose or to formulate the fluid most suitable to his requirements.

Lubricity has several connotations, some of which are in conflict with others. Lubrication, in general, is concerned with minimizing the shearing action of two surfaces between which relative motion exists.

In Fig. 2.1*a*, all surfaces of working parts, no matter how smoothly made, consist—if enlarged sufficiently—of hilly contours. In the absence of lubrication, the hills and valley of mating parts tend to mesh. Were it not for the presence of oxides or fatty acids on the surfaces, the mating faces would actually weld together. The function of lubricants is to minimize the effect of shearing off the hills—or more properly, the asperities—of the surfaces as they move relative to each other. This is done in a number of ways.

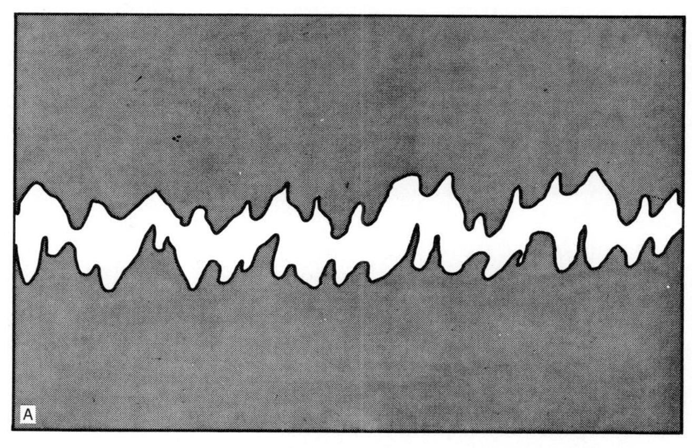

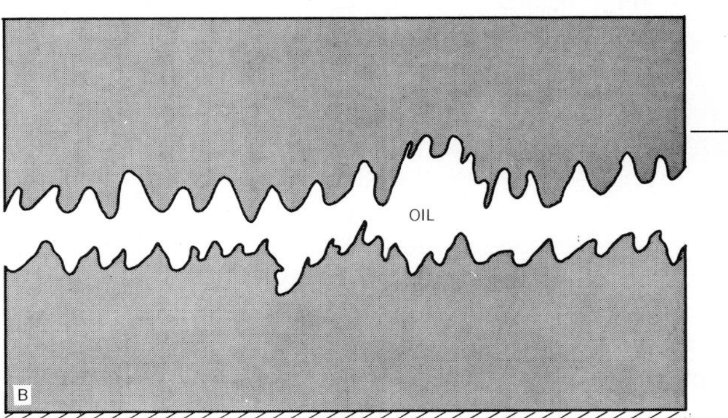

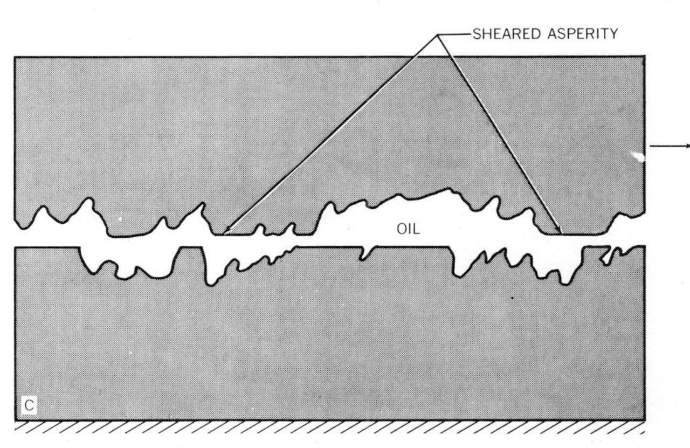

Fig. 2.1. Hydraulic fluid choice must be considered as a boundary layer problem.

Dry lubricants such as lead powder and silver plating interpose a greasy metal between the asperities. The low shear forces required by the greasy metals helps in reducing the sliding forces. Other dry lubricants used are graphite and molybdenum disulphide. Each of these have crystal faces or molecular bonds which are weak. The relative movement then breaks these bonds rather than shears asperities. Properly applied, these are quite effective.

Fluid lubrication, as usually provided by hydraulic fluids, consists of interposing the liquid between the mating faces. If enough fluid can be forced between the two surfaces to separate them completely, Fig. 2.1*b*, hydrodynamic lubrication is achieved. In this condition, the only forces resisting the relative movement of the two surfaces are those needed to shear the fluid. In hydrodynamic lubrication, wear is negligible and the coefficient of friction is very low. Air bearings, Kingsbury thrust bearings and well-designed journal bearings achieve hydrodynamic lubrication.

If the lubricating fluid does not separate the surfaces completely, some shearing of the asperities will occur, Fig. 2.1*c*. The ability of the fluid to maintain some separation, the tendency for the fluid to be dragged between the shearing faces, and the capability of the fluid to form nonwelding surfaces on the newly-sheared faces all contribute to the lubrication effect. The first two of these factors relate strongly to fluid viscosity. The last one depends on the chemical natures of the fluid and the moving surfaces.

The chemical nature of boundary lubrication is described by the chemical polishing theory. Under pressure, two mating metal surfaces will have areas of contact. If relative movement occurs, local high temperatures will develop at the points of contact. At these high temperatures, certain fluids such as the phosphate esters react chemically with the metal to form an alloy having a melting temperature lower than that of the base metal. The low melting alloy will flow away from the pressure and generate a new distribution of forces. Thus, galling or welding is avoided at the cost of relatively high wear. Because this mechanism relates directly to fluid-metal chemistry, one fluid can lubricate only a limited range of metals in this way.

These forms of lubrication are known as boundary layer lubrication. Most practical lubrication problems in power hydraulic systems more nearly approximate boundary layer lubrication than they do hydrodynamic lubrication. Hydraulic fluid choice must be considered as a boundary layer problem. The fluid must have the proper viscosity over the working temperature range to allow it to stay in the working area. It should coat newly sheared surfaces with fatty acids or metal salts in order to avoid welding. It should have low internal shear forces.

Low viscosity gives low shear forces but allows the fluid to be squeezed from between heavily loaded surfaces. A viscous but chemically inert fluid such as a silicone may minimize wear but not prevent galling and seizing caused by pressure welding. A phosphate ester base or a fatty acid base fluid may prevent welding but allow a high wear rate. The lubricating aspect of the choice of a fluid must then be a compromise among the factors of low coefficient of friction, low wear and anti-galling.

Lubricity is of primary concern where moving parts are in contact. Examples of these areas are pump and motor gear teeth and piston-cylinder piston heads and rod ends, valve spools and accumulator pistons. Tests for oil lubricity attempt to relate these requirement areas to standardized testing methods. Running heavily loaded gears with the test oil as a lubricant is a good test for gear pumps but may not have much relevance for piston pumps. Perhaps the most widely used test for oil lubricity is the 4-ball test. In this, three ½-inch diameter steel balls are held in a cup with the test oil. A fourth ball is pressed down onto the other three with a known load and then rotated at a known speed. The torque to rotate the fourth ball can be measured to give the coefficient of friction for the particular composition of balls. After running for a specified period of time at a controlled temperature, the wear scar on the balls is measured. The size of this scar can be related—at least subjectively—to the wear resisting qualities of the oil. If the load were increased until welding occurred, the test could give a measure of the anti-seizing quality of the oil.

Stability of a fluid is no more precise a term than is lubricity. A fluid is stable if its important characteristics change little with use or during storage.

In use, the fluid is subjected to heavy mechanical shear forces. The fluid will be exposed to entrapped air and moisture. It will be in contact with organic materials with which it may combine. The fluid may be subjected to each of these while at high temperature or when under nuclear radiation. While no real fluid is truly stable under all of these conditions, some fluids meet combinations of conditions better than do others.

Heavy shear forces tend to break down fluid molecular bonds and reduce viscosity. If the fluid is not resistant to this tendency, then flow computations and pump designs must account for expected changes in viscosity. Agitation of hydraulic fluids in the presence of oxygen and water accelerates the chemical breakdown of the fluid and the formation of gums and varnishes. If the fluid reacts with active metals such as silver and copper alloys, pump and valve designs must change, or wear and early breakdown will result. Where organic materials such as elastomers or electrical insulations are in contact with the fluid, any reaction must be beneficial or non-existent. As an example of a beneficial reaction, many hydraulic fluids contain additives which promote some swelling in elastomers. This swelling aids the sealing action of O-rings. Obviously, a shrinking action would be very harmful.

High operating temperatures accelerate each of the chemical reactions. Nuclear irradiation tends to promote the growth of polymeric chains and leads to greater viscosity and eventual hardening.

Viscosity When two surfaces which have relative motion are separated by a real fluid, the fluid immediately adjacent to each surface moves with that surface, Fig. 2.2. The velocity varies in proportion to the distance from the moving surface film. Relative motion between adjacent laminae of fluid means that a shearing stress exists between them. Newton's law states that this shearing stress between adjacent

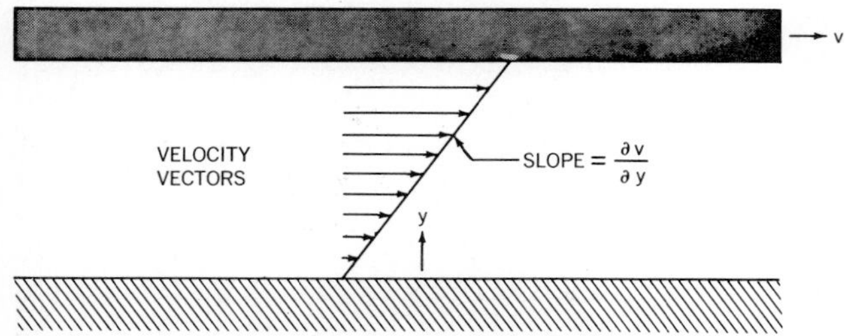

Fig. 2.2. Velocity gradient in fluid between a stationary plate and a parallel moving plate.

laminae of fluid of infinitesimal thickness is proportional to the rate of shear in the direction perpendicular to the motion.

If a very small portion of the velocity gradient is considered, the velocity gradient is a straight line having a slope of $\Delta v / \Delta y$. In accordance with Newton's statement, the shear force along the infinitesimily small distance Δy, which lies in a direction perpendicular to the velocity vector is proportional to the partial derivative of the velocity with respect to the lamina thickness in the y direction.

$$S = \mu R \tag{2.1}$$

Where S = shear stress
$R = \partial v / \partial y$

μ is the constant of proportionality. μ is then the measure of viscosity. It is called the absolute viscosity. μ has the dimensions of mass/length x time. In English units, absolute viscosity is measured in units of lb-sec/in^2 called Reyns. In metric units, viscosity is given in poise units, dyne-sec/cm^2. Both the Reyn and the Poise are very large units. More convenient units are the microreyn (one millionth of a Reyn) and the centipoise (one hundredth of a Poise).

Fluids which obey Newton's law, 2.1, are called "Newtonian." At least during laminar flow conditions most working hydraulic fluids are essentially Newtonian.

Many calculations necessary to hydraulic system analysis involve the use of the ratio of absolute viscosity to the oil mass density. This ratio is called the kinematic viscosity ν (nu).

$$\nu = \frac{\mu}{\rho} \tag{2.2}$$

although the ratio is neither kinematic nor a viscosity. The unit of kinematic viscosity most frequently used in military and aircraft systems is the stoke. This has the dimensions of sq. cm/sec. Because the stoke is a large unit, viscosity is often reported in centistokes (cs = one hundredth of a stoke). The comparable ratio in the English system is the Newt which has the dimensions of square inches per second. Because all of the other parameters of interest to hydraulics engineers are—in the United States, at least—given in English units, this book will use the Newt as the primary measure of viscosity.

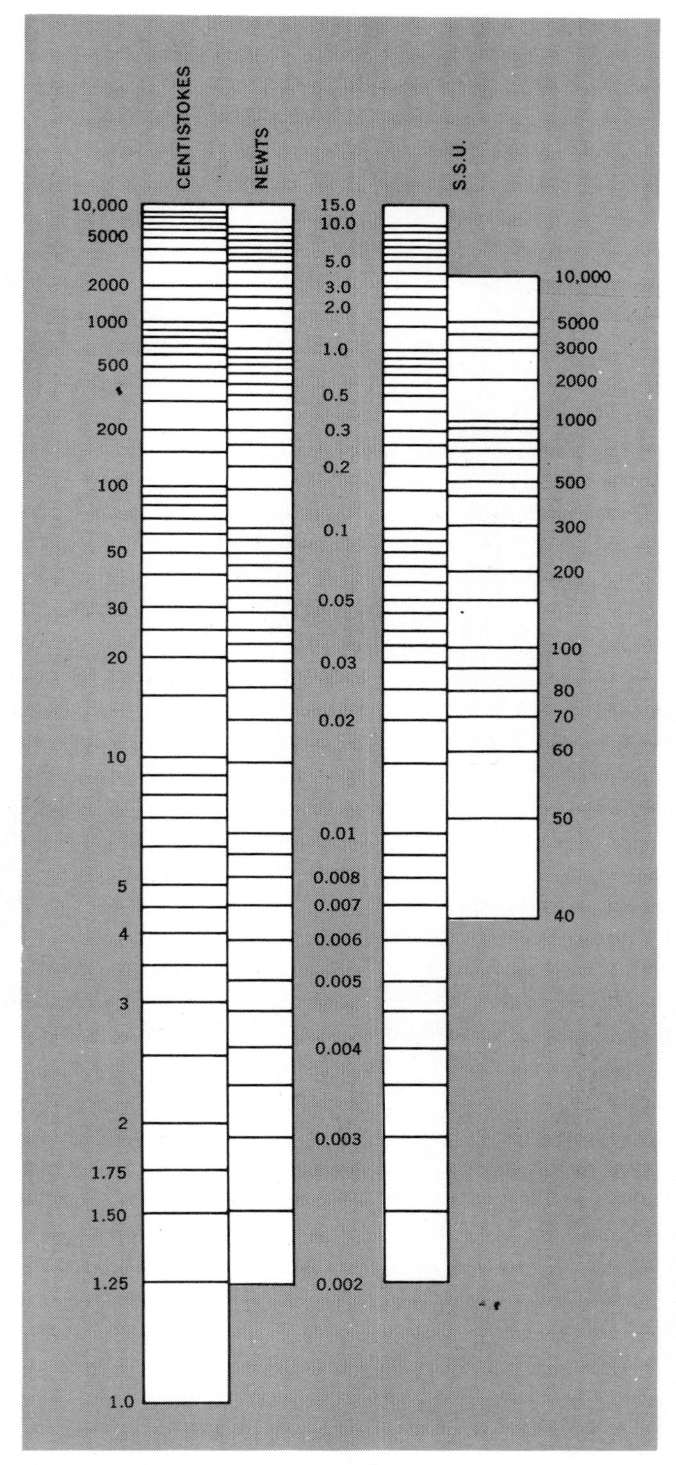

Fig. 2.3. Viscosity conversion chart.

The conversion formula is

$$\nu\,(Newts) = 0.001552\,\nu\,(cs) \tag{2.3}$$

Industrial hydraulic fluid kinematic viscosity is usually given in Saybolt Universal Seconds (SSU). SSU, as determined by ASTM Procedure D445-53T,

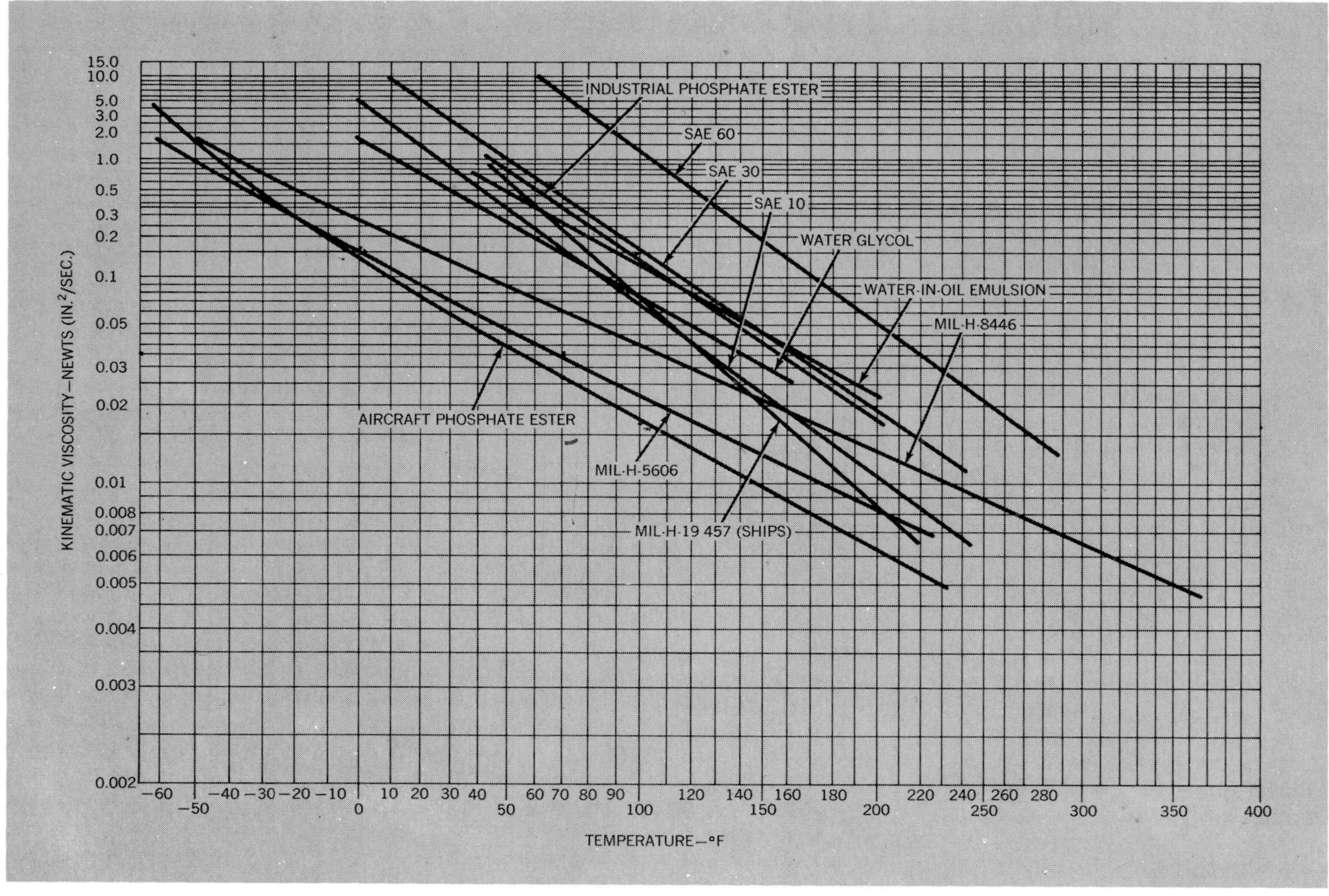

Fig. 2.4. Viscosity-temperature curves for typical industrial and aircraft fluids.

is the time required for 60 milliliters of the oil at 100 degrees F to flow through a standard orifice. Unlike the other units of viscosity, SSU cannot be related to Newton's law. The conversion formulae among Newts, centistokes and SSU are taken from or extrapolated from ASTM Procedures D446-53 and D666-57 or ASTM Special Technical Publication No. 43C.

$$\nu\ (cs) = 0.226t - \frac{195}{T} \quad 32<T<100 \qquad 2.4$$

$$\nu\ (cs) = 0.220t - \frac{135}{T} \quad 100<T \qquad 2.5$$

$$\nu\ (Newts) = 0.00035t - \frac{0.303}{T} \quad 32<T<100 \qquad 2.6$$

$$\nu\ (Newts) = 0.000392t - \frac{0.21}{T} \quad 100<T \qquad 2.7$$

Viscosity-Temperature—The variation of viscosity with temperature is usually shown by a modification of an ASTM (American Society for Testing Materials) chart which plots viscosity either in SSU or CS against temperature. These charts are based on the Walther equation.

$$\log \log (\nu + k) = A + B \log T \qquad 2.8$$
$$k, A, B = \text{Constants}$$
$$T = {}^\circ F$$

By the use of charts based on the Walther equation most petroleum fluids will have straight line plots of viscosity vs. temperature. Fig. 2.4 is a modification of the ASTM chart which gives the viscosity—temperature curves of some typical industrial and aircraft oils in Newts.

The slope of the viscosity—temperature curve of an oil as plotted on the ASTM chart is often used as a measure of the rate of variation of viscosity with temperature. Because of the way the chart is scaled, the slope is measured in inches along the temperature axis. While—in the cartesian coordinate system—this slope is always negative, the minus sign is not usually given.

Probably the most widely used method for expressing the viscosity-temperature relationship is the Viscosity Index (VI). The VI is based on arbitrarily chosen lubrication oils which were once thought to represent the maximum and minimum obtainable viscosity-temperature sensitivities. These were established as zero and 100 on a scale. Fluids are now obtainable which are out of scale on either end of these

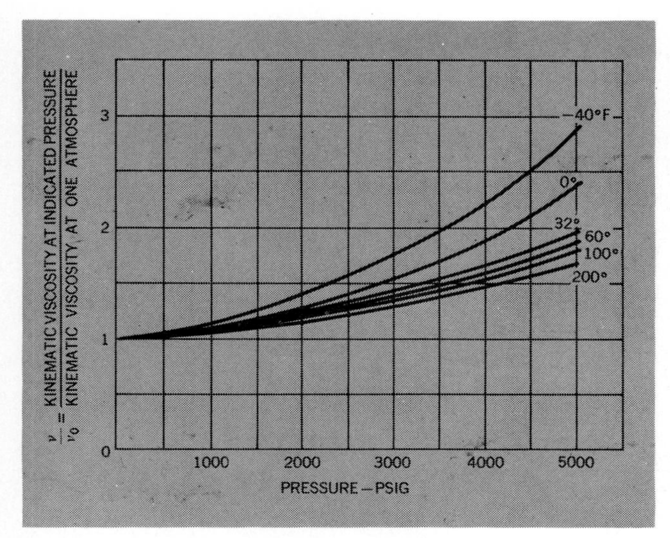

Fig. 2.5. Variation of viscosity with pressure for MIL-H-5606 oil.

limits. The VI of a fluid is obtained by measuring its viscosity at 100°F. Then:

$$VI = \frac{L - V}{L - H} \times 100 \quad 2.9$$

Where L = viscosity at 100°F of OVI fluid which has same vis. at 210° as unknown oil.

H = viscosity at 100°F of 100 VI fluid which has same viscosity at 210° as unknown oil.

V = viscosity at 100°F of fluid under test

Care must be exercised in the use of the VI at high VI values. Two oils having the same viscosity at 100°F could have the same VI, yet they could have widely varying viscosity values at elevated temperatures.

Viscosity-Pressure—The viscosity of hydraulic oils is not much affected by moderate changes in pressure. However, pressurization to 3000 or 4000 psi will increase fluid viscosity sufficiently to require consideration in engineering calculations. Fig. 2.5 shows isothermal changes of viscosity of MIL-H-5606 oil with pressure.

Viscosity-Shear—Mechanical shear causes permanent changes in hydraulic fluids containing the type of complex polymeric chains found in VI improvers such as the methacrylate additive in MIL-H-5606. A piston hydraulic pump imparts heavy shear to fluid passing through it. Shear breakdown is caused by the actual rupturing of polymer molecules. Cavitation, which is the local formation and collapse of vapor bubbles caused by lowering of local pressure below the vapor pressure, also causes permanent viscosity changes. Other energy sources which cause fluid breakdown through shear are sonic energy, very turbulent flow, and very severe pressure gradients.

Compatibility with System Materials—Compatibility of the fluid with the balance of the system covers such phenomena as rusting, miscibility, corrosion, solution of gasses and solids, softening of surfaces, etc. The choice of fluid and system materials should be such as to minimize the bad effects of or to take advantage of these phenomena. A number of compatibility problems have already been discussed under the heading of fluid stability. However, some additional factors bear looking at. Because some deleterious interaction of materials will always take place, the type of breakdown product can affect system operation. As examples, the production of stringy particulate contamination will quickly clog filters. Gums and varnishes can coat lapped valve surfaces and cause sticking and malfunction. Chemical attack of wearing surfaces can either promote lubrication or cause early failures.

Fluid Heat Transfer—Heat is generated in hydraulic systems by pump compression and pump inefficiencies, by pressure loss across relief valves, by friction in transmission lines and by valve losses in the output elements. Local damage to both fluid and the mechanical components could result if the fluid did not carry the heat away to be dissipated or transferred. Properly designed hydraulic systems can operate in high temperature environments. The fluid is often relied upon to carry heat away from seals and servo mechanisms. A fluid having good heat transfer characteristics should have a high specific heat and thermal conductivity. Table 2-1 gives the specific heat and thermal conductivity of some hydraulic fluids.

TABLE 2-1
THERMAL CHARACTERISTICS OF SOME HYDRAULIC FLUIDS

Fluid	Thermal Conductivity BTU/hr/ft²/°F/ft	Specific Heat BTU/lb/°F
Aircraft Phosphate Ester	0.078 at 82°F	0.38 at 73°F
Industrial Phosphate Ester	0.067 at 77°F	0.32
Silicate Ester	0.083 at 77°F	0.44 at 81°F
Industrial Water-Glycol	0.25	0.8
MIL–H–5606	0.077 at 160°F	0.506 at 160°F

Bulk Modulus—One large use of hydraulic power is in control systems. The favorable power to weight ratio and the stiffness of liquid systems make them the frequent choice for the positioning and the rotation of inertial masses. The stiffness of the hydraulic control system is directly related to the lack of compressibility of the oil. The measure of this lack of compressibility is called bulk modulus of elasticity. In the English system, bulk modulus has the units of pounds per square inch. The higher the bulk modulus, the less elastic or the stiffer the fluid. Particularly in control servo-mechanisms handling heavy loads, a high bulk modulus is mandatory for stability.

TABLE 2-2
PROPERTIES OF SOME HYDRAULIC FLUIDS

	MIL-H-5606	Aircraft Phosphate Ester	Industrial Phosphate Ester	Silicate Ester	Industrial Water-Glycol
Flash Point	200 – 225°F			395°F	None
Fire Point	255°F		690°F	450°F	None
Auto Ignition Temp.	475°F	1100°F	over 1200°F	755°F	
Neutralization No. mg KOH/gm	0.2			0.1	
Vapor Pressure	.445 psi at 160°F	.008 psi at 248°F	.002 psi	.009 psi at 300°F	
Coefficient of Expansion	4.6 x 10^{-4}/°F	4.18 x 10^{-4}/°F			
Specific Gravity	0.83 at 68°F	1.065 at 77/°F	1.15 at 60°F	0.93 at 60°F	1.055 at 60°F
Pour Point	-75°F	-85°F	-5°F	-100°F	-60°F
Viscosity Index		238			140
Bulk Modulus	320,000 at 160°F	308,000 at 160°F	387,000 at 70°F	196,000 at 160°F	

Fluid bulk modulus can be determined either by measuring changes in fluid volume as a function of pressure at isothermal conditions[2] or may be determined through measurements of the speed of sound in the fluid under isentropic conditions.[3] The values of bulk modulus may be reported as secant or tangent. The secant bulk modulus is the average or mean of the change in pressure divided by the total change in volume per unit of initial volume. The expression for secant bulk modulus is:

$$\beta_s = \left[\frac{\Delta p}{\Delta V/V_o}\right]_{T,E} \qquad 2.10$$

Where: β_s = secant bulk modulus
V = volume
T = subscript indicating isothermal
E = subscript indicating isentropic

The tangent bulk modulus is the inverse of the fluid compressibility at a specific point. It must, therefore, be measured at a specific quiescent pressure and temperature. The quiescent temperature must be reported. The expression for tangent bulk modulus is:

$$\beta_t = -V_o \left[\frac{\partial p}{\partial V}\right]_{T,E} \qquad 2.11$$

The bulk modulus of hydraulic fluids increases slightly with pressure but decreases sharply with rising temperature. A careful matching of fluid bulk modulus to expected operating conditions is necessary if unstable system functioning is to be avoided. Fig. 2.6 gives values of bulk modulus as a function of temperature for several hydraulic fluids.

Low Volatility—A high vapor pressure (low boiling point) allows the easy formation of bubbles in areas of low local pressure. Typical of such areas in hydraulic systems are pump suction lines and metering ports in servo valves. The formation and subsequent collapse of the bubbles causes local pressure shocks which can erode metal from pump pistons and valve metering lands. This phenomenon is known as cavitation. The use of fluids having low vapor pressures at operating temperature minimizes the occurrence of cavitation.

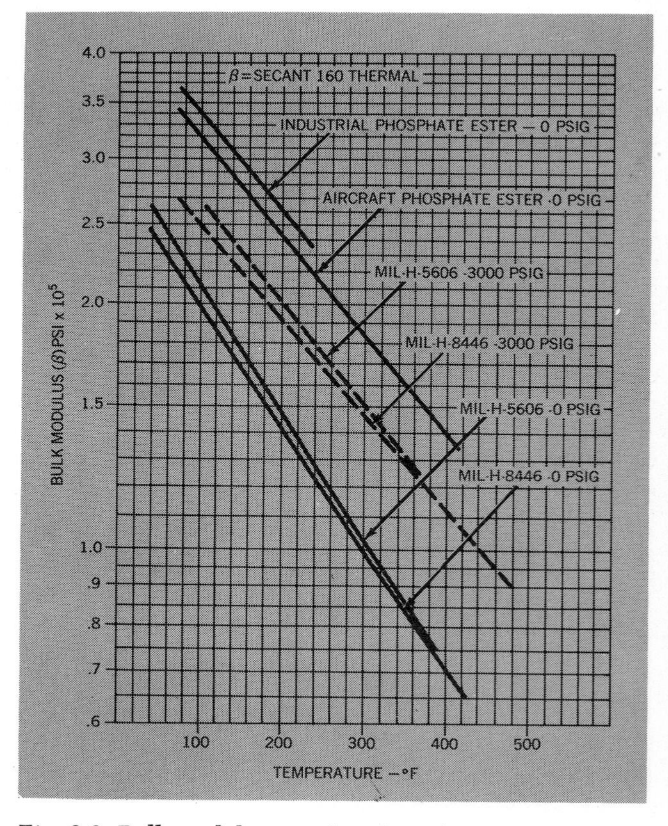

Fig. 2.6. Bulk modulus as a function of temperature.

Actuator piston rods which move in and out past seals usually pull a thin film of oil past the seal. This film of oil is helpful in promoting seal life. Under conditions of low ambient pressure or high ambient temperature, fluids having high vapor pressures would evaporate from this film. If they left gummy or abrasive residues, subsequent inward motion of the piston rod could tear or abrade the seal.

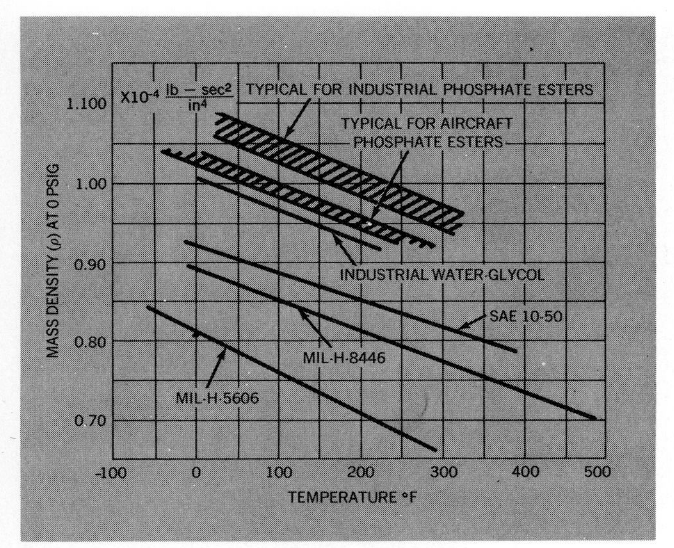

Fig. 2.7. Mass density as a function of temperature.

Cavitation—Although not a separate property of hydraulic oils, cavitation is intimately related to vapor pressure, oil density and the way the system is designed. In the classical theory of cavitation[4,5] during the collapse of the vapor bubbles, the liquid flows radially inward until motion is stopped and rebound occurs. The estimated pressure at rebound is approximately 75,000 psi. The local temperature is estimated in thousands of Kelvin.

A more modern theory of cavitation[6,7] describes the collapse of the vapor bubbles as a flattening in the direction of flow. Often a torus (doughnut shape) is formed just before final collapse. When this torus finally collapses, a microjet of oil ranging in diameter from .00001 inches to .001 inches spurts from the toroidal center at a velocity which can vary from 600 ft/sec to 4000 ft/sec. This microjet has enough energy to tear metal from a surface.

The tendency of a fluid to cavitate is described by the cavitation number N_c[8].

$$N_c = \frac{P - P_v}{Y \rho v^2} \qquad 2.12$$

Where: P = static Pressure
P_v = vapor Pressure
v = local flow velocity
Y = a constant based on the fluid

The smaller the cavitation number, the greater is the tendency to cavitate.

Cavitation damage tends to fall into three categories. When cavitation intensity is severe, metal fracture can occur. However, if the implosive or jet forces are below the yield point of the metal, cavitation can cause corrosion fatigue and stress corrosion. If the cavitation forces are relatively low, protective films, such as surface oxides, can still be removed. This removal allows for corrosive destruction of metals by other mechanisms.

The general conclusions of interest to the hydraulics engineer are that high density, high vapor pressure and high local flow velocities each increase the probability of cavitation damage. Cavitation damage can occur at pressures both slightly above and below the oil vapor pressure. Because vapor pressure increases with temperature, damage rate can very with temperature also. The removal of protective films by low intensity cavitation means that oil additives which attack base metal will accelerate corrosion.

Low Foaming—Vapor or air which is entrained in a fluid sharply reduces the effective bulk modulus of the fluid. The highly compressible gas acts as a soft spring which is in series with the stiffer spring of the oil. Air is frequently introduced into a system or trapped in a system during a filling or ground-cycling operation. Usual bleeding methods do not remove all of this air. If the fluid forms foams, the removal of gasses is much more difficult.

Low Coefficient of Expansion—The coefficient of expansion of a hydraulic fluid is of particular interest to the designer of a system which must operate over a wide range of temperature. In many modern closed hydraulic systems, very little oil is supplied in the reservoir to make up leakage. The primary function of the reservoir is to accommodate thermal expansion of the fluid. Obviously, a system using a fluid with a low coefficient of expansion requires a reservoir which is smaller than would be needed with a fluid which expands greatly with rising temperature.

Low Weight—The importance of weight in flight vehicles is well known. For this reason, many halogenated fluids, which otherwise might be considered as aircraft hydraulic oils, have been rejected. For ease in making calculations involving weight, Fig. 2.7 gives the mass density ρ of several fluids as a function of temperature.

Fire Resistance—No one wishes to use a hydraulic oil which will initiate or support a fire. However, most hydraulic oils will burn under one condition or another. Measuring the tendency of an oil to catch fire or to support combustion is not a simple matter. No single test in use can be used to evaluate all expected operating conditions.

Flammability can be defined as ease of ignition combined with the ability to propagate flame. These two are sometimes in conflict. The term fire resistant is applied to fluids which can be ignited but will not support combustion when the ignition source is withdrawn or which will not cause flame to flash back to the source.

The usual measures of fire resistance are flash point, fire point and Autogenous Ignition Temperature (AIT). These are valuable only as gross measurements of flammability. A number of other tests which measure the flammability of high velocity jets of hydraulic fluid when in contact with flame or when impinging on hot plates have been suggested. None of these have yet been properly standardized.

Flash point of an oil is the temperature at which the liquid gives off sufficient vapors to ignite when approached by a small flame under specified conditions. Test fluid is placed in an open metal cup of standard dimensions. The fluid temperature is raised at a specified rate. At 5°F intervals a small test flame is passed over the cup. The flash point is the fluid temperature when a flash of fire occurs at any point on the surface of the fluid. This procedure is specified in ASTM method D92-57. Another method is given in ASTM procedure D93-58T.

Fire point is determined in a similar manner. After the flash point has been found, the fluid is further heated and the test flame applied at intervals of 5°F. The fire point is the fluid temperature at which the fluid burns continuously for 5 seconds.

The AIT is the temperature at which ignition occurs spontaneously. As described in ASTM procedure D286-58T, a small amount of the fluid is placed on a heated glass surface. The temperature is raised until ignition occurs. AIT is strongly affected by the nature of the glass surface, the amount of liquid on the glass and the size and ventilation nature of the test area.

2.3. FIRE RESISTANT FLUIDS

The largest percentage of hydraulic fluids are petroleum derivatives. These are excellent working fluids but, unfortunately they burn vigorously. Industrial fires, aircraft fires and shipboard fires have each been started by or supported by burning petroleum based hydraulic oils. In order to reduce this fire hazard, a number of fire resistant fluids have been made available. These all tend to fall into one or another of the following types.

1. Emulsions—water in oil
   —oil in water
2. Water-glycol solutions
3. Straight synthetics
4. Blends of synthetics and petroleum oils

In the water-in-oil emulsion, the continuous phase is oil. The water is present in droplets. The viscosities of such fluids vary from about 0.0775 Newts (250 SSU) to about 0.342 Newts (1000 SSU). The larger viscosities are associated with increased water content. These fluids are usable from the freezing range up to about 175°F. Their primary drawbacks are high density and high vapor pressure.

Oil-in-water emulsions contain droplets of oil in a water base. Up to about 30-40% of oil, the viscosity is the same as water. These emulsions are very hard on paints and packings. Some suppliers recommend the use of nitrite elastomers for seals.

Synthetics—Synthetic oils usually used are either chlorinated hydrocarbons or phosphate esters. Each of these are good lubricants, primarily because they react with metals to form weld resisting surface films. These synthetics tend to have low VI values and to get very viscous at low temperature. The phosphate esters, in particular, attack paints, organic coatings and electrical insulation. Special formulations of these materials are needed wherever these fluids are used. Butyl rubber and PTFE seals are necessary. The following table was prepared by Henrickson[10] for typical industrial fire resistant synthetics.

TABLE 2-3

	Chlorinated Hyd	Phosphate Ester	Blend without VI Improver	Blend with VI Improver
Sp. Gr. 60/60	1.42	1.16	1.28	1.13
Vis. Newts 0°F	–	–	–	0.855
50°F	0.719	0.908	0.625	0.137
100°F	0.047	0.072	0.050	0.047
130°F	0.018	0.031	0.021	0.030
210°F	0.004	0.008	0.005	0.013
VI	Low	Low	Low	170
Flash Point	390°F	490°F	400°F	425°F
Fire Point	740°F	600°F	660°F	465°F
Pour Point	10°F	0	10°F	-55°F
AIT	1220°F	1150°F	1200°F	790°F

Water-Glycols—Water-glycol fluids are true solutions, not merely emulsions. A working mixture requires approximately 45% water. These solutions have high VI's and low pour points. The viscosity will rise as water evaporates. These fluids can be used from about −10°F to nearly 180°F. Most available elastomers are not affected and may be used. Metals and alloys containing zinc, cadmium or magnesium should be avoided. Special paints are required.

Good Dielectric—Submerged electrical coils and contacts are sometimes used in solenoids and servo valves in hydraulic systems. The life of these is enhanced if the fluid is a good dielectric.

Toxicity—The fluid should be non-toxic if ingested with food, licked from the hands or breathed in as vapors. The fluid should be tested with animals to assure man's safety. One problem with such a technique is determining how many rats equal a man!

Effect of the fluid on skin and eyes is, perhaps, more important than the effect of ingested oil. Dermatitis studies should be made before any wide scale use of a new fluid be attempted.

The fluid should also be non-allergenic and should have either no odor or a pleasant odor.

Other Properties—Desirably the chosen fluid should be given a distinctive color. It should not have an acid reaction. It should not form harmful solutions or precipitates with water. It should have high surface tension and should not be affected by nuclear irradiation. It should flow readily at the lowest expected operating temperature. Finally, it should be easily available and low in cost.

2.4. ELASTOMERS

A wide variety of rubber-like materials is available to the hydraulic system designer. These elastomers are used in O-rings, diaphragms, and other seals or pressure separators. Some plastic materials are also used for many of the same purposes. In order to differentiate, the following definition of elastomer will be used: "A polymeric material which at room temperature can be stretched to at least twice its original length and upon immediate release of stress will return quickly to approximately its original length."

Because of the wide variation of properties obtainable within each of the several elastomer types of interest to hydraulic system designers, reference should be made to fabricators and molders prior to final decision. By controlling compounding and curing and by adding fillers, plasticizers, accelerators and vulcanizing agents, the fabricator can frequently supply specific desired properties. A list of commonly used polymers originally compiled by Butler[11] is given in Table 2-4.

TABLE 2-4

PROPERTIES OF ELASTOMERS

Material	Usable Temperature Range	Fluids
Buna N Nitrile	-65°F to 275°F	Oils, Refrigerants
Polyacrylate	0°F to 350°F	Oil, Refrigerants
Neoprene	-40°F to 220°F	Oils, Water, Refrigerants
Hypalon	-40°F to 250°F	Oils, Water, Acids
Fluorocarbon	-40°F to 450°F	Oils, Fuels
Butyl - EP	-65°F to 300°F	Air, Phosphate Esters
EPT	-65°F to 300°F	Air, Phosphate Esters
Urethane	-40°F to 212°F	Oils
Silicone	-150°F to 500°F	Oils, Phosphate Esters

Buna N—This is a copolymer of butadiene and acrylonitrile. It is available in a number of polymers. Military Specification O-rings for hydraulic system use conforming to MIL-P-18017 are molded of Buna N. Polymers which contain a high portion of acrylonitrile have excellent resistance to petroleum oils such as MIL-H-5606, good tensile strength and abrasion resistance but have poor low temperature flexibility. Decreasing the acrylonitrile content increases low temperature flexibility but lowers oil resistance.

Butyl—The phosphate-ester hydraulic fluids require the use of butyl elastomers. Buytl is a high molecular weight copolymer of isobutylene with small amounts of isoprene. It is not a good material for use with petroleum based oils. Butyl has low permeability to gases. It is frequently used for gas and vacuum seals.

Neoprene—This chloroprene rubber has good resistance to sunlight, ozone and weathering. At least two versions, Neoprene W and WRT, have shown good high temperature stability.

Silicone—The generally inferior tensile strength and tear resistance of silicone rubbers do not prevent their being used where thermal stability and inertness to reactive fluids are important. Silicones are of real use in both very low (—130°F) and very high (500°F) temperature use in fluid systems.

Polyacrylate—The acrylic rubbers have good oil resistance at temperatures up to 350°F. They are generally of low strength and take a marked compression set. Acrylics have excellent resistance to oxidation, ozone and sunlight.

Viton—This material is a linear co-polymer of vinylidene-fluoride and hexafluoro-propylene. It contains about 65% fluorine. Viton has become nearly a standard material for elastomeric seals for use at elevated temperatures. While having limited elasticity, it has shown good resistance to petroleum oils, synthetic lubricants and silicate esters at working temperatures up to 500°F.

2.5. ELASTOMER CHARACTERISTICS

Fluid Compatibility—Elastomer compounds must be dissimilar chemically to the fluids with which they come in contact. Chemically similar compounds tend to dissolve each other, causing swelling. For fuel and hydraulic oils, the elastomers usually used are nitrile, neoprene, acrylate, fluorocarbon or fluorosilicone. These materials may sometimes undergo drastic volume changes in the fluid but return to near original size and shape when dried. Usually no permanent change in physical properties occurs.

Hardness—This is the physical property most often tested. The Shore "A" durometer, the most frequently used measure, uses a presser foot with an identer point working against a calibrated spring. The durometer measures 100 when pressed firmly on flat glass.

Elastomers act in some ways as if they were highly viscous fluids. The Shore "A" reading could be related to viscosity. A low reading indicates less resistance to pressure, therefore the seal material flows into small openings more easily. A low durometer value is desirable for low pressure seals. Conversely, a high durometer reading indicates a greater resistance to flow or extrusion. Within limits, therefore, the tendency is to use low shore elastomers with low pressures and high shore with high pressures.

The coefficient of friction is lower with high durometer compounds, however, the force required for the same squeeze is greater. The higher force results in greater breakout and running friction. Therefore, to reduce friction either a lower hardness elastomer should be used or the squeeze should be reduced.

Modulus—As used in elastomer technology, this refers to stress at a predetermined elongation—usually 100% of the original length. In general, a high modulus indicates good resistance to extrusion and an ability to recover from overloads.

Compression Set—Set is usually defined as the stress relaxation under a load. The ability of an elastomer to resist compression set is probably its most important physical characteristic if it is to be a good sealant in a hydraulic system. The worst set of characteristics is the combination of high compression set and shrinkage in the fluid. Failure under these conditions is inevitable. In general, compression set is a greater problem with dynamic seals than with static seals. The compression set results in loss of squeeze. This loss is accelerated at high temperature.

Tear Resistance—This is the force per unit thickness required to propagate a nick or cut, or to initiate tearing in a direction normal to the direction of stress. A compound can have a low tear strength and still be a good seal material. A minimum strength of 100 lbs. per inch is desirable in order to assure adequate handling characteristics.

Abrasion Resistance—Abrasion resistance is a measure of the elastomer's resistance to wearing when in contact with a moving surface. Abrasion resistance is a combination of properties—resilience, stiffness, thermal stability, resistance to cutting and tearing—rather than a specific quality. System lubrication can also affect abrasion resistance.

Volume Change—The percent increase or decrease of original volume after the elastomer has been in contact with a fluid is the volume change. As the elastomer swells it loses hardness, has lower abrasion and tear resistance and is more easily extruded under pressure. Up to 50% swell can be tolerated for static seals. 15—20% is the maximum swell normally considered acceptable for dynamic seals.

Swell is desirable as a compensation for compression set and for achieving flexibility at low temperatures.

Shrinkage is accompanied by increased hardness. Shrinkage is, of course, additive to compression set, 3% to 4% shrinkage is the maximum for dynamic seals.

Thermal Reaction—Low temperature effects are usually reversible and may not cause trouble except during protracted operational exposure to the low temperature. Hardness increases at low temperature but decreased flexibility and resiliency and increased brittleness may be more harmful.

Some elastomers become crystalline rather than amorphous at very low (—65°F) temperature. This occurs on the flat cross section of O-rings parallel to sealing surfaces. While appearing as compression set, it can be distinguished by gradual recovery of the original cross-section as temperature increases.

High temperature effects are often non-reversible and, harmful. As the temperature is raised, the elastomer becomes softer and more flexible until the plasticizers are driven out of the compound. The elastomer then can become hard and brittle. Compression set—actual remolding of the seal cross section—can occur short of the embrittlement temperature.

Permeability—This is the ability of an elastomer to allow passage of gasses and liquids. This should not be confused with leakage. Permeability is actually a combination of solubility and diffusivity. Fluids chemically pass into solution on one side of a seal, migrate through to the other side and evaporate. The more a seal is compressed, the lower its permeability.

Coefficient of Thermal Expansion—The measure of expansion in length per unit rise in temperature is the coefficient of thermal expansion. As seen in Table 2-5, elastomers have a coefficient of thermal expansion which is about 7 times that of aluminum and 15 times that of steel. Therefore, high temperature seals require glands having adequate room for expansion. Low squeeze can cause leakage at low temperatures.

TABLE 2-5

LINEAR THERMAL EXPANSION

	In/in/°F x 10^{-6}
Silicone Rubber	90 - 150
Nitrile (40 - 90 Shore) rubber	60 - 110
Neoprene (30 - 80 Shore)	70 - 120
Fluorocarbon elastomer	80 - 100
PTFE	40 - 60
Aluminum	12 - 13
Corrosion Resistant Steel	9 - 11
Mild Steel	6 - 7

Aging—Aging is the sum total of the physical and chemical changes which take place in an elastomer over a given time period. Natural aging may take years before significant changes occur. Aging can be accelerated by exposure to sunlight, ozone and heat. The major effects of aging are brittleness and compression set.

Plastics—High polymeric substances which do not meet the definition of an elastomer are generally considered plastics. They are used in hydraulic systems as back-ups for O-rings, as soft seats for valves and as sealing materials.

PTFE—The single plastic most widely found in hydraulic systems and components is poly tetra fluoroethylene (PTFE). PTFE is a tough, inert, waxy solid. It can be processed only by compacting and sintering.

PTFE is chemically very inert. It has excellent resistance to chemical breakdown at temperatures up to 700°F. It has an extremely low coefficient of friction. PTFE is used in back-up for O-ring seals, for valve seats, for slipper seals over O-rings and as static seals.

One major drawback to the use of PTFE is its tendency to flow under pressure, forming thin feathery films. This tendency to flow can be reduced by the use of filler materials such as graphite, metal wires, glass fibers and asbestos.

FEP—This is similar to PTFE but is composed of fluorinated ethylene-propylene copolymers. While possessing many of PTFE's good properties, FEP can be molded, extruded and cemented.

Polyethylene—Polyethylene is a strong, flexible, waxy plastic having a hard surface. It is highly resistant to chemicals and has very low moisture absorption. When irradiated, it has improved solvent resistant and high temperature strength.

Kel-F—As a polymer of chlorotri-fluoro ethylene, Kel-

TABLE 2-6
PROPERTIES OF ALUMINUM

	A108	355	6061	2014	2024	7075
Density lb/in^3	0.101	0.098	0.098	0.101	0.100	0.101
Modulus of Elasticity (psi)			10 x 10^6	10.6 x 10^6	10.6 x 10^6	10.4 x 10^6
Yield Strength (psi)	14,000		21,000 - T4	40,000 - T4	48,000 - T4	
		25,000 - T6	40,000 - T6	60,000 - T6		72,000 - T6
Ultimate Strength (psi)	21,000		35,000 - T4	62,000 - T4	68,000 - T4	
		35,000 - T6	45,000 - T6	70,000 - T6		150,000 - T6
Elongation-%-2"	2.5	3		20 - T4	19 - T4	
			12 - T6	13 - T6		11 - T6
Thermal Conductivity	1010		1070	840 - T4	840 - T3	840
BTU/ft^2/hr/°F/in						
at 77 °F		1010 - T6		1070 - T6		
Specific Heat	.022	.023	.022	.023	.023	.023
BTU/lb/°F						
Melting Point °F	970-1135	1015-1115	1180-1205	950-1180	935-1180	890-1180
Coefficient of Expansion						
10^{-6}/°F at 100 °F	11.9	12.2	13.1	12.8	12.9	13.1

F possesses properties similar to those of TFE. Kel-F is moldable and transparent. It has low moisture absorption, great chemical inertness and thermal stability.

Other Plastics—Other plastics which are occasionally found associated with power hydraulic systems are polyamide, polypropylene, polyvinyl chloride, polyvinyl fluoride, polyimide and acetate.

2.6. INORGANIC MATERIALS

Metals—A wide range of metals are used in hydraulic systems. The choice of metals is most strongly influenced by the operating temperature. At system temperatures below 275°F, aluminum and its alloys are used except where local stress and lubrication problems dictate other materials. Above 275°F, stainless steel is more frequently used.

Aluminum—Most subsonic aircraft use aluminum alloys as the primary fabrication metal for hydraulic systems. Various casting alloys are used for housings for pumps, valves, filters, and other major components. The most frequently used tubing material is 6061-T6 in tubing specification MIL-T-7031. NASA now specifies that aluminum fittings be machined from 7075. Most AN and MS fittings and other machined aluminum alloy parts are made from 2014 or 2024 alloy in either the T4 (Solution heat treated) or T6 (Solution heat treated and artificially aged) condition. Table 2-6 gives some of the room temperature properties of several aluminum alloys. Table 2-7 shows how certain of the mechanical properties of some alloys vary with increasing temperature.

Ferrous Alloys—Low carbon steel is often used for hydraulic fittings. SAE 4140 and 4340 alloys are frequently used for combined toughness and hardness in uses such as valve bodies and machined parts. Where greater hardness is desired such as in rod end

TABLE 2-7

Alloy	Property	75 °F	300 °F	400 °F	500 °F	600 °F
2014-T6	F_{tu} psi	70K	39K	18K	10K	6.5K
	F_{ty} psi	60K	34K	12K	8K	5K
	Elongation %	13	15	35	45	65
2024-T4	F_{tu}	68K	43K	26K	14K	7K
	F_{ty}	48K	37K	22K	10K	5K
	Elongation%	19	17	22	45	75
6061-T6	F_{tu}	45K	32K	19K	7K	4K
	F_{ty}	40K	30K	16K	5K	2-5K
	Elongation %	17	18	25	55	90
7075-T6	F_{tu}	82K	25K	14K	11K	8.5K
	F_{ty}	72K	21K	12K	8.5K	6.5K
	Elongation %	11	30	60	65	80
355-T6	F_{tu}	43K	32K	12K	8K	4.5K
	F_{ty}	27K	24K	9K	6K	3K
	Elongation %	4.0	2	20	25	50

bearings or valve spools SAE 52100 is often specified.

Corrosion Resistant Alloys—Where the steel element is exposed to the atmosphere or where strength must be retained at elevated temperatures, the so-called "Stainless Steels" are used. Perhaps the single largest use of these corrosion resistant (cres) alloys is for hydraulic lines and connectors. The two cres alloys most widely used for hydraulic tubing are ⅛ hard type 302 or 304 in tubing to Specification MIL-T-6845 and annealed type 321 or 347 in tubing to Specification MIL-T-8808. Tubing of these alloys is easily bent and flared. AN and MS connectors are made of cres types 304, 316 and 347. Type 303 has been used in the past but its broad chemical tolerance has made it unfashionable.

Recent interest in brazed or welded permanent connectors has developed a need for tubing which is

TABLE 2-8
MECHANICAL PROPERTIES OF SOME CORROSION RESISTANT ALLOYS

	302	304	316	321 347	17-7PH	AM350 MIL-S-8840 DA	AM350 MIL-S-8840 SCT	AM355 SCT
F_{ty}	30	30	30	30	140	135	150	165
F_{tu}	75	75	75	75	150-180	165	185	190
Compression Yield Long Term	35	35	35	35	140	142	158	178
Modulus of Elasticity	29×10^6	20	29	29	29	29	29	29
Poissons's Ratio	0.3	0.3	0.3	0.3	0.3			
Weight Density	0.286lb/in^3	0.286	0.286	0.286	0.280	0.202	0.282	0.202
Torsion Modulus						11×10^6	11	11

DA = Double Aged SCT = Subzero Cooled & Tempered

more compatible with thermal joining methods than 302 and 304 and stronger than annealed 321 or 347. When cres alloys type 302, 303, 304 or 316 are heated in the 1200°–1600°F range, the chromium in the metal grain combines with the carbon to precipitate as a chromium carbide. The metal then no longer has good corrosion resistance. 321 and 347 contain stabilizing materials to minimize this effect. However, in the annealed condition the sections required are too thick. Some companies using brazed connectors are specifying tubing of type 321 work hardened to the ¼ hard condition, others are procuring 304 and 316 tubing having an extra low carbon content. No military specifications exists for either of these tube types. The great high-temperature strength of the precipitation hardening cres alloys makes them very attractive.

Carbides—The development of erosion techniques of machining has made the production of complex parts out of such materials as tungsten carbide economically feasible. While tungsten carbide is still being used primarily as inserts to minimize wear or erosive effect, some whole slide valves and bodies have been machined out of solid blocks of it.

Liquid Metals—An unusual use of metals is as hydraulic fluids for high temperature hydraulic systems. Even the very best organic fluids are probably limited to operation at temperatures no higher than 900°F. In order to extend this temperature range upward for environments to be expected in hypersonic flight, the only practical power transmission fluids appear to be liquid metals. Work has been done at General Electric and other places under contract to the Research and Technical Division, Wright-Patterson AFB on a eutectic alloy of sodium and potassium known as NaK-77 and an alloy of sodium, potassium and caesium.

TABLE 2-9
PROPERTIES OF NaK-77

Components of Alloy	Potassium	77.2%
	Sodium	22.8%
Melting Point at 1 atmosphere		9.95°F
Boiling Point at 1 atmosphere		1446.3°F
Specific gravity at 68°F		0.871
at 1400°F		0.693
Absolute Viscosity at 100°F		0.62 cp
at 1200°F		0.14 cp
Kinematic Viscosity at 100°F		0.72 cs
at 1200°F		0.19 cs
Average Heat Capacity		0.2163 cal/gm/°C
Vapor Pressure at 441°F		1.81×10^{-3} mm Hg
at 1341° F		431.85 mm Hg
Bulk Coef. of Thermal Expansion		185×10^{-6} in^3/in^3/°F
Adiabatic Bulk Modulus at 212°F		463,000 psi
at 1000°F		318,000 psi

The data given in Table 2-9 are from a publication "Liquid Metals for Flight Control Systems" from the Armament and Control Products Section, Light Military Electronics Department of General Electric.

REFERENCES

1. Hatton, Roger E., "Introduction to Hydraulic Fluids", Reinhold 1962
2. Ellington, R. T. and Eakin, B. E., "Proceedings" of the National Conference on Industrial Hydraulics, Vol. XI, P163, 1957
3. Peeler, R. L. and Green, J., ASTM "Bulletin" No. 235, 5 January 1959
4. Hickling, R. "Some Physical Effects of Cavity Collapse in Liquids", ASME *Trans. Journal of Basic Engineering* March 1966
5. Plesset, M. S., "The Dynamics of Cavitation Bubbles" ASME "*Trans Journal of Applied Mechanics*" September 1949
6. Robinson, M. J. and Hummitt, F. G., "Detailed Damage Characteristic in a Cavitating Ventor" ASME *Trans Jr. of Basic Engineering* March 1967
7. Ivany, R. D., Hummitt, F. G. and Mitchell, T. M., "Cavitating Bubble Collapse Observations in a Venturi" ASME Trans Jr Basic Eng., September 1966
8. Godfrey, D. and Furby, N. W., "Cavitation of Oils and Hydraulic Fluids" Chevron Res. Co., September 1967
9. McClay, D. and Seah, A. "Cavitation Effects in Poppet Valves" ASMS publication 67-FE-27
10. Henrickson, K. G. "Fire Resistant Fluids" Machine Design November 11, 1965
11. Butler, F., "What You Should Know about O-rings" *Hydraulics* & *Pneumatics* V19, N6 PPS 109-112; June 1966

CHAPTER 2

PROBLEMS

1. Convert the following values of kinematic viscosity from centistokes to Newts

   5 cs, 35 cs, 125 cs, 440 cs, 1300 cs

2. Convert the following values of kinematic viscosity from SSU to Newts

   50 SSU, 75 SSU, 125 SSU, 750 SSU, 2500 SSU

3. What is the absolute viscosity in Reyns of MIL-H-5606 at the following temperatures?

   -20°F, 20°F, 75°F, 150°F, 200°F

4. What is the ASTM slope for the following fluids shown in Figure 2-4?

   MIL-H-5606, MIL-8446, SAE 10, SAE 60

5. A closed system contains 10 gallons of MIL-H-5606 oil at -40°F. The bulk oil temperature is raised to 200°F.

   (a) Assuming there is enough room for expansion so that the pressure would remain constant, what would be the change in volume?

   (b) assuming that no room for expansion is available (i.e. the tubing is infinitely rigid), what would be the rise in pressure?

CHAPTER THREE

Power Generators

3.1. INTRODUCTION

The commonly used sources of power in hydraulic systems are pumps and accumulators. Pumps, in general, are mechanical-hydraulic transducers which convert power from rotating or translating shafts into flow. Accumulators are pneumatic-hydraulic or mechanical-hydraulic transducers which convert power generated in a gaseous medium or energy stored in a gaseous or mechanical spring into flow.

Positive displacement pumps do not generate pressure. They merely provide flow of fluid in a hydraulic system. The nature of the system to which the output of the pump is connected—modified sometimes by mechanism internal to the pump—determines the pressure associated with the output of a pump.

Consider, for example, a pump which has its outlet open to atmosphere. The only output possible is flow. Conversely, if the output of a fixed displacement pump is connected to a blocked line, the pump will still put out nothing but flow. But now, the flow has nowhere to go, and this continuing flow will compress the trapped fluid. Pressure in the blocked line will rise until something breaks. If a relief valve is connected to the blocked line, flow will compress the fluid and pressure in the line will rise until the relief valve setting is reached. After that, excess flow is vented through the relief valve.

Note here that the relief valve establishes the pressure level in the line. The pump merely supplies a flow of fluid.

Several companies build variable delivery pumps in which the output to a blocked or partially blocked line is at a fixed pressure level. These pumps are so designed that the pressure in the system established by resistance to the flow generated by the pump is compared to a reference force. A spring under compression, or some other force generating mechanism, establishes this reference force. The difference between the force developed by system pressure and the reference force is used to modify the output flow of the pump in such a way as to minimize the force difference.

Except for convenience, the devices which measure system pressure and compare it with the reference force need not be packaged with the pumps. The control mechanism is really not a part of the pumping mechanism. In this sense a variable delivery pump is a simple flow generator attached to a device which can command changes in output flow capability.

Similarly, an accumulator connected to atmosphere will discharge oil at atmospheric pressure until it is empty. Only when connected to a system having resistance to flow can pressure be developed.

3.2. PUMPS

Turbine pumps, impeller pumps, and propeller pumps are rarely found in power hydraulic systems. While they are occasionally used for such services as fluid de-aerators, most power systems require positive displacement pumps. At high pressures (2500 to 4000 psi), piston pumps are often preferred to gear or vane pumps.

The efficiency of a pump is related, among other factors, to the resistance of the pump design to leakage flow between fluid at outlet pressure and fluid at inlet pressure. Leakage in a vane pump, Figure 3.1., can occur along the length of the vane and across the vane edges. The leakage path has a theoretical length of zero and a width equal to the width of the vane. The forces which hold the seal against differential pressure stem from centrifugal effects or from elastomers under compression. The corners of the vanes are particularly difficult to seal.

Figures 3.2. and 3.3. show a schematic and a pictorial view of a gear pump respectively. Leakage can occur across the gear tooth contact line and across the sides of the gears. Various techniques of pressure loading the side plates of vane and gear pumps have been tried to reduce this leakage. Recent improvements in the design of gear pumps[1] have resulted in the widespread use of these pumps in earth moving equipment at pressures to 2500 psi. Some units are now available with rating to 3000 psi.

Leakage flow is defined by the Hagan-Poiseuille law.

$$Q = \frac{\pi d^4 \Delta P}{128 \mu l} \quad 3.1$$

Where: Q = flow rate

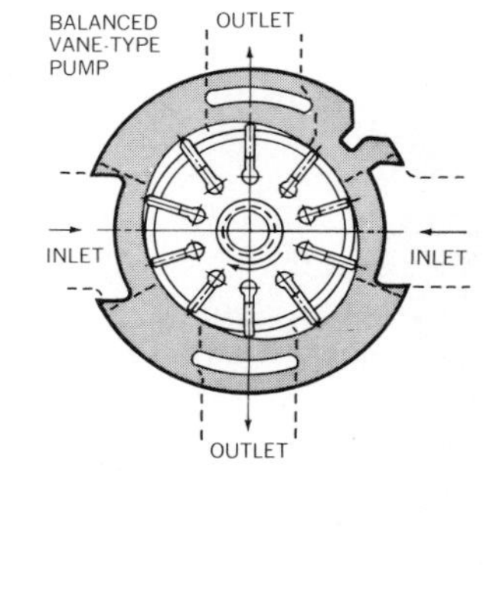

Fig. 3.1. Balanced vane-type pump.

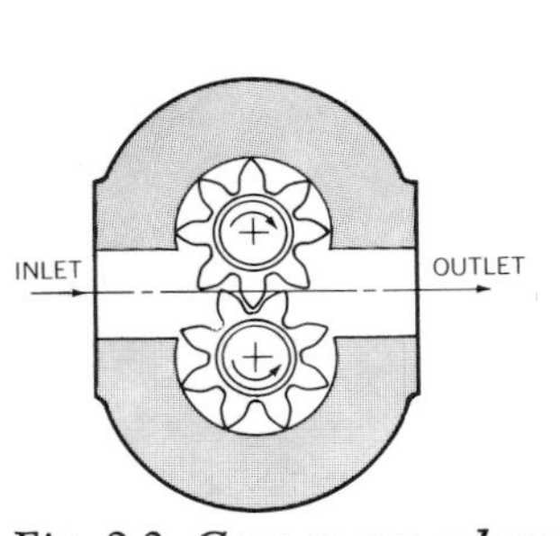

Fig. 3.2. Gear pump schematic.

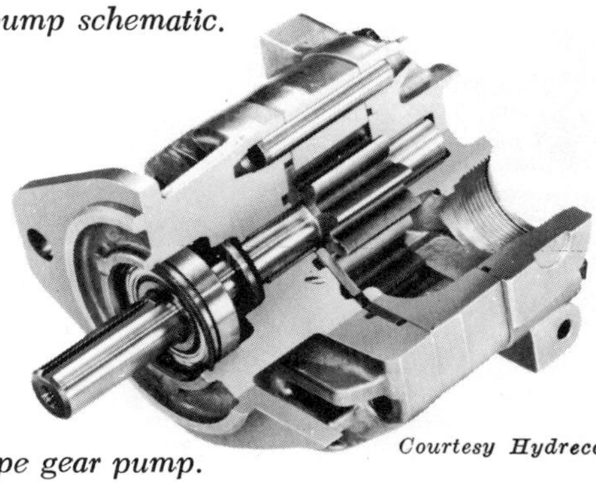

Fig. 3.3. Spur-type gear pump. *Courtesy Hydreco*

As applied to vane and gear pumps, the Hagan-Poiseuille law states that, even when clearances d are made small, a short sealing length l tends toward large leakage flows. Piston pumps minimize this difficulty by relatively deep insertion of a pumping piston into its cavity.

3.3. FIXED DELIVERY PISTON PUMPS

All piston pumps operate by allowing oil to flow into a pumping cavity as a piston retreats and then forcing the oil out into another chamber as the piston advances. Design differences among pumps lie primarily in the methods used to convert rotating input shaft energy into translatory piston energy and in methods of separating inlet from outlet oil.

In one design, Figure 3.4., input shaft rotation rotates the pistons and a piston block. An angle exists between the input shaft axis and the piston block axis. This angle causes the pistons to move in and out of the piston block as the assembly rotates. Valving on the block allows this action to transfer oil from a port connected to the retreating half of the rotation to a port connected to the advancing half of the rotation. The quantity of oil transferred (pumped) per revolution is a function of the diameter of the pistons, the number of pistons, and the angle of the drive.

In another version of this design, Figure 3.5., there is no angle between the input shaft and the piston block. Here, each piston is connected to a pad which slides on a swash plate which is angled with respect to the drive axis. As the piston block rotates, the pad, sliding on the swash plate, moves the piston in and out of the block. Valving is similar to that in the angled pump.

In a different design, Figures 3.6. and 3.7., the drive rotates a swash or cam plate. This plate bears on pads attached to the ends of pistons. The pistons and block do not rotate, and the swash plate forces the pistons into the block. A nutating plate which rides on the piston pads forces the pistons back out as the swash plate retreats. Flow porting and separation are performed on each piston. Part way along the retreating stroke, the piston uncovers an inlet port. During the pumping stroke the piston covers up the inlet port and subsequent motion forces oil past check or one-way valves and into the outlet port.

One of the primary design differences between piston pumps and vane or gear pumps lies in the function of the housing. In a vane or gear pump the housing normally defines the volume available for introducing or expelling oil. In a piston pump much of the housing merely contains the operating mechanism and separates the mechanism from the ambient atmosphere. Oil which leaks past the pistons is trapped in the housing. This oil is often used as the lubricant for the mechanism. Obviously, an outlet for this oil must be provided or it would eventually reach the system operating pressure. A case drain port is then

Fig. 3.4. Bent axis-type axial piston pump. *Courtesy Vickers Inc.*

Courtesy Hydreco

Fig. 3.5. Swash plate-type axial piston pump.

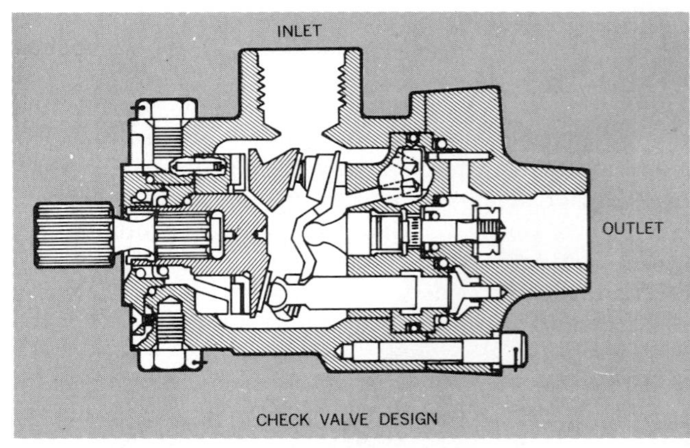

Fig. 3.6. Schematic of checkvalve pump.

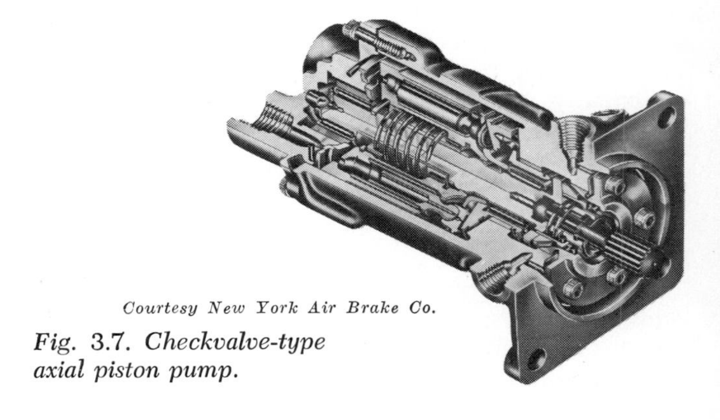

Courtesy New York Air Brake Co.

Fig. 3.7. Checkvalve-type axial piston pump.

provided to allow excess leakage oil to be returned to the sump or reservoir. The check valve pump of Figures 3.6. and 3.7., does not require a case drain.

The case drain oil absorbs most of the heat generated by friction in the mechanism and by windage from the rotating parts. Therefore, this oil is often passed through a heat exchanger and cooled before it returns to the reservoir.

A rotating and pulsating device such as a pump operating against high pressure generates metallic particles due to wear. Many of these particles pass out of the case with the case drain oil. For this reason, many system designers place a filter in the case drain line.

3.4. VARIABLE DELIVERY PUMPS

If means were developed to vary the effective pumping stroke of an axial piston pump, the output flow could then be modified. If this flow modifying device were designed to be responsive to the pressure in the system being supplied, the pump could be made to act—within design limits—as a source of flow at a constant pressure.

This sort of control is achieved by using system pressure in a piston to balance a preloaded spring. The spring preload represents the desired pressure in the system. The difference between the piston force and the spring preload will displace the spring and this displacement is used to modify the flow output of the pump. The controlled flow, when driven into the impedance of the system, tends to hold the system pressure at the desired value.

In bent axis pumps, variable delivery can be achieved by changing the angle of the drive (Figure 3.8). The quantity of fluid pumped per stroke is directly proportional to the *sine* of the angle between the drive shaft axis and the cylinder block axis. As the angle goes to zero, flow is reduced to zero. If the cylinder block is swung past the zero flow point, the flow will reverse as the inlet and outlet ports change function.

Another method of achieving variable volume in a bent axis pump is by varying the position of the porting or vave plate relative to the piston assembly. The valve plate in Figure 3.9. can be rotated to allow more or less fluid to be pumped. For example, were the kidney shaped ports to be at the top and bottom rather than at the sides as shown, the pump would produce no net flow.

Swash plate pumps as shown in Figure 3.5. can be converted to variable delivery pumps by a mechanism which varies the angle of the swash plate. If the swash plate is at a right angle to the drive axis, Figure 3.10., no fluid will be pumped. If the swash plate is tipped, pumping occurs. One company builds such a pump in which the flow is controlled by varying the angle of the swash plate. The pump of another manufacturer uses a fixed angle swash plate and controls flow by means of a variable valve plate.

Figure 3.11. shows a pressure-compensated variable delivery pump in which the fluid is pumped out past check valves. Variable delivery is achieved by varying the effective pumping stroke. Hollow pistons pass through bushings. Until cross-ports in each hollow piston are sealed by the bushing, no pumping action can occur. By moving the bushings axially along the piston, the effective pumping stroke can be varied from zero to full rated stroke.

One of many methods can operate the variable delivery controls an these pumps, for instance: handwheel, electric motor, or servo actuator. In pressure-compensated variable delivery pumps, flow is controlled by a device which senses system pressure, compares system pressure with a reference—usually a compressed spring—and modifies the flow to main-

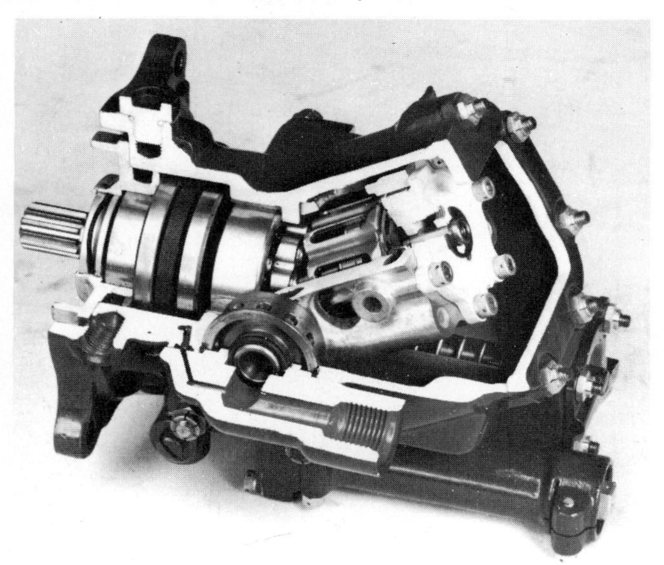

Courtesy Vickers Inc.

Fig. 3.8. Variable delivery bent axis pump.

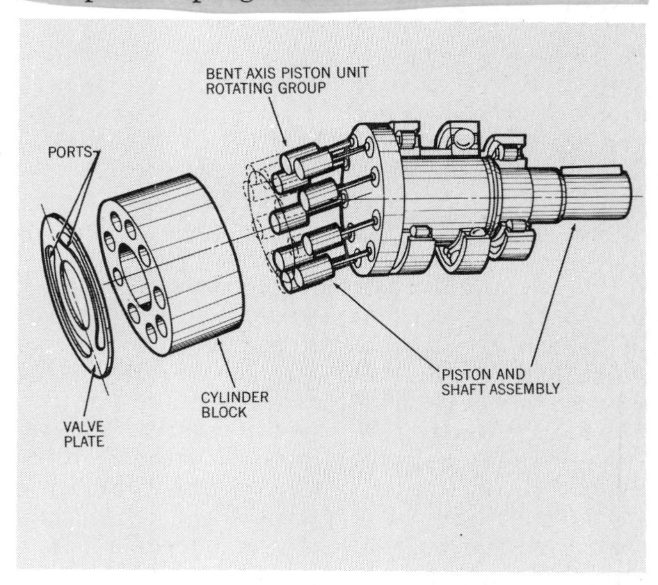

Fig. 3.9. Pump delivery is varied by rotating valve plate.

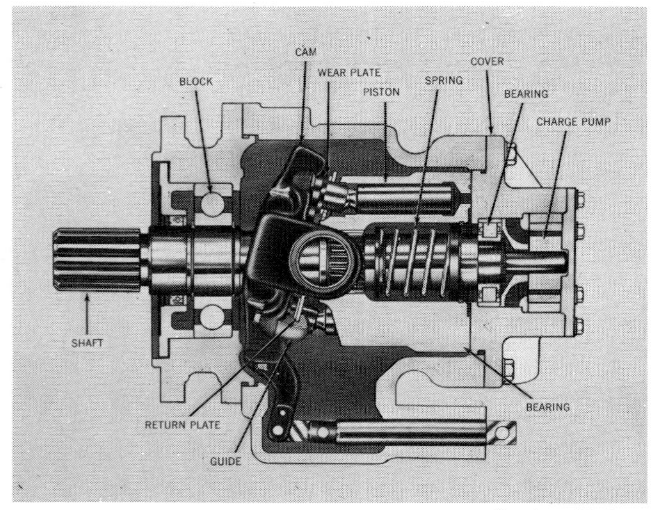

Courtesy Hydreco

Fig. 3.10. Variable delivery swash plate pump.

Fig. 3.11. Pressure-compensated variable delivery pump.

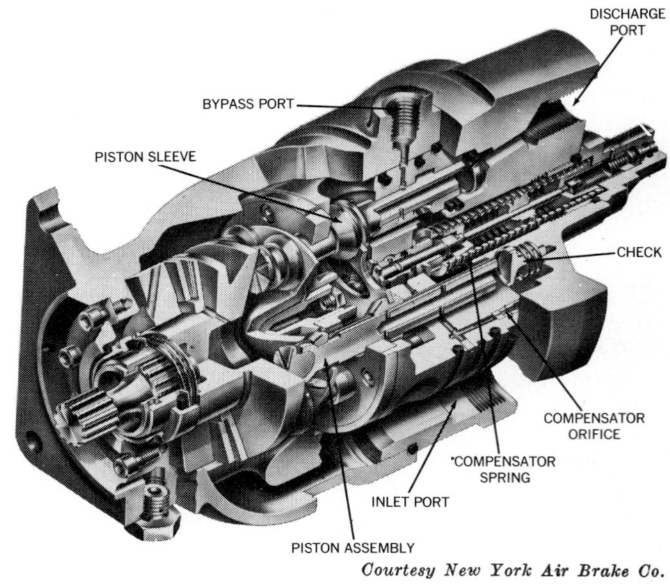

Courtesy New York Air Brake Co.

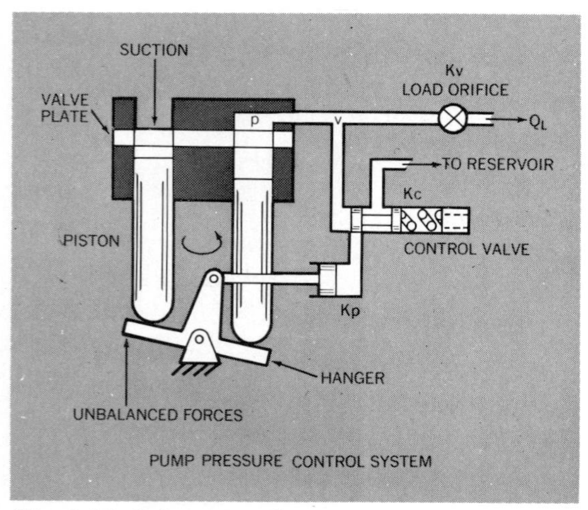

Fig. 3.12. Schematic of pressure compensator.

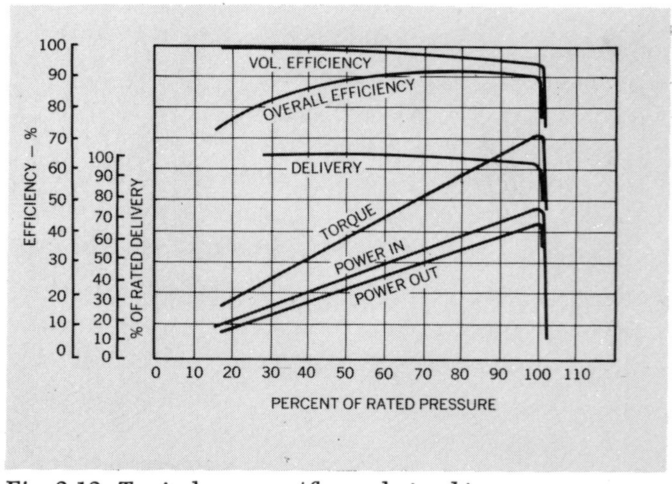

Fig. 3.13. Typical pressure/flow relationship.

tain system pressure at a constant value. In Figure 3.11. system pressure is ported to a small piston on the pump centerline. The force generated by the piston is balanced by springs. Deflection of the springs controls the position of the bushings which regulate flow.

A pressure compensating control technique used by one manufacturer is shown schematically in Figure 3.12. The dynamics of systems used by other manufacturers are similar. Flow from the pumping pistons is converted to pressure at the load. This pressure is imposed on a spring-loaded spool in a compensator valve. The adjustable preload on the spool determines the desired system pressure. If system pressure is lower than desired, the spring in the compensator valve moves the spool to port control pressure to reservoir. Unbalanced forces on the hanger will then move the hanger, increasing flow. Conversely, excessive system pressure will move the valve spool, porting system pressure to a control actuator. This fluid pressure, acting on the control actuator, will move the hanger to reduce flow. In the absence of any pressure, the control system will move the hanger to its full flow position, assuring proper pump start-up.

Figure 3.13. shows typical pressure-flow relationships for pressure compensated variable delivery pumps.

3.5. PUMP/SYSTEM INTERACTION

Frequently, hydraulic system designers choose off-the-shelf pumps with little concern other than supplying sufficient flow at available input power. Early emphasis that positive displacement pumps supply only flow and that pressure is developed by the system suggests that, as a minimum, the dynamic relationship between system and pump should be investigated. In truth, the pump should be chosen in light of several overall requirements and with system detailed design and the nature of the working fluid well in mind.

Positive displacement pumps generate flow. In a fixed delivery pump, provisions must be made to dissipate flow or system pressure will rise until a rupture occurs. The usual means of accomplishing flow control is to place a relief valve in the high pressure line. When the pressure rises above an established amount, the relief valve will vent excess flow back to the reservoir. In such systems, pump flow and relief valve capacity must be carefully matched to assure proper venting.

Flow from a high pressure line through a relief valve to a low pressure element is wasted hydraulic horsepower, which can be calculated from the following relationship:

$$hp = \frac{\Delta P\, Q}{1714} \qquad 3.2$$

Where: Q = *flow in gpm*

This wasted horsepower is converted to heat in the hydraulic system. If not properly removed, the heat can damage the fluid, elastomer seals, and other organic material in the system.

Pressure-compensated variable delivery pumps do not require a relief valve in the high pressure line. The pressure compensation feature eliminates the need for the relief valve. In nearly all working systems, however, at least one is used on a just-in-case basis. The use of a pressure compensator, while avoiding dependence on a relief valve, brings on its own problems. The actuator-spring-spool arrangement in the compensator is a dynamic, damped-mass-spring arrangement. However, when the system calls for a change in setting, the response *cannot* be instantaneous. The speed with which the change is made—and whether oscillations will occur—will be determined by the natural frequency of the compensator

$$f = \frac{1}{2\pi}\sqrt{\frac{SPRING\ RATE}{MASS}} \qquad 3.3$$

as modified by the coulomb friction of the moving metal parts and the viscous friction imposed by the oil viscosity. These response characteristics should either be matched to design system demand rates or the designer should use techniques for supplying oil to the system on a transient basis.

The hydraulic system has a comparable set of response characteristics. The oil in the system possesses mass, is compressible and acts like a stiff spring. The tubing and other pressure-containing elements are made of elastic materials. Internal flow friction provides damping. Therefore, even a pump which reacted instanteneously could not cause an equally instantaneous change in system pressure. In reality, pump operating characteristics cannot be specified unless the dynamic nature of the system is defined.

Positive displacement pumps do not generate smooth, uniform flow. Each piston transfers a discrete quantity of oil from the inlet to the outlet side.

Each slug or quantum of oil transmitted from the low pressure side of a pump and injected into the high pressure system causes a pressure wave to travel down the system. Skaistis[2] and his associates have shown that pumps with an even number of pistons create a pressure ripple which has the same frequency as the pumping frequency (the rate of revolution x the number of pistons) and that pumps having an odd number of pistons create a pressure ripple having a frequency *twice* that of the pumping frequency. However, the ripple frequency is masked by pressure pulses from another source.

In all practical pump designs, the pump must port the fluid to the high pressure side before maximum compression has been achieved. Because system fluid is at a higher pressure than the fluid in the advancing cylinder port, oil will flow from the system into the cylinder. Then, as final compression is achieved, the piston reverses this flow and forces all the oil into the high pressure system. This flow pattern is shown in Figure 3.14a.

Figure 3.14b. shows the reduction in this violent flow reversal which can be achieved by using metering holes or grooves to control the back flow. If the porting is so delayed that a high level of compression of the oil occurs before communication with the high pressure system, the flow pulse can be further reduced as shown in Figure 3.14c. These flow pulses result in very noticeable pressure pulses in the system and occur at the pumping frequency. These pressure pulses are the primary source of hydraulic system noise.

Usually the pumping pulses occur at a frequency too high to excite resonance in hydraulic system components. As an example, a nine-piston pump operating at 2750 rpm will generate pressure pulses at 9 (2750/60) = 412.5 Hz. Rarely are hydraulic system components responsive to such a frequency. Occasionally, however, two or more pumps may be placed in a single system. In such a case, a small difference in drive speed can develop pressure pulse beat frequencies which can severely damage tubing and equipment.

Pumps can be very sensitive to fluid viscosity. Many aircraft have hydraulic systems which must operate at —65 F. Should this same aircraft have to fly at supersonic speeds, the hydraulic oil temperature might reach 400 to 500 F. Few oils are both fluid enough at —65 F and viscous enough at 450 F to be suitable for pumping. With normal design clearances between pumping piston and bore, a fluid having a viscosity lower than 0.0015 Newts will leak badly, give low efficiency, be a poor lubricant, and contribute to pump overheating. On the other end of the thermal scale, a fluid having a viscosity in excess of 20 Newts will not flow into the pumping cavity on the pump suction stroke unless forced in by a supercharge pump or reservoir pressurization.

Pumps are the primary source of energy in the hydraulic system. Unless the pump is designed efficiently, it will waste energy and heat-up the system. Each of these effects can be extremely undesirable. Oil is raised approximately 8 F when compressed from 1 atmosphere to 3000 psi. This heat is recoverable. Non-recoverable heat results from turbulence in the suction line, flow through tortuous internal flow paths, and excess leakage along the pistons. A rise in temperature from the inlet port to the case drain line is often used as a measure of the inefficiency of a pump.

Cavitation is occasionally encountered in pumps which are either poorly designed or poorly matched to the system they drive. True cavitation occurs when the local pressure of a system is lowered below the vapor pressure corresponding to the oil temperature. Bubbles of gas form in the oil. When the pressure rises during a pumping cycle, the bubbles collapse. Local high fluid velocities due to rapid bubble collapse can break metal particles away from metering edges and pumping parts.

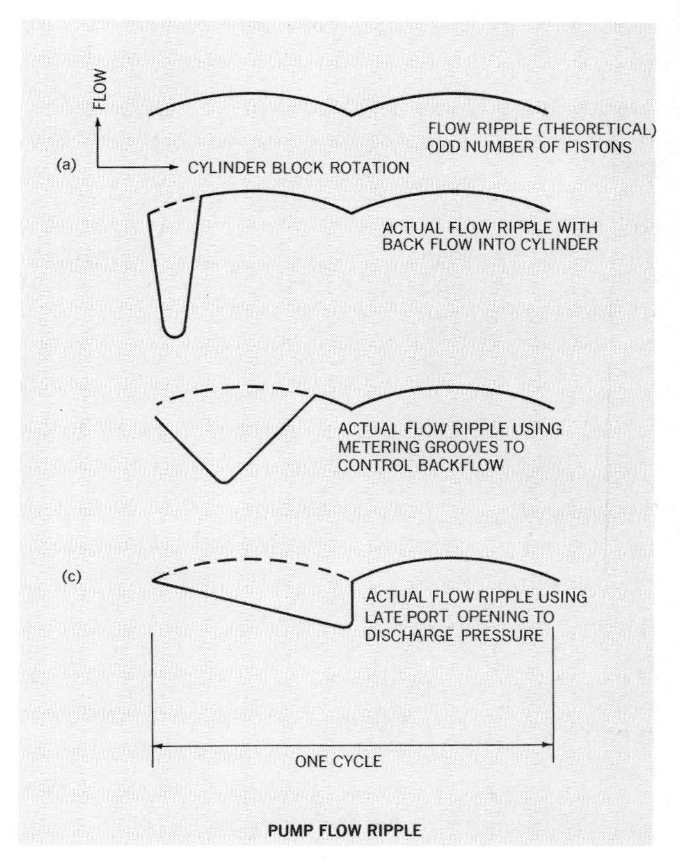

Fig. 3.14. Pump flow ripple patterns.

3.6. PUMP ANALYSIS

Hydraulic positive displacement pump design is still an inexact science. Although many design details of such pumps have yet to be studied, it is possible to establish certain parameters and use them as guides to understand pump operation. Assuming no losses, the flow rate of a pump is

$$Q = dN \qquad 3.4$$

Where: Q = *flow rate*
d = *volumetric displacement per revolution*

and shaft input power is

$$H_s = 2\pi TN \qquad 3.5$$

Where: H_s = *shaft input power*

The power added to the fluid being transferred from a region at pressure P_1 to a region of pressure P_2 is

$$H_o = (P_2 - P_1)\,Q \qquad 3.6$$

Where: H_o = *output power*
Q = *flow rate*

still assuming no losses $H_s = H_o$, the above equations can be combined.

$$T = (P_2 - P_1)\frac{d}{2\pi} \qquad 3.7$$

Of course actual pumps have losses. These include forces involved in the viscous shear of the hydraulic oil in the clearances between pistons and bores, T_v, friction forces in mechanical elements which are directly related to the pressure differential, T_f, and friction forces which are independent of speed and pressure differential, T_c. These forces can be stated as input torque losses.

$$T_a = T_i + T_v + T_f + T_c \quad 3.8$$

Where: T_a = *actual input torque*
T_i = *ideal lossless input torque*
T_v, T_f, T_c = *torques associated with above losses*

and

$$E_T = \frac{T_i}{T_a} \quad 3.9$$

Where: E_T = *torque efficiency*

From a flow standpoint, similar losses can be identified. Among these are viscous leakage flow Q_v which is proportional to the pressure differential, a loss in delivery due to cavitation or to the solution and compression of entrained gasses Q_r, and the leakage flow from high pressure to the case Q_d.

$$Q_P = Q_T - Q_v - Q_r - Q_d \quad 3.10$$

Where: Q_P = *actual flow*
Q_T = *ideal lossless flow*
Q_v, Q_r, Q_d = *flow associated with above losses*

$$E_V = \frac{Q_P}{Q_T} \quad 3.11$$

Where: E_V = *volumetric efficiency*

$$E_O = \frac{H_O}{H_S} \quad 3.12$$

Where: E_O = *overall efficiency*

The basic operating pump equations, as given by Wilson[3], are

$$Q_P = dN - \frac{C_S\, d\,(P_2 - P_1)}{\mu} - Q_r \quad 3.13$$

Where: C_S = *slip coefficient*
$= \frac{K_1 b^3}{d}$
K_1 = *slip constant*

and

$$T_a = \underset{\text{(Pumping Torque)}}{\frac{d\,(P_2 - P_1)}{2\pi}} + \underset{\text{(Viscous Losses)}}{\frac{C_d\, d\, \mu N}{2\pi}} + \underset{\text{(Coulomb Losses)}}{\frac{C_f\, d\,(P_1 - P_2)}{2\pi}} + T_c \quad 3.14$$

Where: C_d = *viscous drag coef.*
$= K_2 N r^2 l w / b$
K_2 = *drag constant*
C_f = *dry friction coefficient*
T_c = *torque loss constant*
l = *length of leakage path*
b = *radial width of leakage path of C_S as defined in 3.13*

The use of these equations involves the determination of the several coefficients. Reference 3 contains some typical values.

3.7. DYNAMIC CHARACTERISTICS OF VARIABLE DELIVERY PRESSURE COMPENSATED PUMPS

Pump pressure control is highly non-linear over its operating range. Therefore, analyses are restricted to small perturbations about a quiescent condition if they are to be linear. The following analysis was made by P. M. Lawhead of Abex Corporation. It is reprinted by permission.

This analysis considers only the idealized pump control circuitry with a simple variable orifice type load. All elements have been linearized by taking small perturbations in the variables about a fixed quiescent value.

Figure 3.12 is a schematic of the pump control under consideration. Delivery from a pump driven at a constant speed is determined by the pump displacement which is varied by changing the hanger angle. The compensator valve ports fluid either to or from a control piston on the hanger, depending upon pressure error about a preset pressure. Thus, if the pressure is high, fluid is passed to the control piston, reducing pump displacement. If pressure is low the valve passes fluid from the control piston to the case, increasing pump displacement.

Flow equations:

$$Q_{\cdot P} = Q_P - Q_L - Q_C - Q_O \quad 3.15$$

Where: $Q_{\cdot P}$ = *compressive flow between the pump and the load due to rate of change of pressure*
Q_P = *output flow of the pump*
Q_L = *leakage flow* $= Q_v + Q_d$
Q_C = *control flow past the compensator valve*
Q_O = *flow past load valve*

$$Q_{\cdot P} = \frac{V}{\beta} P s \quad 3.16$$

Where: V = *volume under compression*
β = *effective bulk modulus of oil*
P = *pump output pressure*
s = *LaPlace operator*

$$Q_O = C_V A_L \sqrt{P_O} \quad 3.17$$

Where: C_V = *loading valve orifice discharge coefficient*
A_L = *loading valve metering area*

in differential form

$$\Delta Q_O = \frac{C_V A_L \Delta P_O}{2\sqrt{P_O}} + C_V \sqrt{P_O}\, \Delta A_L \quad 3.18$$

Where:

$Q_L = C_L P \quad C_L =$ *pump leakage coefficient* 3.19

$Q_C = G_C P \quad G_C =$ *compensator valve gain* 3.20

$Q_P = Nd$ 3.4

Expressing all the equations in their differential form but dropping the differential notation, the pump flow equations can be summarized as follows. (The subscript o on P denotes the variable at its quiescent value.)

$$Q_{\cdot P} = Q_P - Q_o - Q_L - Q_C \qquad 3.15$$

$$= \frac{V}{\beta} P_S \qquad 3.16$$

$$Q_P = Nd \qquad 3.4$$

$$Q_o = C_V \sqrt{P_o}\, A_L + \frac{C_V A_L}{2\sqrt{P_o}} P \qquad 3.18$$

$$Q_L = C_L P \qquad 3.19$$

$$Q_C = G_C P \qquad 3.20$$

Pump displacement is the integral of the flow past the compensator times a constant dependent upon pump geometry. This may be expressed as:

$$d = \frac{G_P Q_C}{s} \qquad 3.21$$

$G_P =$ *pump gain (rate of change of pump displacement per control flow to actuating piston.)*

(This assumes a simple integration with no pressure feedback from hanger dynamics.)

The simultaneous solution to the above equations is expressed in transfer function form

$$\frac{P}{C_V A_L \sqrt{P_o}} = \frac{\frac{1}{N G_P G_C} s}{\frac{s^2}{\omega_n^2} + \frac{2\zeta}{\omega_n} s + 1} \qquad 3.22$$

$$\text{Where: } \omega_n = \sqrt{\frac{\beta N G_P G_C}{V}}$$

$$\zeta = \frac{1}{2}\left(\frac{C_V A_L}{2\sqrt{P_o}} + C_L + G_C\right)\sqrt{\frac{\beta}{V N G_C G_P}}$$

3.8. SYSTEM DAMPING

System damping is primarily a function of three coefficients, two of which will vary considerably, depending on the steady state operating point.

The quantity $\frac{C_V A_L}{2\sqrt{P_o}}$ will vary from a relatively large value when the load valve is fully open (i.e. when A_L is large) to zero at no flow condition.

The pump leakage coefficient C_L is a constant. This term is very small and for all practicable purposes can be neglected.

The compensator valve gain G_C is the primary source of damping for the zero flow mode of operation. This coefficient varies from zero to a large value, depending on the compenator valve position. Assuming zero leakage in the control circuit, G_C will be zero for negative pressure transients. (i.e. pressure drops below the quiescent pressure level). This is true because the control fluid back of the actuating piston is dumped to the pump case rather than back into the output flow. This non-linear damping term is, probably, the predominate source of limit cycling.

Damping for both *plus* and *minus* pressure changes can be accomplished by biasing the compensator valve to one side of null by building a leakage path in the control circuit. This is most usually done with flats across the actuator.

This method of providing pump damping compromises the control system somewhat, but in most cases not seriously. The addition of a leakage path in the control circuit introduces a pressure feedback term to the pump displacement equation. In effect, this changes the steady state response, resulting in a pressure droop from zero to maximum delivery.

In addition, this leakage varies inversely as a function of viscosity, and thus damping can also be a function of oil temperature.

3.9. NATURAL FREQUENCY

The pressure control system natural frequency is determined not only by pump mechanics, (i.e. G_C, G_P, and N), but is also tied into the load characteristics. The effective fluid-bulk-modulus-to-volume-ratio β/V can be varied drastically, depending upon the high pressure line sizing, and compliance. An accumulator between the pump and its load, for example, can easily change the natural frequency by a factor of ten. Entrained air can have a similar effect.

3.10. ILLUSTRATIVE EXAMPLE

Estimate the dynamic characteristics of a variable delivery pump having the following characteristics and parameters:

1. Speed—260 $\frac{\text{rad}}{\text{sec}}$ (2500 rpm)
2. Pressure—300 psi
3. Equivalent fluid bulk modulus—0.18 x 10^6 psi
4. Volume under compression—10 in^3
5. Maximum displacement—0.064 in^3/rad. (0.4 cipr)

A typical pump of this size will have a pump constant G_P of about 1.4 (i.e. the change in pump displacement in in^3/rad. is 1.4 times the change in volume back of the control piston.)

The compensator valve gain constant G_C depends on the steady state operating position of the valve with respect to its null position. For a 3/32-inch diameter metering hole with an 1/8-inch valve displaced to make up .2 in^3/sec., leakage past the stroking piston the compensator valve gain is about .012 in^3/sec/psi.

Solving for the natural frequency and damping ratio at zero flow we obtain the following:

$$\omega_n = \sqrt{\frac{\beta N G_P G_C}{V}} = \sqrt{\frac{(.18 \times 10^6)\ (260)\ (1.4)\ (.012)}{10}}$$

$$= 280 \ rad./sec.$$

$$\zeta = \frac{1}{2} G_C \sqrt{\frac{\beta}{V\ N\ G_P}}$$

$$= \frac{1}{2}(.012) \sqrt{\frac{.18 \times 10^6}{(10)\ (260)\ (.012)\ (1.4)}} = 0.38$$

This damping ratio is computed for zero flow. At maximum flow the term $C_V A_L / 2\sqrt{P_0}$ for a pump of this kind will be about .003 in^5/sec lb which would increase the damping ratio about 25%. The leakage term C_L is of the order 3×10^{-4} in^5/sec lb and can thus be neglected.

A major point to remember is that pump dynamics cannot be separated from system dynamics. The pressure control system natural frequency ω_n is determined not only the pump design characteristics G_C, G_P, and N, but also by the effective bulk modulus β_e and the fluid volume under compression V. The effective bulk modulus includes the fluid bulk modulus, tubing compliance, the amount of entrained air in the fluid, and the size of any accumulators in the pressure part of the system. One pump manufacturer measured a pump control natural frequency of 13 Hz. when the system contained 3 gallons. When the pressure side volume was raised to 10 gallons, the natural frequency dropped to 8 Hz. With a 30-gallon system the pump control natural frequency was 5 Hz.

3.11. DETECTING PUMP FAILURE

Because the pump is the element of a hydraulic system which has the highest rate of relative movement of wearing metal parts, it is a major source of system contamination. When a pump fails, it contaminates the system. Therefore, many system designers are concerned with the detection of incipient failure of pumps. Among pump characteristics which might be used to detect such oncoming failures are:

- Discharge flow
- Case drain flow
- Pump vibration
- Discharge contamination
- Discharge temperature rise
- Case drain temperature rise
- Discharge pressure

A drop in discharge flow could be a good indicator of wear in a fixed delivery pump. However, because a compensated variable delivery pump modifies its output flow in accordance with system demand, a change in output flow could not be detected unless it were compared with some measure of the flow controlling device.

Case drain flow is a good measure of incipient failure. Undue wear of high-pressure leakage surfaces results in increased case drain flow. Even wear between the piston end and slipper shoe or between the shoe and its wear plate can result in increased leakage into the case. Of course, check valve pumps may not have any case drain flow.

A skilled technician can often forecast a pump failure by listening to it. Some laboratory tests indicate that monitoring sound frequencies in the band above 3KHz might allow for automatic detection of incipient failure. One problem is that the vibration characteristics appear to be susceptible to great variation from relatively small effects such as temperature, included air, etc.

Experience relating contamination to pump performance has been accumulated through the requirement for patch tests to be made on all production hydraulic pumps. These patch tests essentially compare the contamination accumulated on inlet, discharge, and case drain filters over a prescribed period of time under fixed operating conditions. By comparing the amount and nature of contamination accumulated on these patches with a standard established for each model of pump, a skilled observer can predict an incipient pump failure or other discrepancy in pump performance. Some device which could compare system contamination against a standard might be usable as a failure warning device. This might be an optical comparison or one based on the change of resistance to flow of a fine screen or filter medium.

The inefficiency of a pump can be determined by the amount of input heat rejected by the pump to the hydraulic fluid.Thus an impending failure which decreases torque or volumetric efficiency of the pump would increase the temperature of the oil. The amount of heat rejected to the fluid can be measured by either the temperature rise from inlet to discharge or from inlet to case drain. However, the temperature rise in the discharge line would be relatively small so that a few degrees error in measurement would be a large percentage of the total heat rejection.

Small and fairly constant flow from the case drain provides a better place to measure heat rejection. However, case drain temperature is affected by ambient temperature, discharge pressure, and pump operating characteristics. With typical temperature rises of 30 to 100 F, a measurement of case drain ΔT modified by environment and established for each pump design, might be a good monitor of incipient failure.

An abnormal change in output pressure, either up or down, is also an indication of some type of pump or system failure. This is easy to measure. This measurement might well be augmented by an analysis of pump output wave form.

3.12. ACCUMULATORS

Hydraulic accumulators store oil under pressure. They do this by letting pressurized oil lift weights, compress springs, or compress gas volumes. Then, if the system pressure lowers below the pressure potential of the energy storing device, the weight drops, the spring extends, or the gas expands. Oil will be expelled from the accumulator under pressure.

Most hydraulic accumulators use a volume of gas—either air or nitrogen—as an energy storage medium. The following discussion relates specifically to gas-oil accumulators of the general designs shown in Figure 3.15. These accumulators supply oil for temporary demands greater than the pump can supply. They average out high frequency demands so that the pump need not respond to load frequencies. They are used to damp pump ripples. Sometimes they are used as a prime source of hydraulic power in small sys-

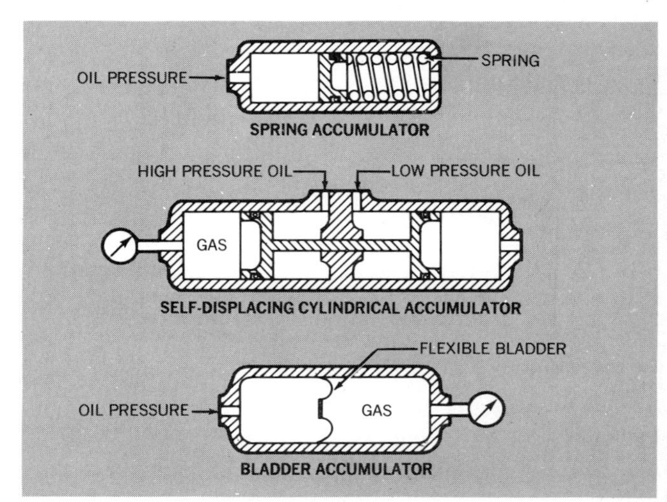

Fig. 3.15. Gas-oil accumulators.

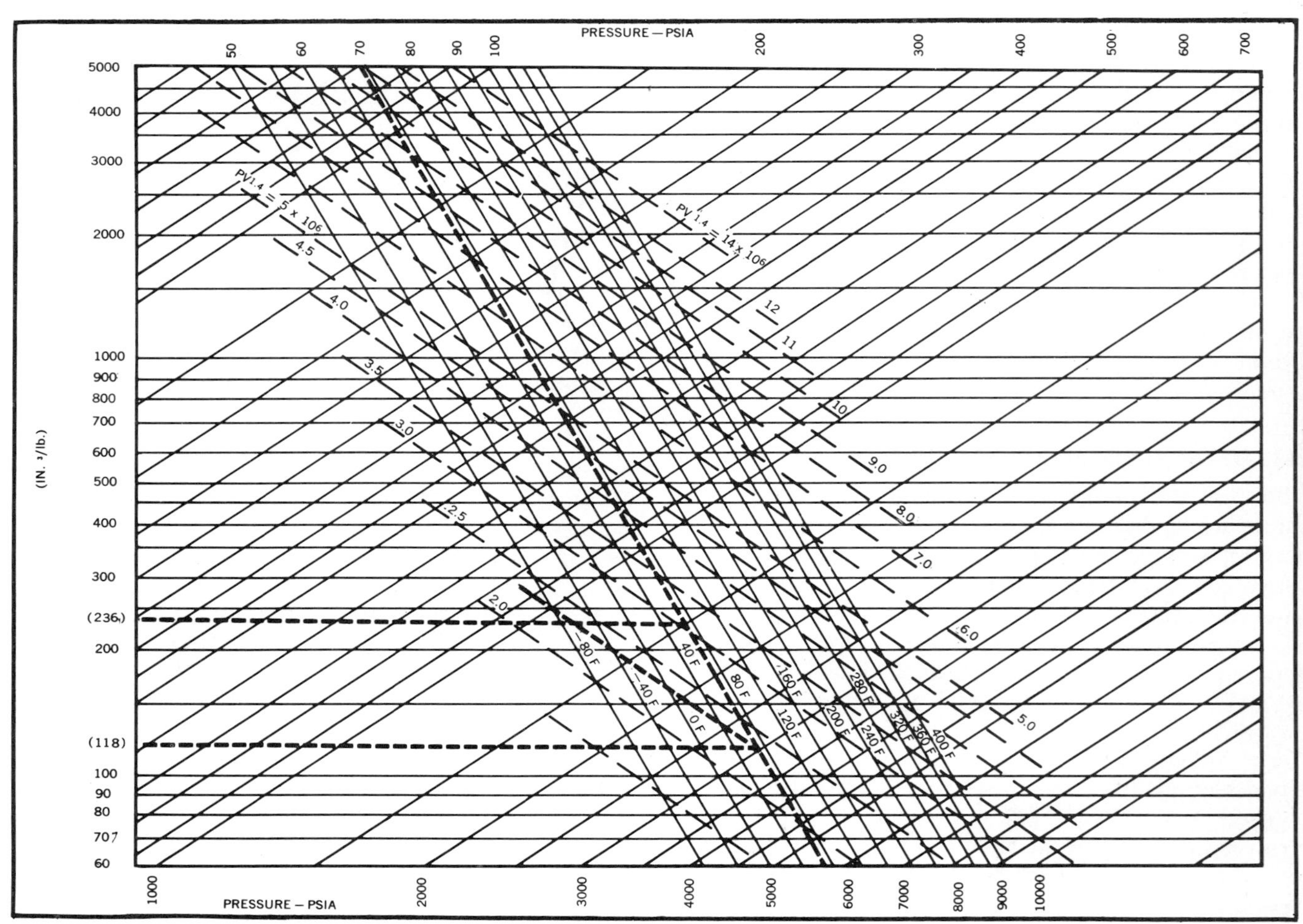

Fig. 3.16. Isothermal and isentropic chart for a perfect gas (K = 1.4).

tems of short operating life such as are sometimes required for ground-to-air or air-to-air missiles.

The key to the understanding of gas-oil accumulators lies in the study of the laws which govern the expansion and compression of the gas. The changes in state of the accumulator gas can follow any number of polytropic curves. These are bounded on one side by slow changes in which the temperature of the gas does not change as the volume changes. Such changes are called isothermal (constant temperature) and are defined by

$$P_o V_o = P_1 V_1 \qquad \Delta T = 0 \qquad 3.23$$

Because isothermal changes involve no change in temperature, no internal energy changes occur in the gas. However, work can be performed by the gas or on the gas in accordance with the general equation.

$$W = \int_{V_1}^{V_2} P\, dV \qquad 3.24$$

In the case of an isothermal change, the work can be determined by setting $PV = K$

Then

$$P = \frac{K}{V} \qquad 3.25$$

$$W = \int_{V_1}^{V_2} \frac{K\, dV}{V} \qquad 3.26$$

This is evaluated to be

$$W = K \Big|_{V_1}^{V_2} ln.\ V = K\ ln.\ \frac{V_2}{V_1} \qquad 3.27$$

In order for the internal energy to remain constant when work is done by the gas, the heat change must balance the work done

$$q = \frac{K\ ln.\ \frac{V_2}{V_1}}{J} \qquad 3.28$$

Where: J = Mechanical equivalent of heat (9340 in-lb/BTU)

The other bounding condition is a reversible adiabatic (isentropic) change which takes place so rapidly that no flow of heat into or out of the gas occurs. The defining relationship is

$$PV^k = K \qquad 3.29$$

The exponent k is the ratio of the specific heat at constant pressure over the specific heat at constant volume. This ratio is not constant over the entire operating range but is sufficiently invariable that, for engineering calculation, it can be assumed a constant. For dry air and dry gaseous nitrogen the ratio is close to 1.4 over the pressure ranges used for most hydraulic systems.

The work performed during an isentropic change is determined by evaluating P from equation 3.29. in terms of V and substituting this value into equation 3.24.

$$W = K \int_{V_1}^{V_2} \frac{dV}{V^{1.4}} \qquad 3.30$$

Upon integration this becomes

$$W = \Big|_{V_1}^{V_2} \frac{-K}{0.4\ V^{0.4}} \qquad 3.31$$

This can also be shown equal to

$$W = -\frac{P_2 V_2 - P_1 V_1}{1 - k} \qquad 3.32$$

Because no heat flow occurs during an isentropic change, the change in internal energy of the gas must equal the work done. The heat change is zero, so the internal energy change is

$$I = q - W = wC_v\ (t_{r2} - t_{r1}) \qquad 3.33$$

Where: w = weight of gas
T_R = Rankine Temperature (°F + 460)
C_v = specific heat at constant volume

Each of these bounding conditions conform to the perfect gas law which assumes a frictionless gas.

$$PV = wRT \qquad 3.34$$

R for nitrogen is 661 in lb/lb/degree F and for air is 640. This equation holds true for any type of change. Any two conditions of a given gas are related by the general equation.

$$\frac{P_1 V_1}{T_1} = \frac{P_2 V_2}{T_2} \qquad 3.35$$

In addition to isothermal and isentropic changes, gas can be expanded at constant pressure. It can undergo temperature change while held at constant volume and it can undergo rapid volume changes in which some heat flow occurs.

When heat flow occurs at constant pressure, ($\Delta P = 0$), the gas volume must vary directly with the absolute temperature (Charles' Law).

$$\frac{V_2}{V_1} = \frac{T_2}{T_1} \qquad 3.36$$

The work done, from equation 3.24 is:

$$W = P\ (V_2 - V_1) \qquad 3.37$$
$$= wR\ (T_2 - T_1) \qquad 3.38$$

The heat absorbed is:

$$q = wC_p\ (T_2 - T_1) \qquad 3.39$$

The change in internal energy is the difference between the work done and heat flow

$$I = q - W = wC_v\ (T_2 - T_1) \qquad 3.40$$

When the gas temperature is changed but the volume held constant ($\Delta V = 0$), no external work is performed. The absolute pressure varies as the absolute temperature.

$$\frac{P_2}{P_1} = \frac{T_2}{T_1} \qquad 3.41$$

The heat absorbed is

$$q = wC_v\,(T_2 - T_1) \qquad 3.42$$

and the change in internal energy is identical with the heat flow.

Actually, most of the gas changes which occur in gas-oil accumulators are polytropic. They occur rapidly, but not so swiftly but that some heat flow takes place. While not precisely true, the change can best be described by

$$PV^n = K \qquad 3.43$$

$$\text{Where: } 1 < n < k$$

Work during a polytropic change is

$$W = \frac{K}{n-1}\left(\frac{V_2^{\,n-1} - V_1^{\,n-1}}{(V_1^{\,n-1})(V_2^{\,n-1})}\right) \qquad 3.44$$

or

$$W = \frac{R}{n-1}\,T_1\left[\left(\frac{P_2}{P_1}\right)^{\frac{n-1}{n}} - 1\right] \qquad 3.45$$

The heat flow is

$$q = \frac{R}{J}\,\frac{n-k}{(k-1)\,(n-1)}\,T_1\left[\left(\frac{P_2}{P_1}\right)^{\frac{n-1}{n}} - 1\right] \qquad 3.46$$

Accumulator gas change calculations can be made by assuming the proper form of the state change and then applying one of the given formulas. This, however, is tedious. Analysis of accumulation problems can be done quickly, within engineering accuracy, by using the curves of Figure 3.16.

Figure 3.16 is a cross plot of isothermal and isentropic changes for a perfect gas having $k = 1.4$. It is, therefore, usable for both air and nitrogen over the range of pressures normally found in hydraulic systems. Consider the following example:

An accumulator having a total volume of 300 in^3 is charged with nitrogen at 1500 psi. The temperature is allowed to stabilize at 75 F. The accumulator is attached to a hydraulic system which is slowly charged to 3000 psi. The 300 in^3 is shared by the gas and the oil. A valve in the system is suddenly opened and a hydraulic motor operated until the oil supply is depleted. At 0.6 in^3/revolution, how many revolutions were made by the hydraulic motor? What was the oil pressure at the time of depletion? How much work was done?

The initial conditions are 300 in^3 of GN_2 at 1500 psi and 75 F. The intercept of the 1500 psia and the (interpolated) 75 F line is located on Figure 3.16. This corresponds to a specific volume of 236 in^3/lb. At 300 in^3, the accumulator holds 300/236 = 1.27 lbs. of nitrogen. The gas is slowly (therefore approximately isothermally) compressed to 3000 psi by the hydraulic system. By following the (interpolated) 75 F line from 1500 psia to 3000 psia, the gas is now found to have a specific volume of 118 in^3/lb. The 1.27 lbs. of gas now occupy 150 in^3 so that the accumulator also holds 150 in^3 of hydraulic oil. When the valve is opened, the oil flows to the motor and the accumulator is emptied rapidly. The expansion can be considered isentropic—or close to isentropic. By following the isentropic slope back to the original specific volume, the final pressure is found to be 1120 psia.

The work is found by integrating the flow over the pressure range. For $PV^{1.4} = K$, the constant K is evaluated by inserting the initial conditions of 3000 psi and 150 cubic inches volume.

$$(3000)\ (150)^{1.4} = 3.3 \times 10^6$$

Then the work can be determined from equation **3.31**

$$W = \frac{3.3 \times 10^6}{2.96} - \frac{3.3 \times 10^6}{3.92} = 2.8 \times 10^5 \text{ in-lbs}$$

or from equation **3.32**

$$W = \frac{(1120)\ (300) - (3000)\ (150)}{1 - 1.4} = 2.8 \times 10^5 \text{ in-lb.}$$

The temperature at the end of the isentropic expansion can be estimated from the chart as being —60 F. If nothing further is done with the system, heat from the ambient air will flow into the accumulator and raise its temperature back to 75 F. Because the accumulator is already full of gas, no expansion is possible and the change will be at constant volume. The heat required can be obtained from **3.42**

$$q = 1.27\ (0.178)\ (60 + 75)$$

$$= 30.3 \text{ BTU}$$

Another approach to an estimate of the work output of the system is to assume that a true isentropic expansion is impossible under the stated conditions. Suppose than an assumption of polytropic expansion is made where

$$PV^{1.15} = K$$

For this system, the constant K can be evaluated from the initial conditions to be 9.9 x 10^5. The work is

$$W = 9.9 \times 10^5 \left. \frac{-1}{0.15\,V^{0.15}} \right|_{150}^{300}$$

$$= 3 \times 10^5 \text{ in-lbs}$$

REFERENCES

1. Eley, James M., "The Modern Gear Pump for Earth Moving Machines", SAE Paper 680250, April 9-10, 1968.
2. Skaistis, S. J.; Kennamer, D. E. and Walrud, J. F., "Quieting of High Performance Hydraulic Systems." *Proceedings* of NCIH, Chicago, Ill., Oct. 22-23, 1964.
3. Wilson, W. E., "Positive Displacement Pumps and Fluid Motors," Pitman Publishing Co., New York, 1950.

CHAPTER 3

PROBLEMS

1. A pump having an output of 3.26 gpm at 3600 rpm is operating with a system having 45 cu. in. of MIL-H-5606 at 80°F under compression. The quiescent pressure is 3000 psi. The $\beta_e = 2.0 \times 10^5$ psi. Assume pump gain $G_p = 1.25$ and compensator gain G_c - 0.010. At zero flow what is the damping ratio ζ ? What is the pump - system natural frequency? If the load is an orifice having an area of 0.05 sq. in. and a coefficient of 0.8, what is the full-flow damping ratio?

2. A fixed delivery pump operating at 1750 rpm delivers 0.6 cipr (cubic inches per rev.). The load requires an average of 3.5 gpm. The balance of the flow goes through a relief valve set at 1500 psig to a reservoir at atmospheric pressure. How much horsepower is wasted?

3. An airplane sitting on the ground at 70°F has its accumulator charged with GN_2 at 1000 psi. The total volume is 260 cu. in. Hours later the airplane is flown and the hydraulic system rapidly pressurized to 3000 psi. In flight, the fluid temperature slowly stabilizes at 150°F. A violent maneuver requires 10 cu. in./sec of oil. The pump can supply only 2 cu. in./sec. (Assume a polytropic expansion of $PV^{1.10}$ = constant.) At the end of 10 seconds what system pressure is available. How much work is done by the accumulator supply?

4. An accumulator having a total volume of 300 cu. in. is charged with GN_2 at 1000 psi. After stabilizing at 100°F the oil side is slowly pressurized with oil at 3000 psi. Later a demand (adiabatic type) of 15 cu. in./sec. is made for 3 seconds. What is the gas temperature at the end of the three seconds? How much work was done?

5. A gas bottle holds 150 cu. in. of gaseous nitrogen (GN_2) at 300 psi. The temperature is 75°F. More gas is slowly forced into the bottle until the pressure becomes 1100 psi. What is the weight of the added gas?

6. A piston accumulator with a 4 in. diameter and a 20 in. stroke is charged with GN_2 at 1200 psi and 70°F. The accumulator is connected to a hydraulic system which operates at 3000 psi. Isothermal bulk modulus of a gas is equal to the pressure. The spring rate of an elastic column is

$$K = \frac{A\beta}{l}$$

If the piston weighs 2.5 lbs. what is the natural frequency of the accumulator, assuming weightless oil?

CHAPTER FOUR

Transmission Lines – Steady Flow

4.1. PRESSURE LOSSES

The Bernoulli equation, 1.18, has an undefined pressure loss P_f due to friction effects. This pressure loss is the sum of a number of pressure losses caused by (1) laminar shearing across adjacent layers of fluid having relative motion. (2) turbulent flow due to piping roughness, (3) turbulent losses caused by changes in flow cross section, (4) losses caused by changes in flow direction or (5) streamlined flow losses caused by orifices or chokes. These various forms of pressure losses in transmission lines will be considered in detail.

In a carefully designed hydraulic system, as much attention will go into the design, analysis and specification of the lines, tubes, hoses, pipes and fittings as into any other portions of the system. The transmission system allows the oil to flow from the power source to the control elements while minimizing energy losses, cost and weight. Given the problem of providing a line to convey oil from one part of a system to another, the designer first seeks the answers to a number of questions. Among them are:

1. What is the average flow requirement?
2. What is the peak flow requirement?
3. Considering the system purpose, should the transmission line be designed for peak flow, average flow or some other value?
4. What is the character of the flow? Will it be steady, pulsating or randomly varying?
5. What is the system pressure?
6. What elements will contribute to variations in pressure?
7. How much energy loss can be accepted in the line?
8. What are the relative importances of losses, weight and cost?
9. What are the temperature requirements?
10. Are flexible connections needed?

A knowledge of the answers to these questions will start the designer on the road to a good system configuration.

4.2. TUBING AND PIPING SIZE

The choice of the size of piping must be based on studies which balance energy loss against cost and weight. Large diameter tubing will conduct the fluid with fewer losses but will weigh more, be harder to bend, flare, weld or braze. It will also cost more.

A commonly used rule-of-the-thumb has been to design for an average flow velocity of 15 feet per second. Tubing is specified by outside diameter. Tubing wall thickness varies in accordance with system operating pressure, tubing material and other factors. Table 4-1 gives available sizes of commonly used tubing for aircraft hydraulics which provide 15 feet per second average flow velocity at the given flow rates. Table 4-2 shows the flow rates which provide flow velocities of 15 fps in steel pipe.

Of course, the rule of 15 feet per second average velocity is arbitrary and has no engineering validity. The tubing size should be chosen after studies of energy losses have been made.

4.3. PRESSURE LOSS IN STRAIGHT PIPES

Loss in pressure between two points in a straight run of tubing or piping can be determined by the Darcy-Weisbach formula

$$\Delta P = f \frac{l}{d} \frac{\rho v^2}{2} \tag{4.1}$$

This formula states that the pressure drop increases with the length of the line, with the mass density of the fluid and with the square of the flow velocity. However, increasing the piping inside diameter reduces the pressure drop. Other factors such as viscosity of the oil, roughness of the tubing and whether flow is laminar or turbulent are accounted for in the "friction factor" f. The proper use of equation 4.1 depends upon the correct determination of the friction factor.

TABLE 4-1
ALUMINUM ALLOY 6061T6 AMS4083

Tube OD	Wall-Inches	ID-Inches	Area-Sq. Inches	Flow-GPM
-4	.035	.180	.0254	1.19
-5	.049	.214	.0360	1.68
-6	.049	.277	.0603	2.82
-8	.065	.370	.1075	5.02
-10	.083	.459	.1656	7.73
-12	.095	.560	.2463	11.50
-16	.124	.762	.4208	19.65
STAINLESS STEEL TYPE 304 MIL-T-6854				
-4	.020	.210	.0346	1.62
-5	.020	.272	.0581	2.71
-6	.028	.319	.0799	3.73
-8	.035	.430	.1452	6.78
-10	.049	.527	.2181	10.19
-12	.049	.652	.3334	15.57
-16	.065	.870	.5944	27.75

TABLE 4-2

Pipe Size	Flow — Schedule 40	Flow — Schedule 80
1/8"	2.66 GPM	1.68 GPM
1/4"	4.85	3.27
3/8"	8.86	6.55
1/2"	14.25	10.90
3/4"	24.90	20.20
1"	40.50	33.50
1-1/4"	70.00	60.00
1-1/2"	95.00	82.50
2"	156.50	137.50
2-1/2"	223.00	198.00
3"	343.00	308.00

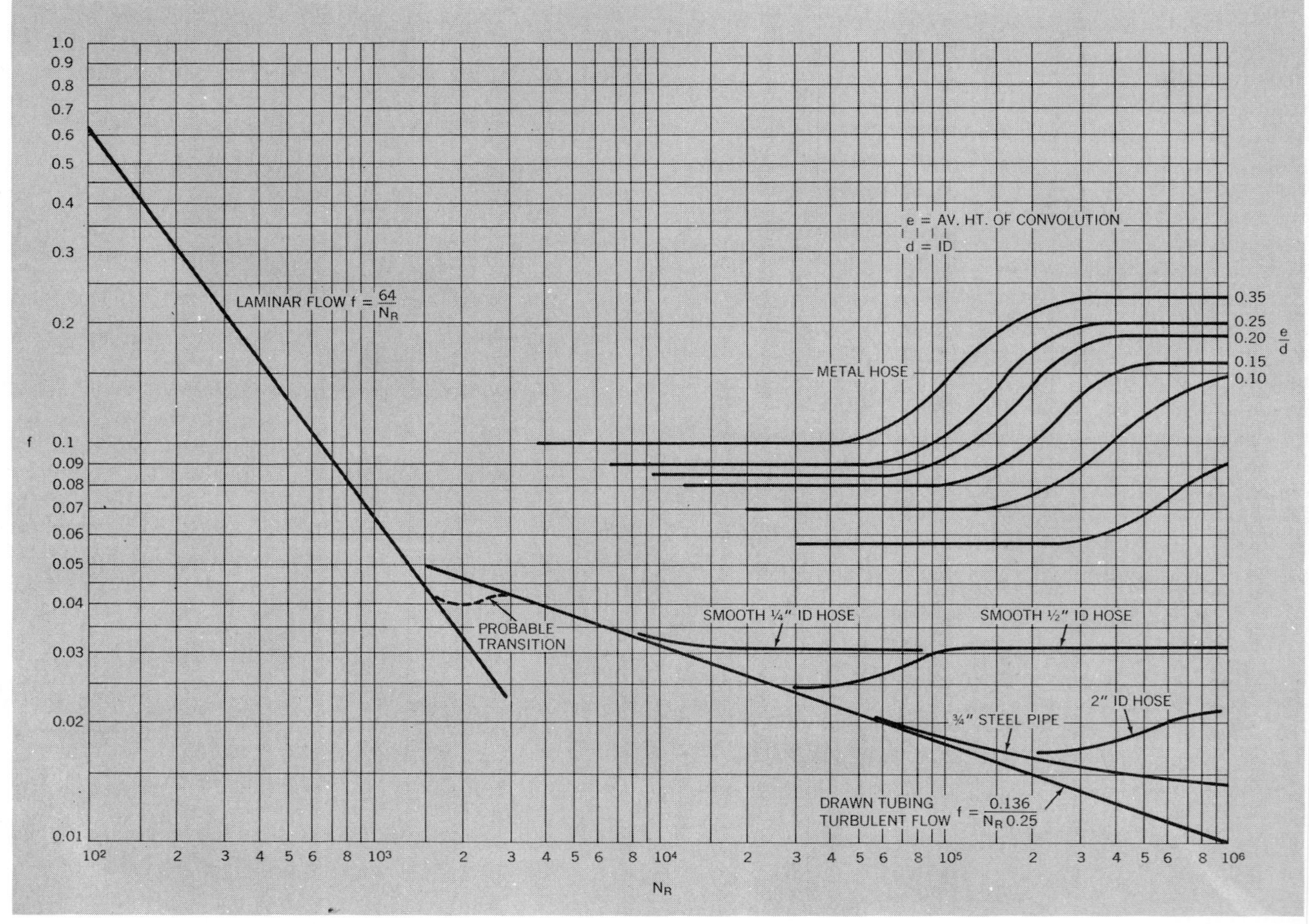

Fig. 4.1. Friction factor vs. Reynolds Number.

Fig. 4.1 shows a plot of friction factor as a function of Reynolds Number. The curve for the laminar flow ($N_R < 1400$) region is markedly different from the curves for the turbulent flow ($N_R > 3000$) region. Also, the internal smoothness of the pipe strongly affects the value of the friction factor in the turbulent region. Over much of the range of Reynolds Number of interest in power hydraulics the values of f are essentially the same for good commercial steel pipe as they are for smooth drawn tubing. Flexible rubber hose deviates from the tubing curve somewhat more than does pipe. Unlined metal hose has, through out the whole turbulent region, a much higher set of values of friction factor than any of the other transmission lines.

For lines having a circular cross-section, flow in the laminar region is usually described by the Hagan-Poiseuille law.

$$Q = \frac{\pi \, d^4 \, \Delta P}{128 \, \mu \, l} \tag{4.2}$$

If 4.2 is solved for ΔP and equations 1.9 and 1.33 combined with it, the Hagan-Poiseuille law can be written

$$\Delta P = \frac{64}{N_R} \frac{l}{d} \frac{\rho v^2}{2} \tag{4.3}$$

By comparing 4.3 with 4.1, the friction factor for laminar flow can be derived.

$$f_{laminar} = \frac{64}{N_R} \tag{4.4}$$

The curve of friction factor in the laminar region shown in Fig. 4.1 is plotted from equation 4.4.

No simple method of relating a circular cross section to other shapes appears to have validity in the laminar region. Fig. 4.2 shows flow and velocity relationships for a number of flow conditions in the laminar region.

The analytic expression for laminar friction factor 4.4 has no counterpart in turbulent flow. The most precise description of the relationship between friction factor and Reynolds Number in the turbulent region is the Prandtl formula for smooth tubing.

$$\frac{1}{\sqrt{f}} = 2 \; log \; (N_R \sqrt{f}) - 0.8 \tag{4.5}$$

This formula is difficult to solve and apply. Because of its greater simplicity the Blasius law is more frequently used for describing turbulent flow.

$$f_{turbulent} = \frac{0.316}{N_R^{0.25}} \qquad \textbf{4.6}$$

The curve for turbulent flow in smooth circular tubes shown in Fig. 4.1 was plotted from equation 4.6.

The usual method for determining the friction factor under turbulent flow conditions for non-circular cross sections is to develop an equivalent mean hydraulic radius (r_m) by dividing the flow cross section in square inches by the wetted perimeter in inches. The mean hydraulic diameter is then

$$dm = 4r_m \qquad \textbf{4.7}$$

The mean hydraulic diameter is substituted for the circular diameter in computing Reynolds Number. The friction factor is then derived from Fig. 4.1 or equation 4.6 as with circular passageways.

Equation 4.1 is commonly used to solve all steady state flow problems in power hydraulics. As an example, a need might exist to determine the pressure drop in a 1-inch OD x .049 wall tube which is 50 feet long. An industrial phosphate ester is being used at a flow rate of 30 gallons per minute. The oil temperature is 100°F.

Reynolds Number must first be computed.
ID = 1.00 — (2) (.049) = .902
Area = 0.639 sq. in.
Flow = 30 gpm = 115.5 cips
Velocity = 115.5/0.639 = 181 ips
From Figure 2-4, Viscosity = 0.14 Newts

$$N_R = \frac{vd}{\nu} = \frac{(181.9)\ (.902)}{0.14} = 1160$$

Flow is laminar
From Fig. 4.1 or equation 4.4, $f = .055$
From Fig. 2.7, the mass density is 9.9 x 10^{-5}

$$\Delta P = f\ \frac{l}{d}\ \frac{\rho v^2}{2} = (.055)\ \frac{(12)\ (50)}{.902}$$

$$\frac{(9.9 \times 10^{-5})\ (181)^2}{2} = 59.5\ psi$$

However, when either flow velocity or flow rate is unknown or if the tubing inside diameter must be determined, equation 4.1 cannot be applied directly. This is because the friction factor depends upon Reynolds Number. The computation of Reynolds Number requires a knowledge of flow rate, line diameter and fluid kinematic viscosity. The usual procedure when faced with such a problem is to assume a reasonable value of the unknown and solve the problem. The extent of error in the assumption can easily be judged and a re-trial made. By continued iteration the calculation can be made as accurate as may be desired. As an example of this technique, consider the problem just solved except that the pressure drop allowable is 6.0 psi and the proper size of tubing must be determined.

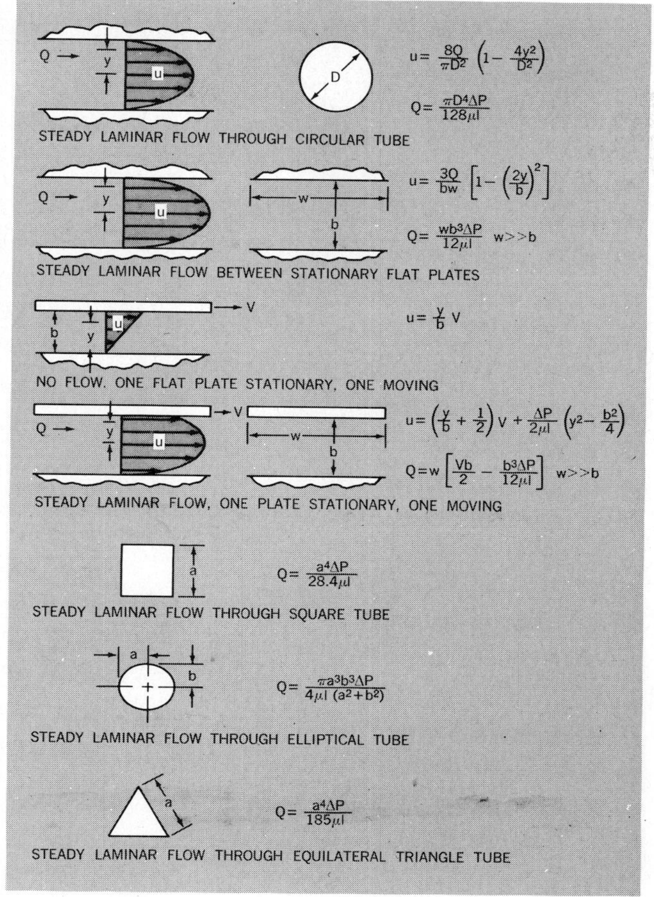

Fig. 4.2. Flow and velocity relationships.

An initial assumption of 1.00 ID is made. The area would then be 0.7854 sq in. The velocity would be

$$v_1 = 115.5/0.7854 = 147\ ips$$

$$N_{R_1} = \frac{(147)\ (1.00)}{0.14} = 1050$$

For $N_R = 1050$ $\qquad f = .061$

$$\Delta P_1 = (.061)\ \frac{(600)}{(1.00)}\ \frac{(9.9\ x\ 10^{-5})\ (147)^2}{2} = 3.28$$

Because this is less than the allowable pressure drop, the line may be uneconomically large. The next trial might assume a line ID of 0.8 inches. This yields an area of 0.503 sq in.

$$v_2 = 115.5/0.503 = 230\ ips$$

$$N_{R_2} = \frac{(230)\ (0.8)}{0.14} = 1310$$

$$f_2 = .0489$$

$$\Delta P_2 = (.0489)\ \frac{(600)}{(0.8)}\ \frac{(9.9\ x\ 10^{-5})\ \ (230)^2}{2} = 9.6$$

This is too high and the assumption of ID = 0.8 is uneconomically small. A new assumption of ID

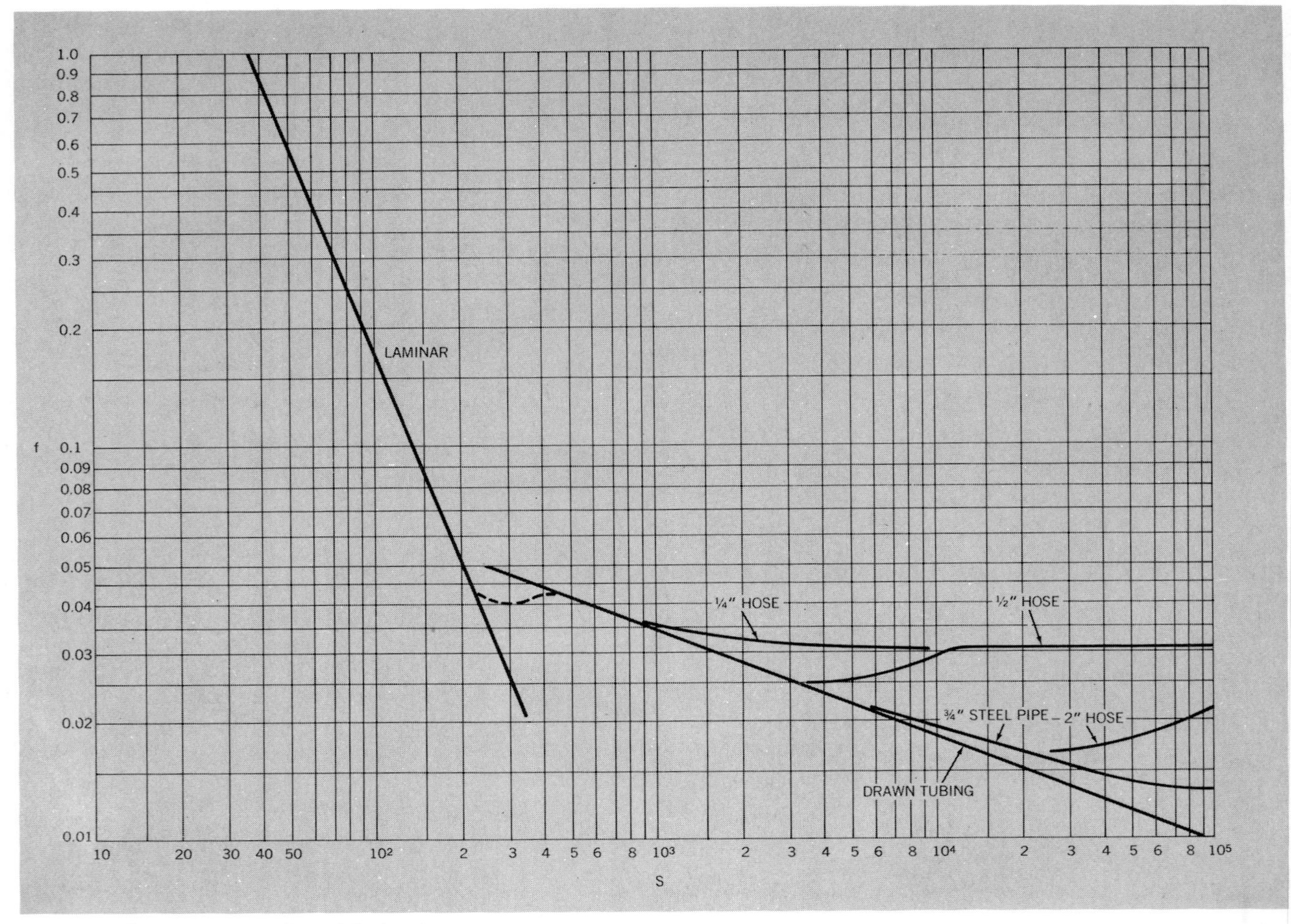

Fig. 4.3. Plot of constant S vs. friction factor.

between 0.8 and 1.0 can be made and the process repeated until the computed pressure drop is sufficiently close to the allowable 6 psi.

A more direct—though not necessarily simpler—procedure for this class of problems is to use some dimensionless ratios which were first suggested by Johnson[1]. Among these ratios, the following can have real usefulness.

$$f = \frac{\pi^2}{8} \frac{\Delta P}{l} \frac{d^5}{\rho Q^2} \qquad 4.8$$

$$S = \frac{1}{\nu}\left[\frac{\Delta P}{l} \frac{d^3}{\rho}\right]^{1/2} \qquad 4.9$$

$$T = \frac{1}{\nu}\left[\frac{\Delta P}{l} \frac{Q^3}{\rho}\right]^{1/5} \qquad 4.10$$

These can also be used to develop the following relationships.

$$S = N_R \left[\frac{f}{2}\right]^{1/2} \qquad 4.11$$

$$T = \frac{N_R}{4}\left[8\pi^3 f\right]^{1/5} \qquad 4.12$$

Figs. 4.3 and 4.4 are plots of S and T against friction factor using the relationships of equations 4.4 and 4.6 for smooth tubing.

The usefulness of these relationships can be shown by solving the problem which previously required iteration. Because viscosity, mass density and flow rate are known and the pressure drop specified, the T number can be calculated.

$$T = \frac{1}{0.14}\left[\frac{60}{600} \bullet \frac{(115.5)^3}{9.9 \; x \; 10^{-5}}\right]^{1/5}$$

$$T = 491$$

From Fig. 4.4, the friction factor can be evaluated as $f = .056$ and the problem solved by direct substitution into equation 4.3 or 4.8. Problems in which the flow rate or velocity are unknown can be solved by determining the S number and using that to evaluate the friction factor.

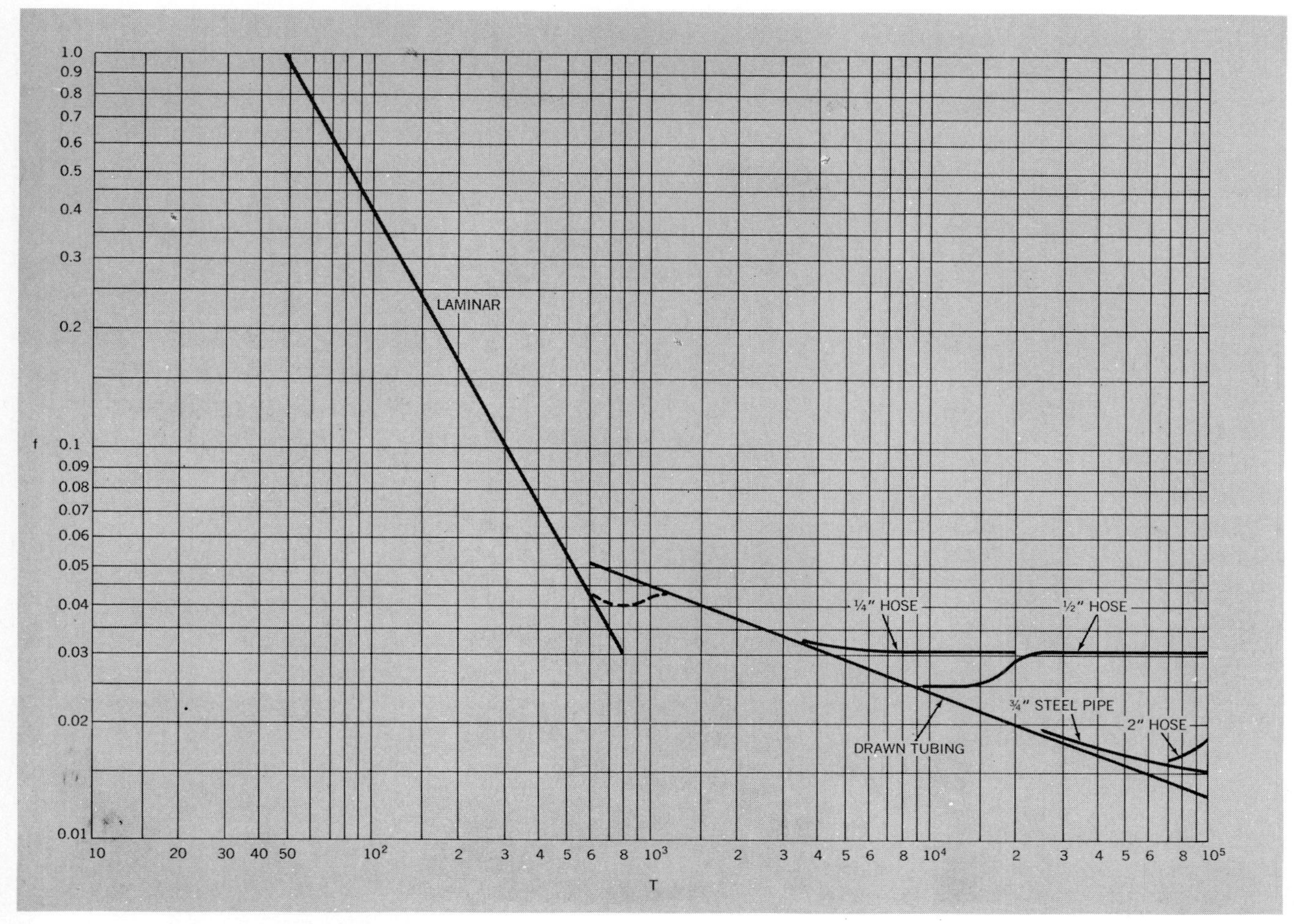

Fig. 4.4. Plot of constant T vs. friction factor.

4.4. TUBING BENDS

Not all tubing runs in straight lines. Where a bend occurs, the pressure loss will be greater than the equivalent length of straight tubing. The effect of these bends can be determined by computing the pressure drop in the bend.

$$\Delta P = \left[f\frac{l}{d} + cK_B \right] \frac{\rho v^2}{2} \qquad 4.13$$

Where K_B = correction factor for bend angle
c = resistant coefficient for 90° bends

Figs. 4.5 and 4.6 taken from the *Aerospace Fluid Component Designer's Handbook*[2] give values for K_B and c.

Assume, as an example, that the pressure drop needs to be known for a 75° bend in a ½" x .035 tube. The bend has a radius of 1.25 inches. 10 gpm of MIL-H-5606 at 80°F is flowing.
The ID is .500 — .070 = .430
Inside area = 0.145″

gpm = 38.5 cips $\qquad v = 38.5/0.145 = 265$ ips
From Fig. 2.4, $\nu = 0.03$ Newts (in²/sec)
From Fig. 2.7 $\rho = 7.7 \times 10^{-5}$
$N_R = (205)\,(.430)/0.03 = 3800$
From Fig. 4.1, $f = 0.04$
R/d = 1.25/.430 = 2.9
From Fig. 4.5, $K_B \approx 1.0$
From Fig. 4.6, $c = 0.88$

$$l = (2.5\pi)\,\frac{(75)}{(360)} = 1.64 \text{ inches}$$

The pressure drop computation is

$$\Delta P = \left[(.04)\,\frac{(1.64)}{(.430)} + 0.88 \right] \frac{(7.7 \times 10^{-5})\,(265)^2}{2} = 6.16 \; psi$$

4.5. ORIFICES

Nearly every use of orifices in power hydraulics involves *submerged* orifices in which the orifice outlet is an oil-filled tube or pipe. The basic equation for a

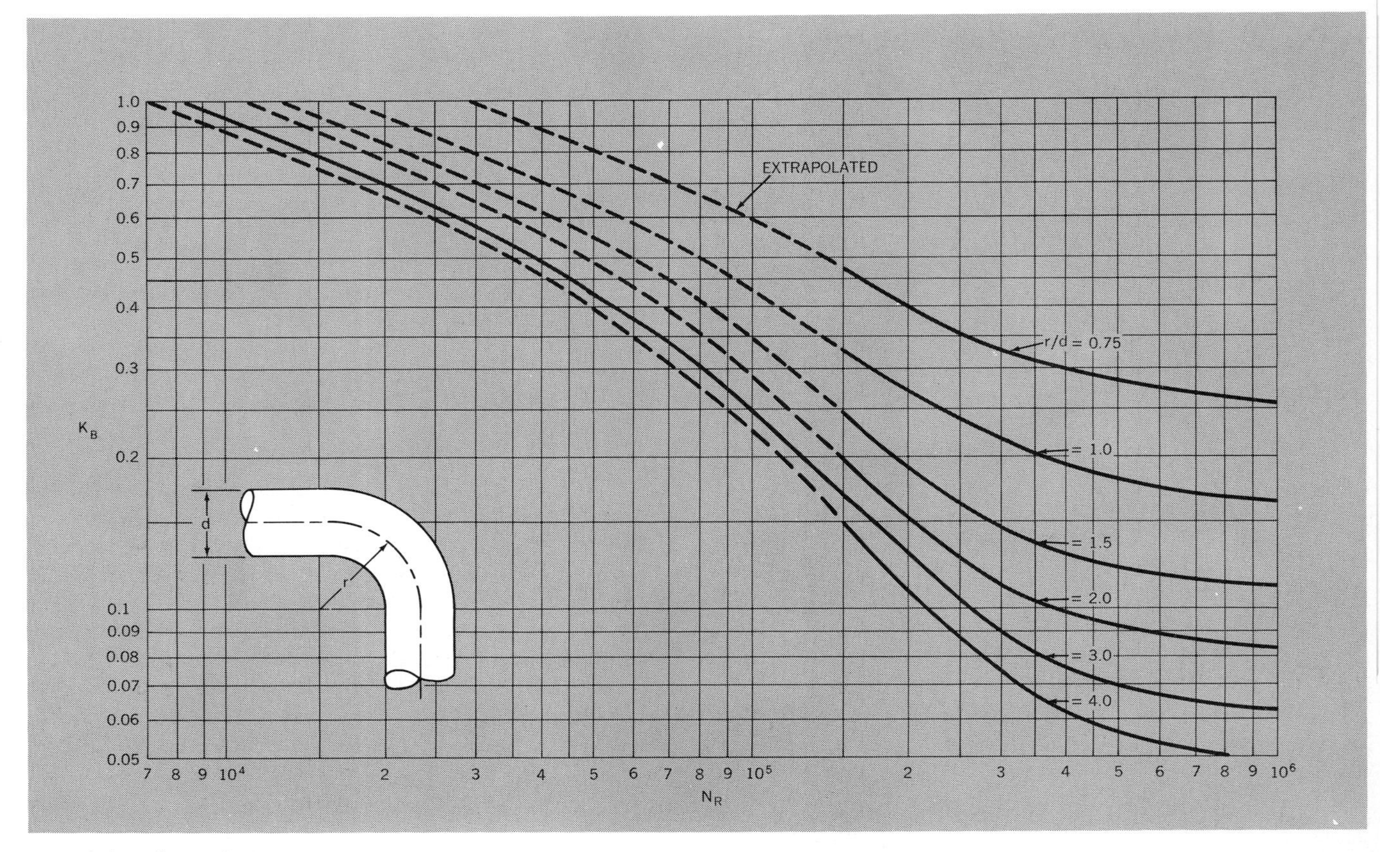

Fig. 4.5. Values of K_B

circular orifice was developed in the first chapter.

$$Q = cA\sqrt{\frac{2\Delta P}{\rho}} \qquad 1.16$$

The addition of the coefficient c is necessary to account for (1) the reduction of actual discharge area as the fluid jet continues to contract after it leaves the orifice and (2) the fluid friction developed as the flow streamlines first converge from the upstream area to the orifice area and the friction developed as the issuing jet reacts turbulently with the fluid in the downstream line. The value of orifice coefficient varies with the ratio of diameters and with Reynolds Number of the upstream flow conditions. Fig. 4.7 gives values of the orifice discharge coefficient over the range of Reynolds Number of interest to power hydraulics. For very low Reynolds Numbers ($N_R \leq 25$), Hodgson[3] gives the following:

$$c = k\sqrt{N_R} \qquad 4.14$$

Equation 1.16 can be further refined by considering the velocity pressure conditions in the orifice and in the upstream passage. If the dimensions of the orifice are small then all of the factors in Bernoulli's equation 1.18 will be essentially constant except the velocity pressure $\frac{\rho v^2}{2}$. Because velocity and pressure are convertible, the following equation must be valid as a modification of equation 1.18.

$$P_u + \frac{\rho v_u^2}{2} = P + \frac{\rho v^2}{2} \qquad 4.15$$

where: u signifies upstream

$$v^2 = v_u^2 + \frac{2}{\rho}(P_u - P) \qquad 4.16$$

Making the substitution $Q = Av$ and applying the orifice coefficient to cover the friction and area losses

$$\frac{Q^2}{A^2} = c^2\left[\frac{Q^2}{A_u^2} + \frac{2}{\rho}(P_u - P)\right] \qquad 4.17$$

Solving for Q

$$Q = \frac{cA\sqrt{\frac{2(P_u - P)}{\rho}}}{\sqrt{1 - c^2\frac{(A^2)}{(A_u^2)}}} \qquad 4.18$$

This is a slightly more accurate, though less wieldy, equation than is 1.18.

When orifices must be placed in series, Fig. 4.8, the analysis is made by using the principle of flow continuity. Unless oil is stored in the system, the flow out must equal the flow in.

$$Q = Q_1 = Q_2 \qquad 4.19$$

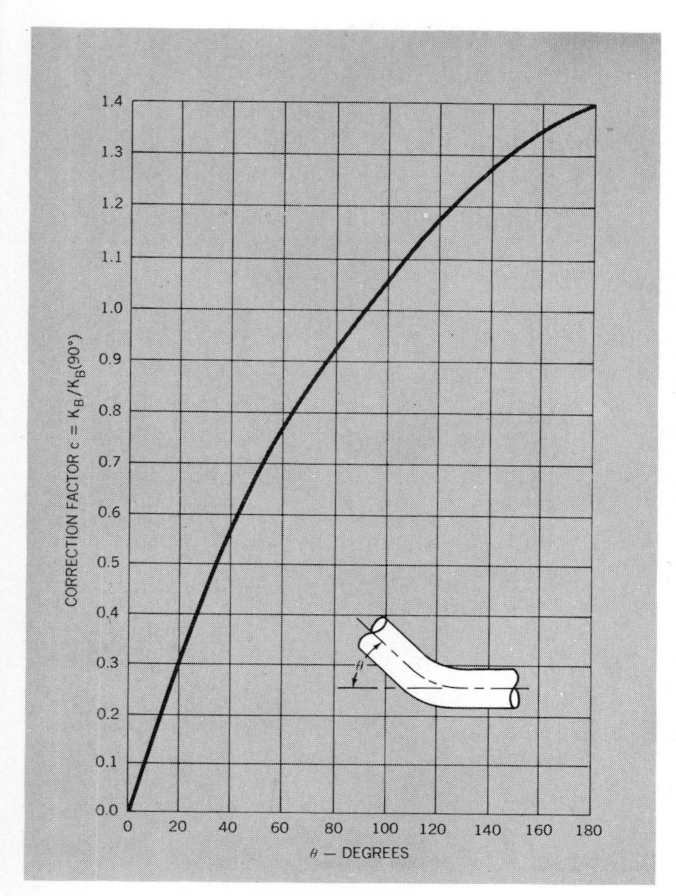

Fig. 4.6. Values of correction factor c.

$$P_0 - P_1 = \frac{\rho}{2}\left[\frac{Q_1}{c_1A_1}\right]^2 \qquad 4.20$$

$$P_1 - P_2 = \frac{\rho}{2}\left[\frac{Q_2}{c_2A_2}\right]^2 \qquad 4.21$$

$$P_0 - P_2 = Q^2\left[\frac{\rho}{2c_1^2A_1^2} + \frac{\rho}{2c_2^2A_2^2}\right] \qquad 4.22$$

$$Q = \sqrt{\frac{2(P_0 - P_2)}{\rho}}\sqrt{\frac{c_1^2A_1^2c_2^2A_2^2}{c_1^2A_1^2 + c_2^2A_2^2}} \qquad 4.23$$

Orifices in parallel are shown by similar reasoning to obey the equation

$$Q = (c_1A_1 + c_2A_2)\sqrt{\frac{2(P_0 - P_1)}{\rho}} \qquad 4.24$$

A word of caution must be inserted concerning the calculation of pressure drop in sharp edged submerged orifices. The equations presented here and those given in nearly all texts are derived from the consideration of flow issuing from a liquid container and going into an open gaseous space (free air). In submerged jets, the pressure relationships hold true only for (1) an upstream pressure measured several tube diameters upstream of the orifice and (2) a downstream pressure measured at the vena contracta. Unless some mechanism, such as a high flow demand, exists to extend the vena contracta pressure conditions downstream, the high flow velocity at the orifice will rapidly decay and the downstream pressure will build up nearly to the upstream levels.

Fortunately, the downstream conditions in most control valves meet the stated requirements quite well. The work done in a motor and the large flow capacity of motors and actuating cylinders provide a good drain for pressure side metering lands of control valves. Return side control lands usually have downstream conditions which approximate free air pressure. Therefore, as will be more fully developed in a later chapter, most control valves can be analyzed as if they were orifices.

If, however, an orifice is relatively small compared to the upstream area and the downstream flow demand is not great, pressure recovery will occur quickly. Under these conditions, the pressure loss approximates that of a sudden contraction followed by a sudden expansion. Where orifices are in series, the greater pressure drop should be taken by the upstream orifice and the orifices should be physically located close together. If the upstream orifice were the larger, particularly if the spacing between orifices were in excess of several tube diameters, pressure recovery could occur prior to the fluid's meeting the second orifice.

As the downstream pressure of an orifice is reduced, the flow rate will increase in accordance with 1.16. At some point, in this procedure, the pressure at the vena contracta will be reduced to the vapor pressure of the fluid. Further reduction in downstream pressure will then not result in further increases in flow rate. Under these conditions, gas bubbles will form in the fluid surrounding the vena contracta. If, at the time of this occurrence, the downstream conditions allow for pressure recovery above the vapor pressure, the gas bubbles will collapse. This formation of gas bubbles and the subsequent collapsing of the bubbles is cavitation.

As will be described in Chapter 8, many control valves act as if they were orifices. Cavitation in the valve can rapidly erode and destroy the metering lands of control valves. Because cavitation is a complex phenomenon, remedial action when confronted with valve cavitation may involve changing or modifying the fluid as well as making mechanical changes. Any mechanical change which reduces the pressure drop across an individual orifice will tend to eliminate cavitation. The most straight-forward method of doing this is to install valving in series so that the pressure drop across any one metering orifice will be too small for cavitation to occur.

If the downstream pressure is also below the vapor pressure of the fluid, the bubbles will persist. The downstream flow then consists of a mixture of oil and gas bubbles. This is known as two-phase flow. Because density of a two-phase fluid is lower than the density of the pure oil, the quantity flow is

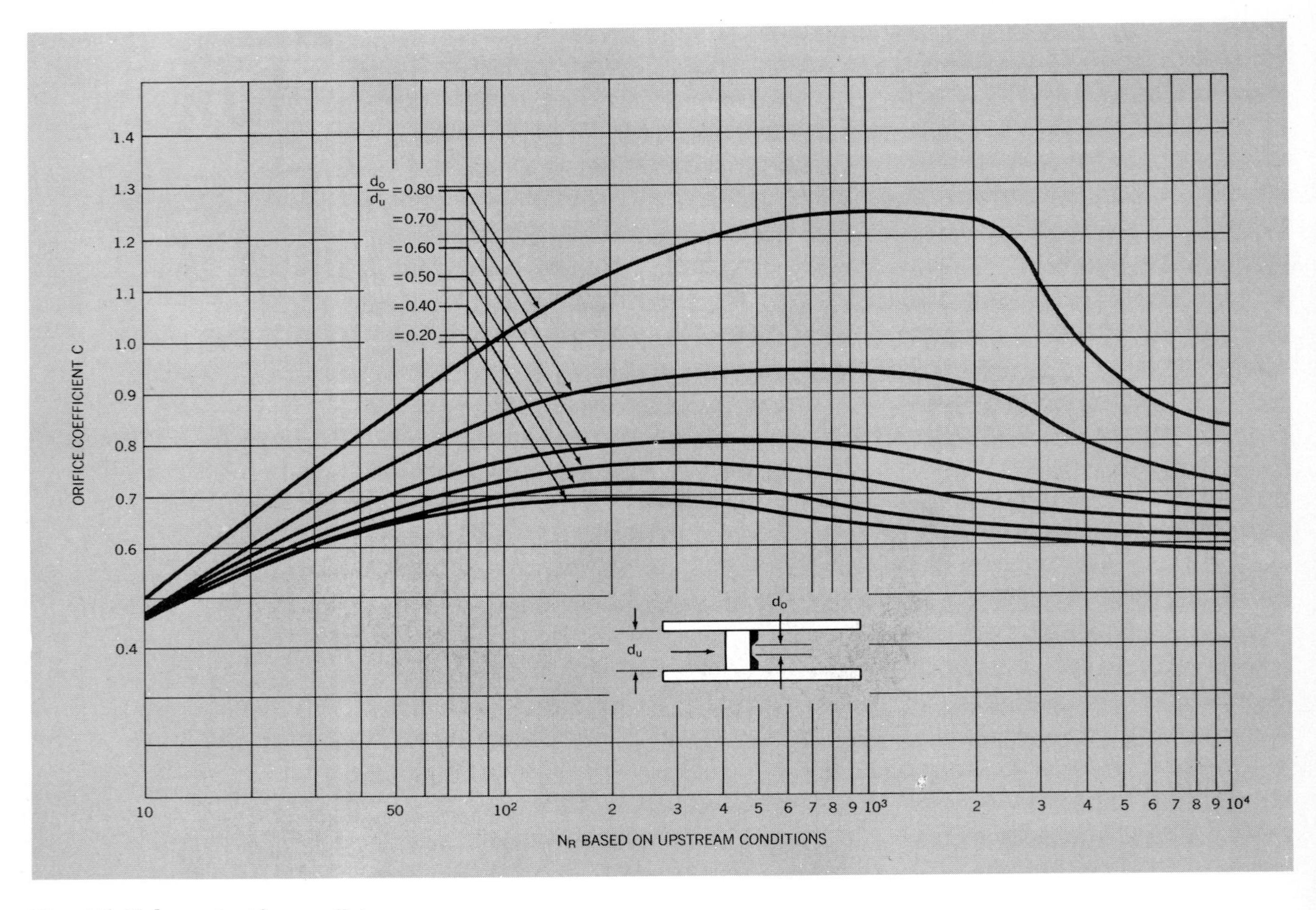

Fig. 4.7. Values of orifice coefficient.

reduced. Steady state flow of oils which are flashing to gas is unusual in power hydraulics. Hanssen[4] and Ziegler[5] have discussed the details of the design of valving for these conditions.

4.6. EXPANSION

When hydraulic oil flows from a small tube into a larger diameter tube or into a large chamber, part or all of the velocity energy is converted into pressure energy. The efficiency of this conversion is affected by the entrance conditions. The pressure change is accomplished by losses due to flow turbulence. These losses can be computed.

$$P = K_e \left[1 - \frac{(d_i)^2}{(d_o)^2} \right]^2 \frac{\rho v_i^2}{2} \qquad 4.25$$

Where:
K_e = inlet coefficient
d_i = inlet diameter
d_o = outlet diameter
v_i = inlet velocity

For convenience in computation, equation 4.25 is plotted in Fig. 4.9. The values of the inlet coefficient, K_e, are given in Table 4.4.

When the flow path is increased slowly the losses can be computed from equation 4.25 using a coefficient obtained from Fig. 4.10.

4.7. CONTRACTION

The various references differ somewhat concerning the proper computation of losses which occur during

TABLE 4-3

$\frac{d}{D}$	K
0.421	0.135
0.632	0.118
0.834	0.106

TABLE 4-4

Well Rounded	0.04
Slightly Rounded	0.23
Sharp Edged	0.485 – 0.56
Inward Projecting Pipe	0.62 – 0.93

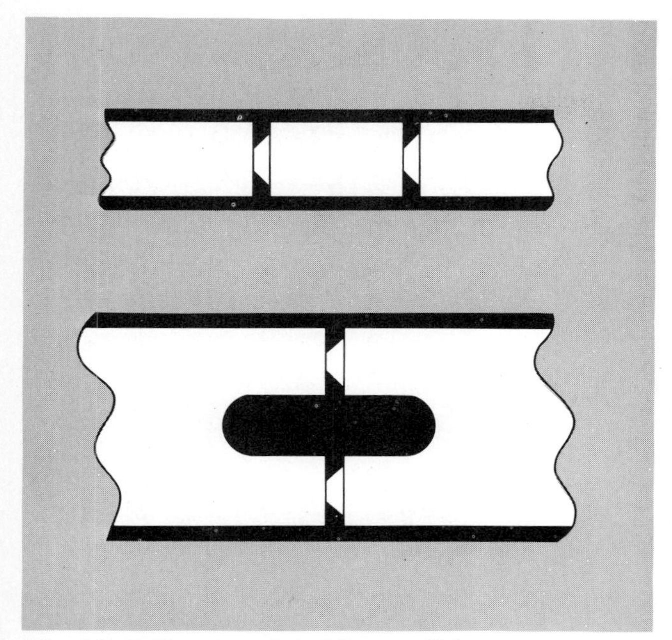

Fig. 4.8. Orifices in series and in parallel.

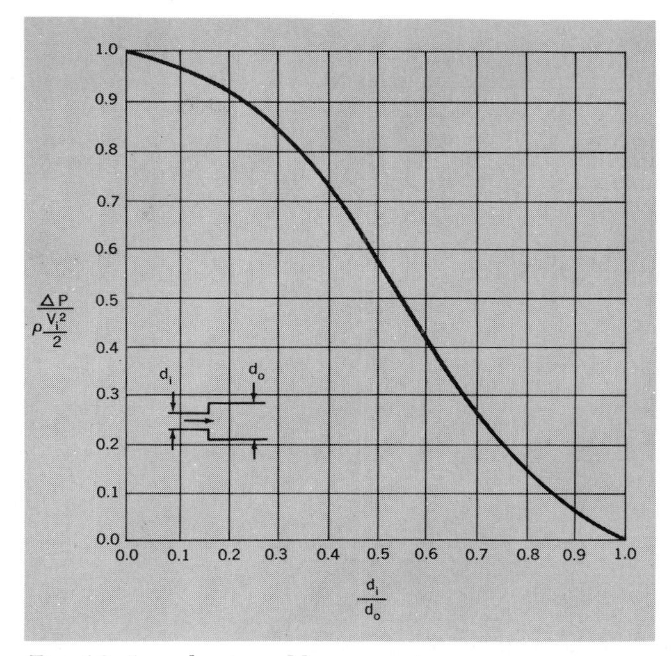

Fig. 4.9. Loss due to sudden expansion.

a contraction. Equation 4.26 appears to have validity at values of Reynolds Number greater than 5000. At $N_R < 5000$, an arbitrary increase in the computed pressure drop should be applied.

$$P = \frac{\left[1 - \frac{(d_o)^2}{(d_i)}\right] \rho \, V_i^2}{2} \qquad 4.26$$

This curve is also plotted in Fig. 4.11.

The best—though admittedly inadequate—expression for the loss in a gradual contraction seems to be

$$P = 0.04 \frac{\rho \, V_o^2}{2} \qquad 4.27$$

4.8. ANNULAR ORIFICES

Leakage paths in piston pumps, partially opened plug valves or needle valves, and quiescent leakage of some spool valves are each examples of annular orifices. The velocity profile across any longitudinal section of an annular orifice is assumed to be parabolic, Fig. 4.12. This is an extension of the flow analysis between two parallel plates. If a fluid particle in the lower half of the stream be considered, the motion of the particle in a direction to the right can only occur as a result of a differential pressure which shall be termed dp. The movement of the particle is resisted top and bottom by shear forces. Because the relative velocity between adjacent laminae of fluid decreases as the center of the flow stream is approached, the shear force on the upper surface of the particle is smaller than the shear force on the lower surface of the particle by the differential df. The slope of the force curve across

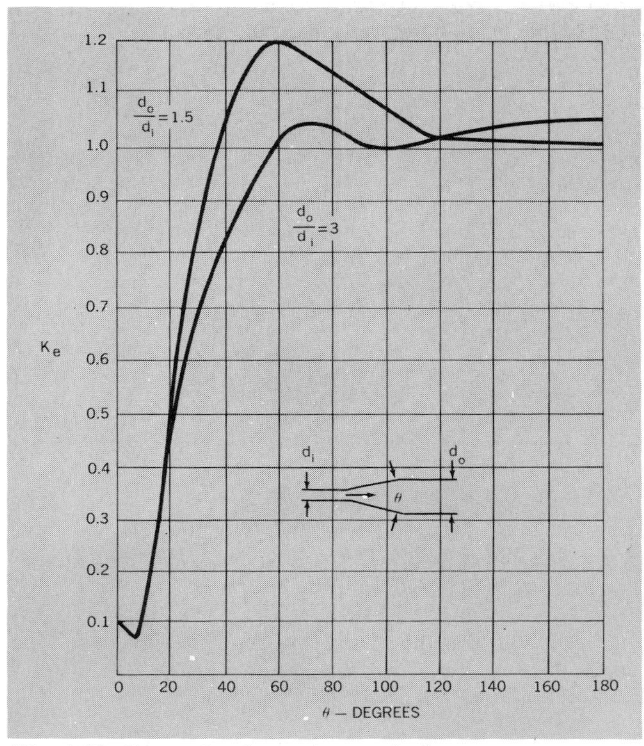

Fig. 4.10. Correction factor for gradual expansion.

the particle is

$$\frac{dy}{dx} = \frac{df}{dp} \qquad 4.28$$

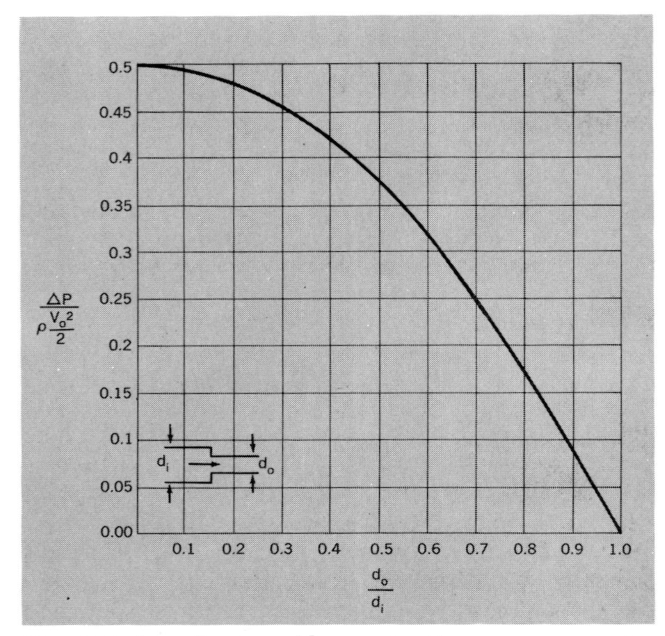

Fig. 4.11. Loss due to sudden contraction.

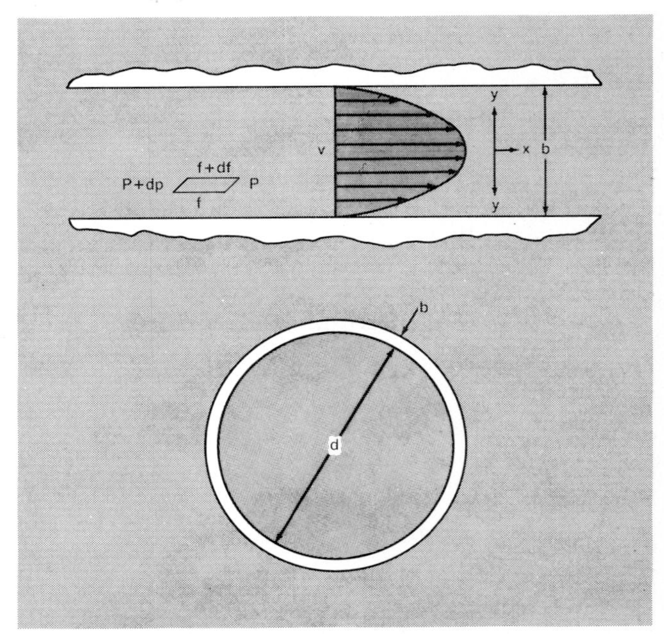

Fig. 4.12. Concentric annular orifice.

or

$$\frac{dp}{dx} = \frac{df}{dy} \qquad 4.29$$

where f is the shearing force due to viscosity. The Newtonian equation, 2.1, for viscosity can also be written

$$f = \mu \frac{dv}{dy} \qquad 4.30$$

Substituting this into 4.29

$$\frac{dp}{dx} = \mu \frac{d^2v}{dy^2} \qquad 4.31$$

integrating

$$\frac{dv}{dy} = \frac{y}{\mu}\frac{dp}{dx} + c_1 \qquad 4.32$$

If the integration be taken from the midpoint of the flow path to the walls, then

$$\frac{dv}{dy} = 0 \text{ when } y = 0, \text{ therefore } c_1 = 0$$

so equation 4.32 becomes

$$\frac{dv}{dy} = \frac{y}{\mu}\frac{dp}{dx} \qquad 4.33$$

Integrating again

$$v = \frac{y^2}{2\mu}\frac{dp}{dx} + c_2 \qquad 4.34$$

at the wall ($y = \frac{b}{2}$) the velocity is zero, so

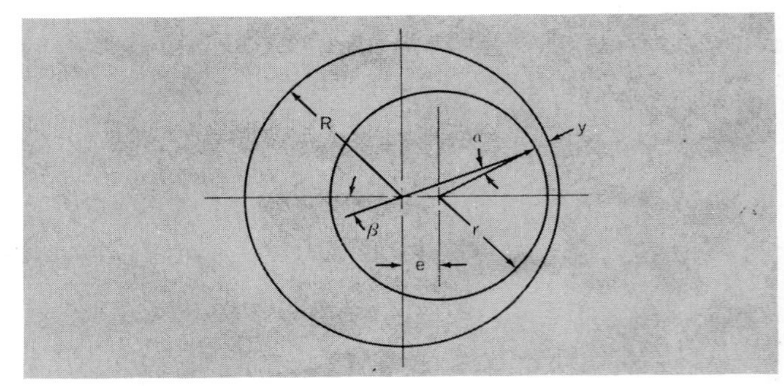

Fig. 4.13. Eccentric annular orifice.

$$c_2 = -\left[\frac{b^2}{8\mu}\frac{dp}{dx}\right] \qquad 4.35$$

therefore

$$v = \frac{1}{2\mu}\frac{dp}{dx}\left[(y^2 - \frac{b^2}{4}\right] \qquad 4.36$$

The average velocity is

$$v_{ave.} = \frac{\int_o^{b/2} y\,dy}{\frac{b}{2}} \qquad 4.37$$

$$v_{ave.} = \frac{b^2}{12\mu}\frac{dP}{dx} \qquad 4.38$$

If the clearance b is assumed to be of a constant value over a length l in the x direction, the pressure gradient will be uniform. That is

$$\frac{dp}{dx} = -\frac{\Delta P}{l} \qquad 4.39$$

Equation 4.39 then becomes

$$v_{ave.} = \frac{b^2 \, \Delta P}{12 \, \mu l} \tag{4.40}$$

Multiplying 4.40 by b, the radial dimension, and by the circumferential length πd (actually the circumferential length is $\pi(d+b)$ but this introduces needless complexity) yields the classical equation for flow through an annular orifice formed by concentric members having circular cross sections.

$$Q = \frac{\pi \, d \, b^3 \, \Delta P}{12 \, \mu \, l} \tag{4.41}$$

In this equation, the flow increases as the cube of the clearance and decreases only directly with length. This means that in reducing leakage through such a flow path, careful attention to fits and clearances yields proportionately greater reductions in leakage losses than mere lengthening of overlap or increase in leakage path length.

This analysis has, so far, assumed that the annular orifice is formed by concentric members. Because of machining irregularities, side loads, improper positioning and other effects, some degree of eccentricity must always be expected, Fig. 4.13. A piston of radius r is eccentric in a bore of radius R by an amount e. The clearance y at any point is

$$y = R - (r \cos \alpha + e \cos \beta) \tag{4.42}$$

For very small clearances, α is very small and $\cos \alpha \approx 1$. The equation can then be written

$$y = R - (r + e \cos \beta) \tag{4.43}$$

Going back to equation 4.41 for the concentric condition, substitute y, the eccentric clearance, for b, the concentric clearance For an incremental circumferential distance $R d\beta$, the incremental flow is

$$dQ = c \, y^3 \, R \, d\beta \tag{4.44}$$

Where: $c = \dfrac{\Delta P}{12 \mu l}$

substituting 4.43 in 4.44

$$dQ = c[R - (r + e \cos \beta)]^3 \, R \, d\beta \tag{4.45}$$

Reverting to earlier terminology where $R - r = b$ then

$$Q = \int_0^{2\pi} c(b - e \cos \beta)^3 \, R \, d\beta \tag{4.46}$$

Calling the relative eccentricity $\dfrac{e}{b} = \epsilon$, then

$$Q = \int_0^{2\pi} cb^3(1 - \epsilon \cos \beta)^3 \, R \, d\beta \tag{4.47}$$

integrating

$$Q = 2\pi \, Rb^3 c \left[1 + \frac{3\epsilon^2}{2} \right]$$

or

$$Q = \frac{\pi \, d \, b^3 \, \Delta P}{12 \, \mu \, l} \left[1 + \frac{3\epsilon^2}{2} \right] \tag{4.48}$$

This analysis states that when the eccentricity is maximum, $e = b$ and $\epsilon = 1$. The flow is then 2½ times greater than would be indicated by the use of equation 4.41. In practical terms, equation 4.48 states that the use of needle valves or other devices dependent upon annular orifices as precise metering orifices is dangerous unless extreme care to assume concentricity is taken. Even a slight eccentricity will make a significant change in flow conditions.

4.9. CONNECTORS

Considering the widespread use of tubing and piping connectors (commonly called fittings), surprisingly few data on pressure losses in them are available. Some experimental information was developed by Campbell[6] for AN flared connectors in military systems using MIL-H-5606 oil. These studies showed little measurable difference between the pressure loss developed in straight-through flow in unions, tees and crosses and 90-degree flow through elbows, tees and crosses. With the possible exception of threaded pipe connections, the data from Campbell are probably reasonably applicable to most SAE and bite-type connectors. Equation 4.49 is an empirical relationship derived from Campbell's plots

$$\Delta P = (1.81 - 0.01t) Q^n \tag{4.49}$$

where:
- Q = flow in gpm
- T = temperature in °F
- n = factor from Table 4-5

Straight through flow in threaded pipe connectors is best considered as a problem involving a rapid expansion and a rapid contraction. Knowing the internal diameters of the pipe and the connector, equations 4.25 and 4.26 can be applied sequentially. If the flow is bent, then some reasonable modification of equation 4.13 should be applied in addition. The following values of cK_B can be used in equation 4.13 to determine the pressure drop in some typical steel pipe fittings.

4.10. FLEXIBLE HOSE

Both straight and curved runs of flexible elastomer or PTFE lined hose can be handled as if they were rigid tubing. As shown in Fig. 4.1, such hose

TABLE 4-5

Connector Size (in 16ths)	n
-4	2.08
-6	1.22
-8	.84
-10	.63
-12	.50
-16	.35
-20	.26
-24	.20
-28	.17
-32	.14

acts like tubes or pipes until Reynolds Number exceeds about 10,000. The friction factor for sizes between ¼″ and 2″ ID can be obtained from Fig. 4.1 by interpolation for the higher values of N_R. Hose bends can be analyzed by the use of equation 4.13. Some further data on hose-coupling assemblies are available from Pollman[7]. Some feeling for the effect of couplings can be obtained from the following table which was developed from data on smooth PTFE lined hose which was supplied by Resistoflex Corporation, Roseland, New Jersey. The measurements were taken with MIL-L-7808 lubricating oil at 50 psi and 90 $\pm 10°F$. Under these conditions the oil viscosity was 0.0248 Newts and the mass density about 8.1 x 10^{-5}lb-sec^2/in.^{4}.

Metal hose which does not contain a smooth liner presents a more difficult problem. The relative roughness of the internal bellows surface sharply affects the value of the friction factor. This relative roughness is the e/d ratio shown on Fig. 4.1. These curves were taken from the work of Daniels and Cleveland[8]. The variation in friction factor due to line bends in flexible metal hose is the same as is determined by equation 4.13.

4.11. LINE ASSEMBLIES

For any hydraulic system in which the flow is uniform or steady, the pressure loss for a complete transmission line can be computed or estimated. Because pressure losses in series are additive, the pressure drop in a line assembly is the sum of the individual pressure drops along the line. The designer must remember that the pressure available at the valve or motor is the supply pressure minus the losses in both the pressure line and the return line.

For those rare cases where two lines are operating in parallel, the total pressure drop is

$$\Delta P_T = \frac{\Delta P_1 \, \Delta P_2 \, Q_T}{\Delta P_1 Q_2 + \Delta P_2 Q_1} \qquad 4.50$$

where: T is subscript for total and 1, 2 are subscripts for individual flow paths

$$Q_T = Q_1 + Q_2$$

TABLE 4-6

cK B	Fitting
2.2	Return Bend
1.8	90° Flow in Tee
0.9	90° Ell
0.4	45° Ell

TABLE 4-7

Hose Sizes	Min. ID	Flow Rate	Per Foot Hose Assembly	Per Foot Hose Lengths
-4	.212	3.85 cips (1 GPM)	4.00 psi	0.42 psi
-6	.298	3.85	0.42	0.30
		7.70	1.00	0.80
		9.63	1.58	1.10
-8	.391	7.70 (2 GPM)	0.28	0.18
		15.40	0.70	0.53
		22.10	1.03	1.11
-10	.485	7.70	0.10	0.08
		15.40	0.40	0.24
		22.10	0.85	0.48
		30.80	1.40	0.80
-12	.602	7.70	0.05	0.03
		15.40	0.10	0.09
		22.10	0.23	0.19
		30.80	0.45	0.31
		38.50	0.75	0.45
		44.20	1.10	0.62
		53.90	1.48	0.79
-16	.852	15.40 (4 GPM)	0.04	0.03
		22.10	0.05	0.04
		30.80	0.13	0.05
		38.50	0.25	0.06
		44.20	0.40	0.08
		53.90	0.55	0.10

References

1. Johnson, S. P. "A Survey of Flow Calculation Methods" Preprinted Papers and Program, ASME Summer Meeting, June 1934, pg 98.
2. RPL-TDR-64-25 "Aerospace Fluid Component Designer's Handbook" AF Rocket Propulsion Laboratory, Edwards, California, May 1964
3. Hodgson, ASME Transactions, V51, 42, 1929
4. Hanssen, A. J. "Accurate Valve Sizing for Flashing Liquids" *Control Engineering* V8 n2, February 1961, pp 87-90
5. Ziegler, J. G. "Sizing Valves for Flashing Liquids" *ISA Journal,* V9 n3, March 1957 pp 92-95
6. Campbell, J. E. "Investigation of the Fundamental Characteristics of High Performance Hydraulic Systems" USAF Technical Report 5997 W-P AFB June 1950
7. Pollman, F. "Calculating Pressure Drop in Hydraulic Hose" *Hydraulics and Pneumatics,* February 1968
8. Daniels, C. M. and Cleveland, T. R., "Determining Pressure Drop in Flexible Metal Hose" *Machine Design,* November 25, 1965

CHAPTER 4

PROBLEMS

1. 1.5 gpm of industrial phosphate ester at 80°F is flowing through a 5/8-inch *OD* x 0.035-inch wall tube. What is Reynolds Number? Is flow laminar or turbulent? What is the value of the friction factor?

2. 30 gpm of MIL-H-5606 at 100°F is flowing through a 1-inch *OD* x 0.049-inch wall tube. What is Reynolds Number? Is flow laminar or turbulent? What is the value of friction factor? If the line is 50 feet long, what is the pressure loss due to flow?

3. A flow of 20 gpm of an aircraft phosphate ester at 70°F is required. If the pressure drop due to flow must be limited to 2 psi per foot of line, what is the inside diameter of tubing which should be used?

4. A hydraulic system uses SAE 10 oil at 100°F. The piping is 3/4-inch schedule 80 pipe (approximate *ID* is 0.79 inch). The allowable pressure drop is 5 psi per foot of pipe. What flow rate will give this pressure drop?

5. Compare the pressure loss in a section of 1-inch x 0.049-inch wall tubing bent to an angle of 45° on a neutral axis radius of 2 inches with the loss in a straight section of the same size and length. $N_R = 5 \times 10^5$.

6. 1 gpm of MIL-H-5606 at 75°F is flowing through an 0.25-inch diameter orifice. The upstream tube is 1/2-inch x 0.035-inch wall. What is ΔP across the orifice?

7. 1.5 gpm of aircraft phosphate ester at 80°F is flowing in a line 5/8-inch x 0.035-inch. If a 0.050-inch diameter orifice is placed in the line, what is ΔP ?

8. A flexible hose has an end connector which screws into a 3/4-inch pipe tap having a minimum *ID* of 0.921 inches. The connector *ID* is 0.656 inch for a length of 2.88 inches. Where the connector terminates in the hose, the *ID* suddenly expands to 0.750 inch. What is the pressure drop in this end connector if 25 gpm of SAE 10 oil at 80°F is flowing through it?

9. A pump piston 0.281 inch in diameter is inserted 0.3 inch into a bore 0.283 inch in diameter. The piston axis is 0.001 eccentric with the bore axis. The fluid is MIL-H-5606 at 120°F. With a differential pressure of 2950 psi across the piston, what is the rate of leakage?

CHAPTER FIVE

Transmission Lines – Unsteady Flow

5.1. CAUSES OF UNSTEADY FLOW

Opening and shutting of valves, the ripple created by the pump, and the operation of the several actuators all cause flow to vary. Under these conditions the oil is accelerated in each direction. Because the oil has mass it tends to resist any change in velocity. Because it is compressible it can store and release energy like a spring.

Energy losses with a compressible fluid under unsteady flow are due to complex impedance rather than pure resistance. Perhaps the simplest way to understand complex hydraulic impedance is by electrical analogy. Hydraulic transmission lines, particularly under laminar flow conditions, have characteristics which are analogous to electrical resistance, capacitance and inductance. For instance, the Hagan-Poiseuille law

$$Q = \frac{\pi \, d^4 \, \Delta \, P}{128 \, \mu \, l} \qquad 5.1$$

can be written

$$P = QR_H \qquad 5.2$$

$$\text{Where: } R_H = \frac{128 \, \mu \, l}{\pi \, d^4}$$

Equation 5.2 has the same form as Ohm's law $E = IR$, so that the expression

$$R_H = \frac{128 \, \mu \, l}{\pi \, d^4} \qquad 5.3$$

can be termed the hydraulic resistance of the line for laminar flow.

When the flow is turbulent the analogy becomes less precise.

$$R_H = \frac{f \, l \, \rho}{2 \, d \, A^2} \qquad 5.4$$

where $\rho = R_H Q^2$

An expression for hydraulic inductance can similarly be developed. Consider a tube of length l and interior area A. A fluid having a mass density ρ is flowing through the tube at rate Q. If the fluid were being accelerated, the force creating the acceleration would be $F = Ma$. But

$$M = A \, l \, \rho$$

and

$$a = \frac{dv}{dt} = \frac{1}{A} \frac{dQ}{dt}$$

and

$$F = PA$$

By combining these relationships and solving for P

$$P = \frac{\rho l}{A} \frac{dQ}{dt} \qquad 5.5$$

Equation 5.5 is analogous to the expression for voltage drop across an inductance

$$E = L \frac{di}{dt}$$

so

$$L_H = \frac{\rho l}{A} \qquad 5.6$$

Hydraulic capacitance can be derived from the general expression for secant bulk modulus

$$\beta = \frac{dP}{dV/V_o} \qquad 5.7$$

or

$$\beta \, dV = V_o \, dP$$

Taking time derivatives

$$\beta \frac{dV}{dt} = V_o \frac{dP}{dt}$$

However

$$Q = \frac{dV}{dt} \quad \text{and} \quad V_o = Al$$

so

$$Q = \frac{Al}{\beta} \frac{dP}{dt} \qquad 5.8$$

This is analogous to

$$i = C \frac{de}{dt}$$

so

$$C_H = \frac{Al}{\beta} \qquad 5.9$$

In problems involving unsteady flow, the elasticity of the tubing acts as if it were compressibility of the oil. The effect of the tubing elasticity is most easily introduced by developing an equivalent fluid-tubing bulk modulus of elasticity.

$$\beta_t = \frac{t \, E}{D \left[1 - \frac{\mu}{2} \right]} \qquad 5.10$$

Where:
- β_t = tubing bulk modulus
- t = tubing wall thickness
- E = Young's modulus
- D = tubing O. D.
- μ = Poisson's ratio

$$\beta_e = \frac{\beta \, \beta_t}{\beta + \beta_t} \qquad 5.11$$

Where
- β_e = Equivalent bulk modulus
- β = Fluid bulk modulus

5.2. EFFECTIVE BULK MODULUS OF GAS-OIL MIXTURE

Dissolved gas has little or no effect on the bulk modulus of hydraulic oil. Entrained air acts like a soft spring in series with the oil column compliance. To these springs must be added the expansion, with rising pressure, of the fluid's container. The spring rate of a fluid column is

$$K = \frac{A\,\beta}{l} \qquad 5.12$$

If, in addition to compressing the oil under increasing pressure, air is compressed and the container expanded, β_e is no longer the fluid bulk modulus but an *effective* bulk modulus which takes into account all of the volume changes which occur.

Assume that an actuator piston exerts an increasing pressure on a gas-oil mixture in a cylinder. The change in volume as seen by the piston is

$$\Delta V_e = -\Delta V_g - \Delta V_o + \Delta V_c \qquad 5.13$$

Where:
g, o, c, e are subscripts referring to gas, oil, cylinder and total volumes.

From equation 5.7

$$\Delta V_e = \frac{V_e\,\Delta P}{\beta_e} \qquad \Delta V_o = \frac{V_o\,\Delta P}{\beta_o}$$

therefore

$$\frac{1}{\beta_e} = \frac{\Delta V_e}{V_e\,\Delta P} \qquad \frac{1}{\beta_o} = \frac{\Delta V_o}{V_o\,\Delta P} \qquad 5.14$$

The bulk modulus of the gas can be determined from equation 5.7 and from the change of state equations. For isothermal changes of state

$$P_{go}\,V_{go} = P_{gt}\,V_{gt}$$

Where: $P_{gt} = P_{go} + \Delta P \quad , \quad V_{gt} = V_{go} - \Delta V$

$$P_{go}\,V_{go} - (P_{go} + \Delta P)\,(V_{go} - \Delta V) = o$$

$$P_{go}\,V_{go} - (P_{go}\,V_{go} - P_{go}\,\Delta V + V_{go}\,\Delta P - \Delta P\,\Delta V) = o$$

casting out $\Delta P \Delta V$ as infinitesimal and solving

$$\beta_{g_T} = \frac{V_{go}\,\Delta P}{\Delta P} = P_{go} \qquad 5.15$$

By similar reasoning the adiabatic bulk modulus

$$\beta_{g_E} = \frac{C_p}{C_v}\,P_{go} \qquad 5.16$$

The radial displacement of a cylinder is

$$\Delta r_1 = \Delta P\,\frac{r_1}{E}\left[\frac{r_2^2 + r_1^2}{r_2^2 - r_1^2} + \mu\right] \qquad 5.17$$

The volume is $V_c = \pi r_1^2\,l$ and the change in volume is

$$\Delta V_c = \pi\,(r_1 + \Delta r_1)^2\,l - r_1^2\,l$$

$$= \pi\,(r_1^2 + 2r_1\,\Delta r_1 + \Delta r_1^2 - r_1^2)\,l$$

Eliminating and casting out the infinitesimal

$$\Delta V_c = 2\pi\,r_1\,\Delta r_1\,l \qquad 5.18$$

$$\frac{1}{\beta_c} = \frac{\Delta V_c}{V_c\,\Delta P} = \frac{2\pi\,r_1\,\Delta r_1\,l}{\pi\,r_1^2\,l\,\Delta P}$$

$$= \frac{2\,\Delta r_1}{r_1\,\Delta P} = \frac{2}{E}\left[\frac{r_2^2 + r_1^2}{r_2^2 - r_1^2} + \mu\right] \qquad 5.19$$

Where the container is a thin walled tube rather than a thick walled cylinder, the thin wall equation would be used similarly to derive the relationship already given in equation 5.10.

Because the volume change is proportional to the inverse of the bulk modulus, the total effective bulk modulus of the system is obtained by summation of inverse values. Based on equation 5.7.

$$\frac{1}{\beta_e} = \frac{V_g}{V_e}\left[\frac{1}{\beta_g}\right] + \frac{V_o}{V_e}\left[\frac{1}{\beta_o}\right] + \frac{1}{\beta_c} \qquad 5.20$$

The effect of entrained air on the effective bulk modulus can be demonstrated by assuming some typical values. Assume that MIL-H-5606 oil at 80F is in a steel cylinder under a pressure of 100 psi. The cylinder has an ID of 1.5 inches and a wall thickness of 0.1 inches. β_o under these conditions is 225,000 psi. β_g is 100 psi. Taking E for steel to be 30 million psi and Poisson's ratio to be 0.333. $\beta_c = 1.825 \times 10^6$. The bulk modulus, as a function of gas volume ratio is

$$\frac{1}{\beta_e} = \frac{V_g/V_e}{100} + \frac{1 - V_g/V_e}{225{,}000} + \frac{1}{1.825 \times 10^6}$$

The effect of varying air volume ratios, under these conditions, can be seen from the following table.

V_g/V_e	β_e
0	2.02×10^5
.005	1.84×10^4
.01	9.5×10^3
.02	5.8×10^3
.03	3.28×10^3
.05	1.98×10^3
.10	10^3

For this example, a not unusual case, 1% of air by volume entrained in the oil reduces the effective bulk modulus to only 4.6% of that of pure oil. At 5% air by volume, the bulk modulus is 0.98% of that of oil and at 10% the bulk modulus is down to 0.495% of that of clear oil. Where high bulk modulus is needed entrained air must be rigidly excluded from the working area.

5.3. EFFECTIVE BULK MODULUS OF SYSTEMS

The effective bulk modulus of elasticity of a complete transmission system can be computed from the basic relationship of equation 5.7.

$$\frac{1}{\beta_e} = \frac{\Delta V_t}{V_t \Delta P}$$

In Figure 5.1, the system contains hose, tubing, oil and gas. The total effective, bulk modulus is obtained by summing the reciprocals.

$$\frac{1}{\beta_e} = \frac{1}{\beta_t} + \frac{V}{V_e}\left[\frac{1}{\beta}\right] + \frac{1}{\beta_H} + \frac{V_g}{V_e}\left[\frac{1}{\beta_g}\right] \quad 5.21$$

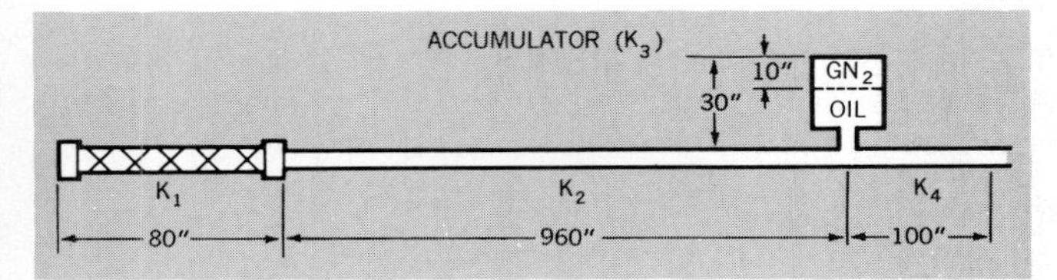

Fig. 5.1. Typical system for computing effective bulk modulus

To illustrate the use of this equation assume that, in Figure 5.1 the tube and hose each have an inside area of 0.7 in² and the accumulator has an ID of 10″ (Area = 78.5 in²). The β of tube plus oil as computed by equation (18) is 250 KPSI. β_H of the hose plus fluid is 100 KPSI. At the working pressure of 3000 psi, the adiabatic bulk modulus of GN_2 is 1.4 (3000) = 4200 psi

$$V_g = (10)\ (78.5) = 785\ in^3$$

$$V_e = 1140\ (0.7) + (20)\ (78.5) + 785 = 3155\ in^3$$

$$V_H = 80\ (0.7 = 56\ in^2$$

$$V_{oil} = 1060\ (0.7) + 20\ (78.5) = 2312\ in^3$$

$$\frac{1}{\beta_e} = \frac{785}{3155}\left[\frac{1}{4200}\right] + \frac{56}{3155}\left[\frac{1}{100{,}000}\right] + \frac{2312}{3155}\left[\frac{1}{250{,}000}\right]$$

$$\beta_e \approx 1.685 \times 10^4\ psi$$

With the expanded concept of *effective* bulk modulus in mind, Equation 5.9 can now be rewritten

$$C_H = \frac{A\ l}{\beta_e} \quad 5.22$$

If the flow is assumed to vary sinusoidally, equation 5.2 must be modified to

$$P = QZ_o \quad 5.23$$

Where Z_o, the characteristic tubing impedance is

$$Z_o = \sqrt{\frac{R_H + j\ w\ L_H}{G_H + j\ w\ C_H}} \quad 5.24$$

where G_H = conductance (leakage) to ground

If the line resistance and conductance are ignored, 5.24 becomes

$$Z_o = \sqrt{\frac{\rho\beta_e}{A^2}} \quad 5.25$$

A pressure wave will travel along an oil-filled tube at the speed of sound. This is called celerity

$$c = \sqrt{\frac{\beta}{\rho}} \quad 5.26$$

For MIL-H-5606 at 75°F and 3000 PSI, $c = 4480$ ft/sec. The wave length λ (lambda) is

$$\lambda = \frac{c}{f} \quad or \quad \lambda = \frac{2\ \pi\ c}{\omega} \quad 5.27$$

From well-developed transmission line theory, Z_s, the source impedance as shown in Figure 5.2 is related to the line and load impedances

$$Z_s = Z_o \frac{Z_R \cosh j\omega\ l/c + Z_o \sinh j\omega\ l/c}{Z_o \cosh j\omega\ l/c + Z_R \sinh j\omega\ l/c}$$

But: $\cosh j\ \theta = \cos\ \theta$ and $\sinh j\ \theta = j\ \sin\ \theta$ so

$$Z_s = Z_o \frac{Z_R \cos\ \omega\ l/c + j\ Z_o \sin\ \omega\ l/c}{Z_o \cos\ \omega\ l/c + j\ Z_R \sin\ \omega\ l/c} \quad 5.28$$

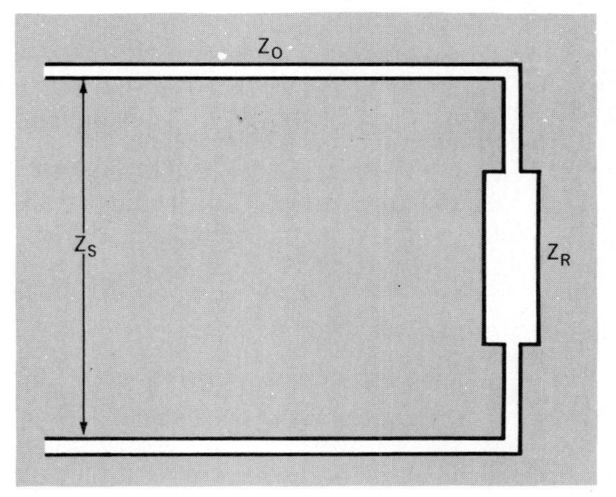

Fig. 5.2. Source impedance is related to line and load impedance

5.4. OPEN-END LINE

If the transmission line is open to the atmosphere at the load end, the load impedance $Z_R = 0$. The analysis can be made by assuming that a wave reflects from the open end so that pressure propagates both ways at once. While this is not a realizable condition, it does simplify the computation. Equation 5.28 for $Z_R = 0$ becomes

$$Z_s = j\ Z_o \tan\ \omega\ l/c \quad 5.29$$

The input impedance is zero when $l = 0, \frac{\lambda}{2}, \lambda, \frac{3\lambda}{2}$ . . . These are the conditions of resonance.

5.5. CLOSED-END LINE

If the power source is driving into a closed end tube, Z_R the load impedance, becomes infinite. This cannot be readily handled in equation 5.28 unless both top and bottom are divided by Z_R,

$$Z_s = j\, Z_o \cot \omega\, l/c \qquad 5.30$$

Resonance occurs when

$$\omega\, l/c = \frac{\pi}{2}, \frac{3\pi}{2}, \frac{5\pi}{2} \ldots\ldots$$

The fundamental frequency (lowest frequency producing resonance) is

$$f_t = \frac{c}{4l} \quad \text{Hertz} \qquad 5.31$$

The load pressure P_R is related to the inlet pressure P_S

$$\frac{P_R}{P_S} = \frac{1}{\cos \omega\, l/c} \qquad 5.32$$

Equation 5.32 states that the ratio P_R/P_S varies from unity when $\frac{\omega l}{c} = 0, \pi, 2\pi \ldots$ to infinity when resonance occurs. While fluid resistance will not allow a rise to infinity, this relationship will allow for computation of forbidden line lengths when the nature of the input is known.

As an example, consider a nine piston pump operating at 3600 rpm driving into a hydraulic system which contains MIL-H-5606 oil at 75F. The control valve at the end of the line is closed. $c = 4480\,(12) = 53{,}800$ in/sec. Resonance occurs at odd multiples of $\pi/2$

$$\omega = 9 \left[\frac{3600}{60}\right] (2\pi) = 3390 \text{ radians/second}$$

$$l_{res} = \frac{\pi/2\ (53{,}800)}{3390} = 24.90 \text{ inches}$$

$$= 2.075 \text{ feet}$$

Therefore, a length of 2.075 feet or an odd multiple of that length should be avoided.

5.6. LINE WITH AN END CHAMBER

Assume that the line terminates in an end chamber. If the chamber has zero volume, the condition is the same as for a closed end tube. If the chamber has infinite volume, the condition is the same as an open end tube.

If the chamber cross-sectional area is more than 10 times the chamber length, the effect of hydraulic inductance is small and the hydraulic resistance is negligible. The chamber only adds capacitance to the system. The combined capacitance of the chamber and line is

$$C = \frac{V_c}{\beta_e} \qquad 5.34$$

Where: C = Combined capacitance
V_e = Volume of chamber

then

$$Z_R = -j\frac{1}{\omega C} = -j\frac{\beta_e}{\omega V_c} \qquad 5.35$$

$$Z_s = j\, Z_o \frac{Z_o \sin \omega\, l/c - \frac{\beta_e}{\omega V_c} \cos \omega\, l/c}{Z_o \cos \omega\, l/c + \frac{\beta_e}{\omega V_c} \sin \omega\, l/c} \qquad 5.35$$

Equation 5.35 can be simplified:

$$Z_s = j\, Z_o \tan(\omega\, l/c + \phi) \qquad 5.36$$

Where: $\phi = \tan^{-1} \frac{\beta_e Z_o}{\omega V_c}$

If V_c = infinity, equation 5.36 becomes equation 5.29.

If $V_c = 0$, equation 5.36 becomes equation 5.30.

Resonance for equation 5.36 occurs at

$$f_{to} = (n\pi - \phi)\frac{c}{2\pi l} \qquad 5.37$$

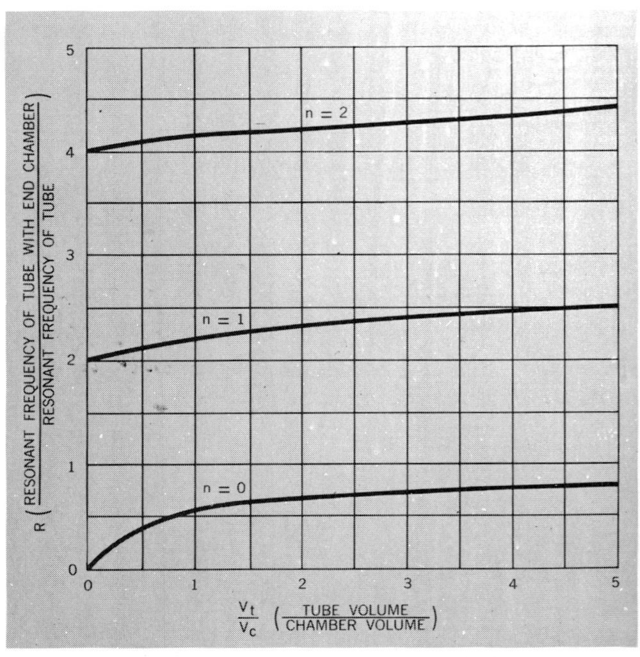

Fig. 5.3. Relationships for n = 0, 1, and 2

Compare this to resonance of a closed end tube.

$$R = \frac{f_{tc}}{f_t} = 2\left[n - \frac{\phi}{\pi}\right]$$

letting

$$b = -\frac{2\ V_t}{\pi\ V_c}$$

then

$$R = \frac{2\pi n - 2\tan^{-1} b/R}{\pi} \qquad 5.38$$

Based on equation 5.38, relationships for various values of n can be shown between R, the resonance ratio, and V_t/V_c, which is proportional to b. Curves of these relationships for the range $n = 0$ to $n = 3$ are shown in Figure 5.3. The resonant frequency of a line terminating in a chamber can be computed by:

- Computing the resonant frequency of the tube using equation 5.31.
- Computing V_t/V_c.
- Finding the appropriate value of R from Figure 5.3. Usually this will be the fundamental or $N = 0$ value.
- Computing the resonant frequency of the tube and chamber.

$$f_{tc} = [R(n)]f_t \qquad 5.39$$

As an example of the use of equation 5.39 in a practical problem, consider a 1" x .049 hydraulic line containing MIL-H-5606 fluid at 75°F. The line is 25 feet long and terminates in an actuator having a volume of 120 cubic inches. When the actuator valve is wide open and the actuator is in the position which exposes maximum volume to the line, what is the natural frequency of the transmission system?

The tube ID is

$1 - (2)(0.049) = .902$ inches

Tubing inside area $= 0.64$ sq. in.

Line frequency $= \frac{4480\ (12)}{4\ (25)\ (12)} = 44.8$ Hz

$$\frac{V_t}{V_c} = \frac{192}{120} = 1.6$$

From Figure 5.3 R = 0.65

$f_{tc} = 0.65\ (44.8) = 29$ Hz

A variation of this problem is the general question of the fundamental resonant frequency of a tube of inside diameter d and of length l which termines in a chamber of volume V_c. This approach would be needed, for instance, if the chamber volume and tube diameter were known and the length of the tube needed to be determined.

The relationship is

$$l = \frac{2\ R\ V_c}{d^2} \tan \frac{\pi\ R}{2} \qquad 5.40$$

Figure 5.4 shows $l/(V_c/d^2)$ plotted against R for the fundamental resonance of $n = 0$.

Use of equation 5.40 is illustrated by the following example. Assume that a line 10 feet long containing MIL-H-5606 at 75°F terminates in an actuator having a volume of 30 cubic inches. The line must not respond to an excitation of 80 Hz. What line inside diameter must be avoided?

Line frequency $= \frac{4480\ (12)}{4\ (10)\ (12)} = 112$ Hz

$V_c = 30$ cubic inches

$$\frac{l}{V_c} = \frac{120}{30} = 4$$

$$R = \frac{80}{112} = 0.714$$

From Figure 5.4

$$\frac{l/V_c}{d^2} = 3\ ,\quad 4d^2 = 3\ ,\quad d = 0.865''$$

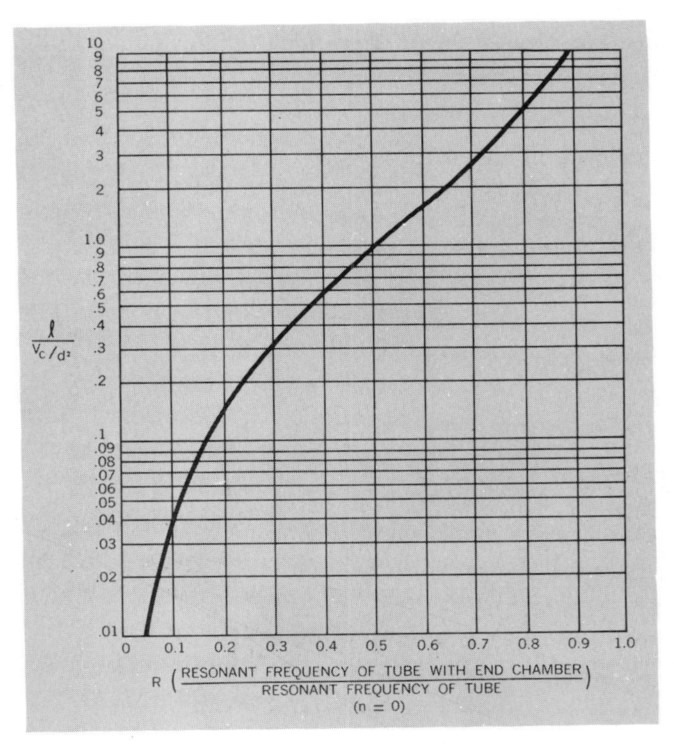

Fig. 5. 4. Relationship of tube length to resonance

5.7. LINE WITH A FIXED ORIFICE

The foregoing analyses assumed that the lines were either wide open or fully closed at the end. In the two basic cases considered, the source impedance is an imaginary vector (because the real component, the line resistance, was ignored) on a complex plane. The resonant condition occurs when the vector length is zero. In any practical situation, some damping will always be supplied by friction of the oil on the tube and shearing of adjacent laminae of oil.

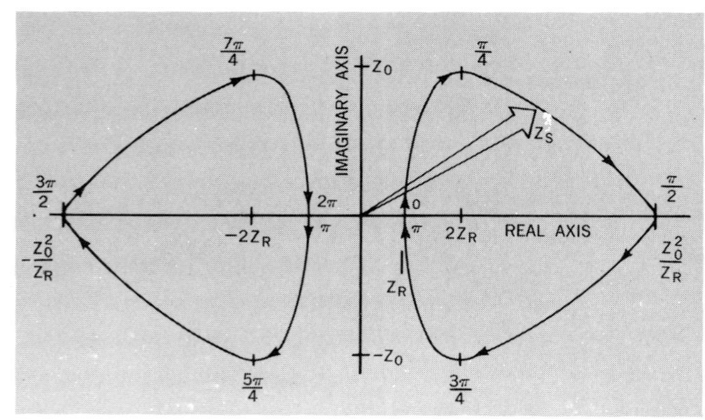

Fig. 5.5. Plot of vector Z_R on complex plane as $\omega l/c$ goes from zero to π

A similar analysis can be made for a line which ends in a fixed orifice which vents to a low pressure area. Because flow always occurs through the orifice, the load impedance Z_R always has a real value. Therefore, the condition of zero source impedance does not occur. The condition of *minimum* damping may be of interest however. Because of the second order relationship between flow through an orifice and the pressure differential across it, the relationship becomes complex unless very small changes in flow are considered. For incremental flow changes, the slope of the pressure/flow curve may be considered a straight line and the load impedance can be expressed as a first order relationship.

$$Z_R = \frac{P}{Q} \qquad 5.41$$

The expression for the source impedance is again equation 5.28. This expression can be evaluated for vales of $\omega\ l/c$ increasing from zero. Figure 5.5 shows a plot of the vector Z_R on the complex plane as $\omega\ l/c$ goes from zero to π. As can be seen, the condition of minimum damping occurs at

$$\omega\ l/c = o,\ \pi,\ 2\pi,\ 3\pi\ .\ .\ .\ .\ .\ n\pi$$

when $Z_s = Z_R$.

5.8. LINE WITH A MODULATING VALVE

The description of a line which ends in a variable orifice such as a servo valve becomes more complex. The load impedance Z_R cannot be described unless a fixed position of the valve is assumed. In this case the problem reverts to the case of a line ending in a fixed orifice. However, the complex relationships between valve position and the pressure across the valve can be developed.

By the same reasoning used earlier, based on incremental flow changes, the flow into a line-plus-valve system from a constant pressure source can be described by

$$Q_1 = c_1Y + c_2P_2 \qquad 5.42$$

Where:
Q_1 = flow into system
Y = valve displacement
P_2 = load pressure
c_1, c_2 = constants

The flow out of the system is

$$Q_2 = Q_1 \cosh j\omega\ l/c \qquad 5.43$$

The load pressure is

$$P_2 = Z_oQ_1 \sinh j\omega\ l/c$$

These can be combined into the following relationships

$$\frac{Q_1}{Y} = \frac{c_1}{\cos \omega\ l/c + c_2\ Z_o\ j \sin \omega\ l/c} \qquad 5.45$$

$$\frac{P_2}{Y} = -\ Z_oc_1 \frac{j \sin \omega\ l/c}{\cos \omega\ l/c + c_2\ Z_o\ j \sin \omega\ l/c} \qquad 5.46$$

5.9. PULSE DAMPING

One of the most difficult problems encountered in developing hydraulic systems is the attenuation of pressure pulses. The two most severe sources of harmful pressure pulses are the pumps and suddenly closing valves. An understanding of ways to analyze pressure pulses and to compensate for them can be derived from an extenuation of the studies on transmission lines.

5.10. PUMP RIPPLE

Pressure ripple can be harmful and may need to be suppressed. One excellent technique is to attach a closed end side branch to the transmission line at the trouble area. If the side branch has zero input impedance, all of the ripple will pass into the branch and none will be passed along the transmission line. The design conditions for the side branch stem from equations (40) and (41). When combined they give

$$l = \frac{\lambda}{4} \qquad 5.47$$

Therefore, a side branch having a length or an even multiple of one quarter wave length will provide pulse damping for a ripple of a specific frequency. Usually, a line and chamber are designed using Figures 5.3 and 5.4 to give a side branch having the same resonant frequency as a line which is quarter of a wave length long.

Another method which has been used quite extensively is the Quincke tube shown in Figure 5.6.

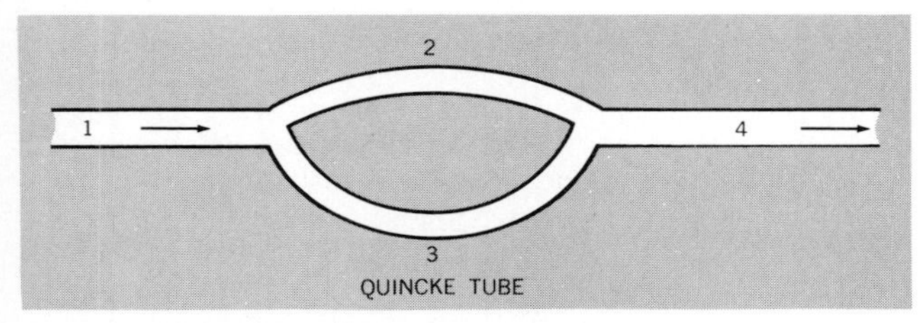

Fig. 5.6. Quincke tube

The theory of the Quincke tube is given by Stewart[1]. The tube is so given that Area (1) = Area (4) = Area (2) plus Area (3). If

$$l_3 = l_2 + \frac{(2n + 1)\,\lambda}{2} \qquad 5.48$$

then excess pressure at the junction of (2), (3) and (4) is zero. A wave or pressure ripple traveling from left to right cannot be transmitted to line (4).

5.11. ACCUMULATORS AS PULSE DAMPERS

Many system designers install accumulators in their systems "just to damp out pump ripple." The techniques for analyzing the use of gas-oil accumulators for such damping have been developed. Equation 5.30 is as applicable to a gas accumulator as to a hydraulic line. To be an effective damping agent, the accumulator should be designed to have zero input impedance. This will occur if

$$\omega \frac{l}{c} = \frac{\pi}{2}, \frac{3\pi}{2}, \frac{5\pi}{2} \ldots\ldots$$

or if, in accordance with equation 5.31:

$$f_t = \frac{c}{4l}$$

Assume that the accumulator is charged with dry nitrogen and that the system operating pressure is 3000 psi. The operating temperature is 80°F.

At 80°F and 3000 psi the specific volume of GN_2 is 125 in³lb so

$$w = 1/125 = 0.008 \text{ lb/in}^3$$

$$\rho = 0.008/386 = 2.08 \times 10^5 \text{ lb-sec}^2/\text{in}^4$$

The isothermal bulk modulus of a gas is equal to the pressure. The polytropic bulk modulus is equal to the pressure times the polytropic exponent. The adiabatic bulk modulus is then 1.4 times the pressure. Then, for the resonant condition

$$\frac{\omega l}{c} = \omega l \sqrt{\frac{2.08 \times 10^{-5}}{4.20 \times 10^{-3}}} = \frac{\pi}{2}, \frac{3\pi}{2}, \frac{5\pi}{2} \ldots\ldots$$

If a 9 piston pump operating at 3600 rpm is considered, ω = 3390 rad/sec

$$l = \frac{\frac{n\pi}{2}}{(3390)\ (7.07 \times 10^{-4})}$$

$$= 0.66'' ,\ 1.98'' ,\ 3.31'' \ldots\ldots$$

Accumulators having gas columns of these lengths are easy to design and install. Such accumulators are very effective in damping pump ripple and in reducing system noise. Where the mass of piston accumulators must be considered

$$\omega = \sqrt{\frac{A\beta_g}{l_g M_p}}$$

5.12. QUICK-CLOSING VALVE

All of the previous discussions assumed that the disturbances in the flow were periodic in nature and that they can be described mathematically as sinusoids. While these assumptions are obviously not always true, the analyses are sufficiently close for engineering evaluation.

Another major class of disturbances are transient rather than periodic. These are usually the result of a rapid opening or rapid closing of a valve. Of the two, the rapid closing of a valve is the more severe. This is because the velocity of flow is converted to a pressure rise above quiescent. This pressure rise is reflected back and forth along the tube at a velocity determined by the speed of sound until the damping in the system attenuates the pressure.

A hydraulic system, under the condition of a rapidly closing valve, acts very like a weight on the end of a spring. The weight, if displaced from its neutral position and released, will bob up and down. In the hydraulic system, pressure surges will bounce back and forth. These phenomena only die down as a result of internal friction or other damping. In most systems this attenuation is quite slow.

The analysis of the pressure surge can be done in two ways. The system can be considered analogous to a weight suspended between two springs. This is called a "lumped parameter" approach because all of the mass is lumped into one spot. In the "distributed parameter" method, the system is treated as if it were an infinite number of small masses and springs all interacting. Each of these methods will be described for a simple transmission line ending in a valve which has been rapidly closed.

5.13: LUMPED PARAMETER ANALYSIS

Figure 5.7a shows the simple transmission system and 5.7b, the mass spring system which is analogous to it. Assuming that the system has some damping, the differential equation which describes this system is

$$F(t) = M\frac{d^2x}{dt^2} + B\frac{dx}{dt} + Kx \qquad 5.49$$

Where: B = damping coefficient

A solution of the form $Ae^{-\sigma t}\ sin\ \omega t$ will be assumed. This solution is an exponentially damped sinusoid. Because no forcing function $F(t)$ exists and because a line velocity v_o exists when the valve is closed, equation 5.49 becomes

$$O = M\frac{d^2x}{dt^2} + B\frac{dx}{dt} + Kx \qquad 5.50$$

$$v\ (t = o) = v_o$$

Equation 5.50 can be solved by LaPlace techniques to give a mid-point displacement as a function of time.

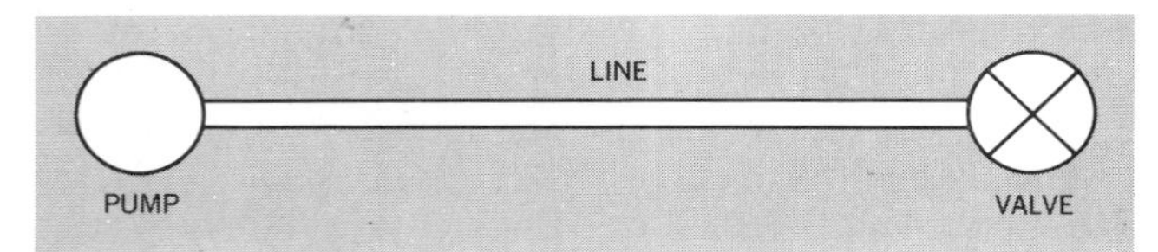

Fig. 5.7a. Simple transmission system.

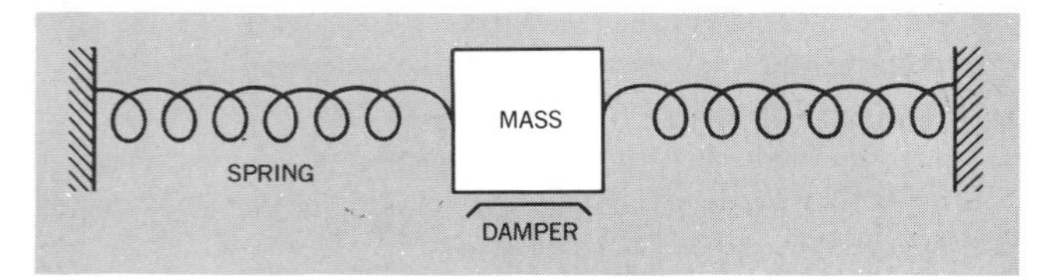

Fig. 5.7b. Mass spring system analogous to simply transmission system.

Taking the inverse LaPlace transform

$$x(t) = \left[\frac{-v_o}{\sqrt{\frac{K}{M} - \left[\frac{B}{2M}\right]^2}}\, e^{-\frac{B}{2M}t}\right]\left[sin\sqrt{\frac{K}{M} - \left[\frac{B}{2M}\right]^2}\ t\right] \qquad 5.51$$

Now, substituting equation 5.12 into $F = Kx$ and $P = FA$, yields

$$P = \frac{\beta\ x}{l} \qquad 5.52$$

so

$$P(t) = -\frac{\dfrac{\beta v_o}{l\sqrt{\frac{K}{M} - \left[\frac{B}{2M}\right]^2}}\, e^{-\frac{B}{2M}t}}{sin\sqrt{\frac{K}{M} - \left[\frac{B}{2M}\right]^2}\ t} \qquad 5.53$$

This states that the pressure will have a maximum of

$$\frac{\beta v_o}{l\sqrt{\frac{K}{M} - \left[\frac{B}{2M}\right]^2}} \text{ psi above base pressure}$$

and the frequency of oscillation of the pressure wave is

$$\omega = \sqrt{\frac{K}{M} - \left[\frac{B}{2M}\right]^2}$$

and the exponential damping rate will depend upon $B/2M$.

5.14: DISTRIBUTED PARAMETER ANALYSIS

This analysis is predicted upon the classical wave equation which is developed in several texts, among them Thomson[2].

$$\frac{\partial^2 P}{\partial t^2} = c^2\frac{\partial^2 P}{\partial x^2} \qquad 5.54$$

This equation, when restrained by the effects of a suddenly closed valve, can be shown to yield the following pressure series.

$$P(t) = P_o + \frac{4\rho c V_o}{\pi}\left[sin\frac{\pi ct}{2l} + \frac{1}{3}sin\frac{3\pi ct}{2l} + \frac{1}{5}sin\frac{5\pi ct}{2l}\ \ldots\ldots\right] \qquad 5.55$$

This is a rectangular wave of amplitude $\pm 4\rho\ c\ v_o/\pi$ about P_o with a frequency of $c/4l$ Hz.

As an example of how to use equation 5.55 consider a hydraulic line 1356 inches long. The tubing is 1″ x .075 wall so the ID is 0.87″ and the inside area is .595 sq. in. For the fluid being used, $\rho = 8.9 \times 10^{-5}$, $\beta_e = 8.9 \times 10^4$ and the kinematic viscosity $\nu = .0023$ in²/sec. The flow rate when the valve was closed is 75 gpm. For this line v_o then is 480 in/sec. $P_o = 3000$ psi. The celerity of sound $c = \sqrt{\frac{\beta}{\rho}} = 3.18 \times 10^4$ in/sec. The pressure fluctuation is

$$\Delta P = \frac{4\rho c\, v_o}{\pi} = \frac{4\ (8.9 \times 10^{-5})\ (3.18 \times 10^{4})\ (480)}{\pi}$$

$$= 1710\ psi$$

The oscillation frequency is $\frac{c}{4l} = \frac{3.18 \times 10^{4}}{4\ (1356)}$

$= 5.84$ Hz.

This pressure wave is plotted in Figure 5.8a.

Unfortunately, this analysis does not include any damping or pressure attentuation due to conversion of part of the energy in the pressure pulses to heat. Intuitively, one realizes that the pressure pulses will not continue forever. They will, in time, disappear. Several studies of the attenuation of pulses have been made[3,4]. Perhaps the easiest to apply is the statement of Brown[3] for the attenuation of the peak pressure downstream due to a unit impulse at the upstream position. Though not completely valid for this analysis, Brown's attenuation represents the easiest one to apply.

$$P_d = e^{-\frac{lv}{r_i^2 c}}\ P_u \qquad 5.56$$

Where r_i = inside radius of tube

For the values of the example, this becomes, for each reflection of pressure

$$P_d = e^{-0.00524}\ P_u = \frac{P_u}{1.0053}$$

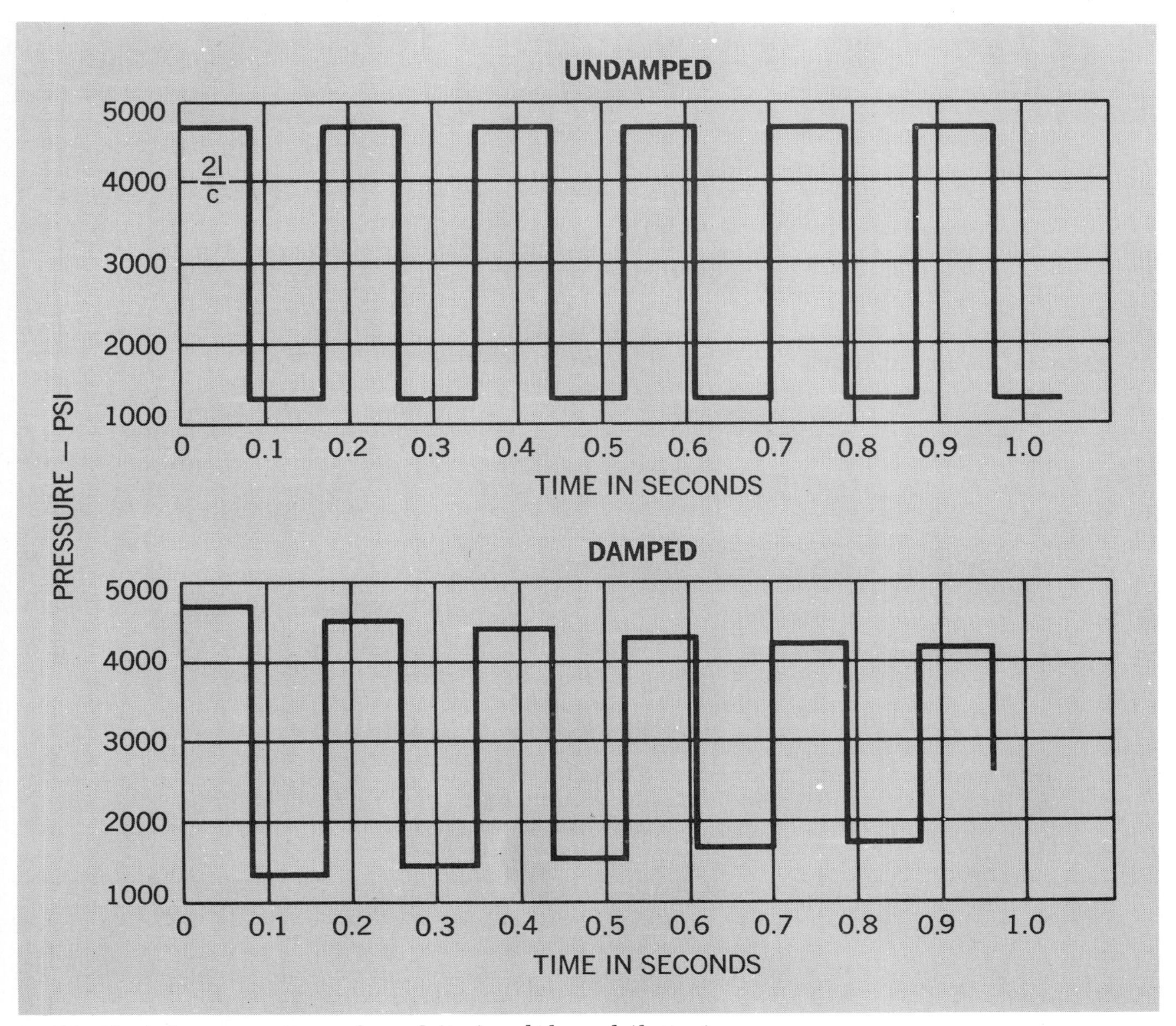

Fig. 5.8. Classical pressure wave undamped (top) and damped (bottom).

The pressure for each cycle can be computed in a sequential manner.

$P_o = 3000$ psi
$P_1 = 3000 + 1710 = 4710$
$P_2 = 3000 - \left(\frac{1710}{1.0053}\right) = 1299$
$P_3 = 3000 + \left(\frac{1701}{1.0053}\right) = 4692$
$P_4 = 3000 - (1692/1.0053) = 1317$
$P_5 = 3000 + (1683/1.0053) = 4674$
$P_6 = 3000 - (1674/1.0053) = 1335$
$P_7 = 3000 + (1665/1.0053) = 4656$
$P_8 = 3000 - (1656/1.0053) = 1353$
$P_9 = 3000 + (1647/1.0053) = 4638$
$P_{10} = 3000 - (1638/1.0053) = 1371$
$P_{11} = 3000 + (1629/1.0053) = 4620$
$P_{12} = 3000 - (1620/1.0053) = 1389$
$P_{13} = 3000 + (1611/1.0053) = 4602$

Because the pressure reverses every $2l/c$ seconds, the pressure peaks will trace out an envelope which is an exponential curve.

$$P(t) = P_o \pm e^{-\frac{vt}{2 r_i^2}} \Delta P_1 \qquad 5.58$$

In this instance, the envelope curve is $P(t) = 3000 \pm 1710e^{-.00608t}$. This curve suggests that the pressure pulses will damp 10% of their initial value in 380 seconds. This damped curve is plotted in Figure 5.8b.

5.15. SLOWLY CLOSING VALVE

If the valve does not close instantaneously but does take less time than $2l/c$, the action is still considered instantaneous. If the time is greater than $2l/c$ but still short enough to generate pressure transients, then other methods must be used to determine the pressure levels.

An iterative method has been proposed by Paynter[5]. The following relationships are used.

Y_o = *Maximum valve opening*
Y = *Actual valve opening*
$Y(t)$ = *Valve opening as a function of time*
$y = Y/Y_o$

$$K = \frac{Z_o Q_o}{2 P_o}$$

Where: Z_o = line impedance $= \frac{\rho c}{A}$
Q_o = *Initial flow*
P_o = *Initial pressure*

$m = Ky$
$i = t/2T$

Where: T = transit time of surge $= l/c$

$$A_i = \frac{2 (m_{i-1} - m_i)}{1 + m_i}$$

$$Y_i = \frac{1 - m_{i-1}}{1 + m_i}$$

h = *Pressure increment* $\Delta P/P_o$
$h_i = A_i - Y_i h_{i-1}$

The time t to close the valve is divided up into i increments each $2T$ long. A table is then set up, listing in order i, y, m_i, m_{i-1}, $m_{i-1}-m_i$, $1-m_{i-1}$, $1+m_{i-1}$, A_i, Y_i, h_i, *and* ΔP. A step-by-step solution of the value of the pressure increments is then possible.

Using the same example as for the instantaneously closing valve

$$Z_o = \frac{\rho c}{A} = \frac{(8.9 \times 10^{-5})\ (3.18 \times 10^4)}{0.595} = 4.75$$

$Q_o = 75$ gpm $= 289$ in^3/sec. , $P_o = 3000$

$$K = \frac{(4.75)\ (289)}{2\ (3000)} = 0.23$$

Assume the valve closes in 0.512 seconds. This is 6 times $2l/c$, see table I

Another technique is basically graphical and has been developed by King[6]. This method is based on the following relationships

$$p_{1t1} - p_{2t2} = \frac{c V_o \rho}{P_o} (v_{1t1} - v_{2t2})$$

$$p_{2t2} - p_{1t3} = - \frac{c V_o \rho}{P_o} (v_{2t2} - v_{1t3})$$

TABLE I—VALVE CLOSING TIMES

i	y	m_i	m_{i-1}	$m_{i-1}-m_i$	$1-m_{i-1}$	$1+m_{i-1}$	A_i	Y_i	h_i	ΔP
0	1	.23			1	1.23	0			
1	.83	.19	.23	.04	.77	1.19	.067	.65	.067	200 psi
2	.75	.15	.19		.81	1.15	.068	.71	.024	72
3	.50	.12	.15		.85	1.12	.071	.76	.043	129
4	.33	.08	.12		.87	1.08	.074	.80	.040	120
5	.17	.04	.08		.92	1.04	.077	.89	.041	123
6			.04	.04	.96	1.0	.08	.96	.041	123

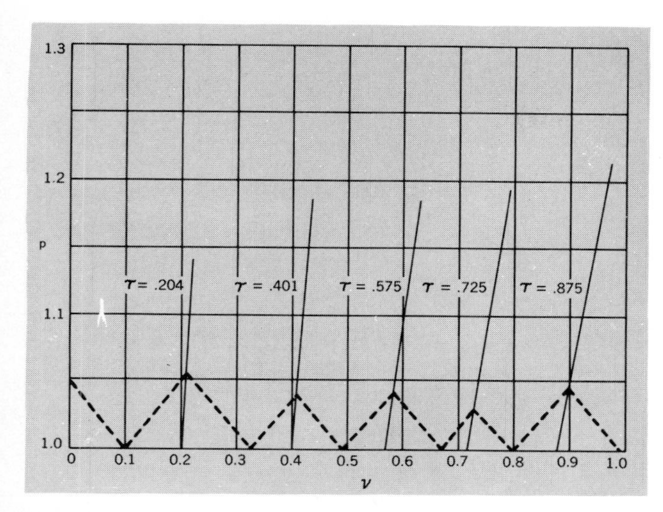

Fig. 5.9. Plot of pressure variations up to point of complete valve closure.

Where: $p = \frac{P}{P_o}$

$v = \frac{V}{V_o}$

Assuming that the valve acts as an orifice

$$v = \frac{V}{V_o} = \tau \sqrt{p}$$

Where: $\tau = \frac{c_d \ A_v}{(c_d \ A_v)_o}$

As the valve closes $1 > \tau > 0$. The valve closing time is divided into a series of increments, preferably equal to $2l/c$. A value of τ is computed for each time interval termination. A graph is then set up for ρ versus v. On this graph are plotted parabolas of $\tau =$ constant for each of the computed values of τ. Starting at $\tau = 1$, ($\rho = 1$, $v = 1$) draw a line having a slope of $-\frac{cv_o\rho}{p_o}$. The intercept of this line with the next τ line will give the ρ ratio for that time. From this intercept, another line with a positive slope $\frac{cv_o\rho}{p_o}$ is drawn. The intercept of this second line with the $\rho = 1$ abscissa is the starting point of a new reflection process. This is repeated until an intercept with the $v = 0$ ordinate occurs. From this point on the system will oscillate with damping as in the previous examples.

To illustrate this technique, the same example used previously will be solved so that a comparison may be made. All of the previous conditions will hold, the valve closing rate is assumed to be linear with time. Further, as the flow is shut off, c_d, the coefficient of discharge will be assumed to increase from 0.6 to 0.78 in steps of 0.03. The following table may now be established.

Time	A_v/A_{vo}	c_d/c_{do}	τ
0	1	1.0	1.0
.0853	.834	1.05	0.875
.1707	.667	1.10	0.725
.2560	.500	1.15	0.575
.3413	.334	1.20	0.401
.4266	.107	1.25	0.204
.512	0	1.30	0.000

The plot of the pressure variations up to the point of complete valve closure is shown in Figure 5.9. From this the following table of pressure ratios and actual pressures was derived.

Time	p	P
0	1.0	3000
.0853	1.055	3150
.1707	1.02	3090
.2560	1.058	3165
.3413	1.03	3065
.4266	1.07	3210
.512	1.03	3085

REFERENCES

1. Stewart, G. W., "The Theory of the Herschel — Quincke Tube", *Physical Review* vol n4 April 1928.
2. Thomson, W. T., "LaPlace Transformation Theory and Engineering Applications," Prentice-Hall 1950.
3. Brown, F. T., "The Transient Response of Fluid Lines," *Journal of Basic Engineering,* December 1962 pp 547 - 553.
4. Oldenburger, R. and Hoodson, R. E., "Simplification of Hydraulic Line Dynamics by Use of Infinite Products," *Journal of Basic Engineering,* March 1964, pp 1 - 8.
5. Paynter, H. M., "Fluid Transients in Engineering Systems," Sec. 20, Handbook of Fluid Dynamics," McGraw-Hill, 1961.
6. King, W. H. and Brater, E. F., "Handbook of Hydraulics," McGraw-Hill 1963 5th Edition.

CHAPTER 5

PROBLEMS

1. Tubing of Type 304 corrosion resistant steel is 1/2-inch *OD* x 0.020-inch wall. If it contains hydraulic oil having a bulk modulus of 220,000 what is (1) the tubing bulk modulus and (2) the equivalent bulk modulus of the tubing and oil?

2. A 1-inch x 0.049-inch Type 304 corrosion resistant steel tube contains MIL-H-5606 oil at 100°F and 250 psi. The oil contains 1% by volume of entrained air. What is the effective bulk modulus?

3. 3 feet of -12 hose (*ID* = 0.625-inch) is attached to a 10 foot run of 3/4-inch x 0.035-inch cres Type 304 tubing. The lines contain an aircraft phosphate ester at 175 psi and 120°F. The oil contains 0.8% by volume of entrained air. What is the effective bulk modulus of the transmission system? $\beta_H = 4.5 \times 10^4$ psi.

4. Compute the celerity of sound for (1) MIL-H-5606 at 80°F and 0 psig, (2) an industrial phosphate ester at 100°F and 0 psig, (3) MIL-H-8446 at 170°F and 3000 psig, (4) aircraft phosphate ester at 220°F and 0 psig.

5. A 9-piston pump is driven at 4500 rpm, pumping aircraft phosphate ester at 125°F. If there is a closed valve at the end of the line, calculate the first three resonant line lengths.

6. An 11-piston pump, operating at 3600 rpm, pumps MIL-H-8446 at 200°F and 3000 psi. The valve at the end of the line is closed. What are the first three resonant line lengths?

7. A line 1/2 inch x 0.035 inch is filled with MIL-H-5606 at 100°F. The line is 20 feet long and terminates in an actuator. In the stalled position, the actuator has a column of 8 cubic inches. What are the first three resonant frequencies of the line and chamber?

8. A 3000 psi system operating at 125°F has a 9 piston pump driven at 5650 rpm. How long should the GN_2 column in the accumulator be in order to damp the pump pressure ripple?

9. An infinitely rigid line 1 inch x 0.079 inch, 55 feet long carries 12 gpm of MIL-H-5606 at 3000 psi and 110°F. The valve on the end of the line is suddenly shut. What are the amplitude and frequency of the pressure cycles which result? What is the pressure decay per transit of the pressure pulse?

CHAPTER SIX

Transmission Lines –Components

6.1. RESERVOIRS

A reservoir provides oil to make up for system leakage, allows space for the expansion or contraction of the oil with changes of temperature, enables gas bubbles to rise out of the oil and dirt particles to sink to the bottom, and allows room for heat dissipation or for the inclusion of thermal control devices. In some specialized systems a reservoir is also a low-level pressure control system.

To get the greatest usefulness from a reservoir, the outlet should be near the low point of the tank. The actual lowest point is often not used so that a sump will exist. Any dirt can then settle into the sump and not be recirculated in the system. The location of the inlet is not so critical. It should, however, be below the surface of the normal oil level so that air bubbles will not be trapped in the oil by return fluid falling into the reservoir. To enable dirt to settle out and air bubbles to rise, the flow velocity through the reservoir should be low. The inlet and outlet lines should be as large as possible and the reservoir itself as big as system economics allow.

A few small additions can greatly increase the usefulness of a reservoir, Fig. 6.1. A cover will sharply reduce the amount of dirt, moisture and other contamination which gets into the oil. If the oil must pass through a large area, fine mesh screen before reaching the outlet, gross contamination can be screened out before going to the pump suction line. As will be discussed in Chapter 7, the screen will also remove many of the air bubbles in the oil by means of the surface tension of the oil.

Table 2-2 gives coefficients of expansion for some hydraulic fluids. Typically the coefficients range between 4 and 5 x $10^{-4}/°F$. This means that the volume of the oil will increase 4 - 5% for each 100°F rise in temperature. Where the range of oil temperature is sizable, the reservoir tank must be big enough to account for this change in volume.

The sizing of a reservoir must, as a minimum, provide for

1. Leakage make-up fluid.
2. Normal volume variations in the system.
3. Thermal volume variations.

Additional space will be needed if heaters or cooling coils must be placed in the reservoir. Topping all these is the overall need for a maximum size in order to ease the removal of dirt and air bubbles.

The type of circuit affects reservoir size. For example, when a hydraulic system is pressurized, oil can flow into an accumulator. Unless a self-balancing accumulator is used, this oil will be stored in the accumulator until the system is shut off. The oil volume will then be returned to the reservoir. The operation of differential-area cylinders will also create temporary storage of oil in the system. This oil can be returned to the reservoir upon reverse cycling.

Where contamination from dirt or water is an unusually severe problem—as with sensitive hydraulic servos—or where high operating temperatures, severe radiative conditions or other difficult environments exist, a completely closed hydraulic system may be necessary. Typical of such requirements are military airplane or submarine hydraulic systems, control servos in nuclear power plants and hydraulic power on space vehicles. These closed hydraulic systems use an airless reservoir in which no contact between the oil and environment is allowed.

The airless reservoir, Fig. 6.2, is essentially a spring-loaded, low-pressure accumulator. The spring establishes a system reservoir pressure which is above atmospheric pressure. This can be very helpful in assuring that the pump does not cavitate.

Fig. 6.3 shows a reservoir which uses an initial pressurization from a constant pressure gas source

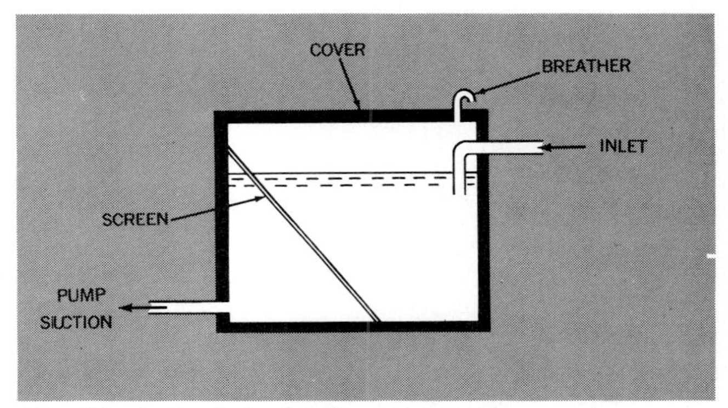

Fig. 6.1. A simple hydraulic reservoir.

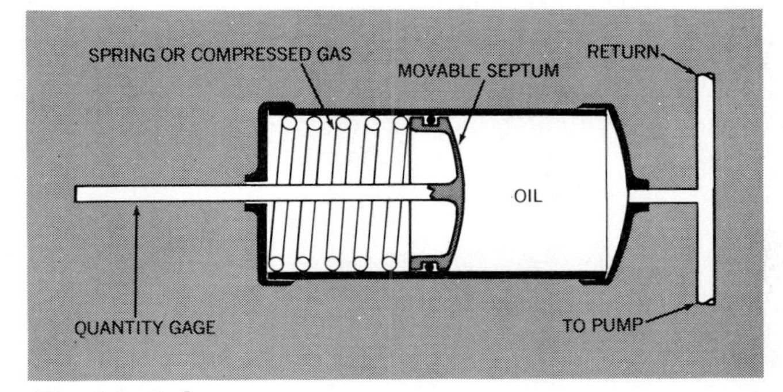

Fig. 6.2. Airless reservoir.

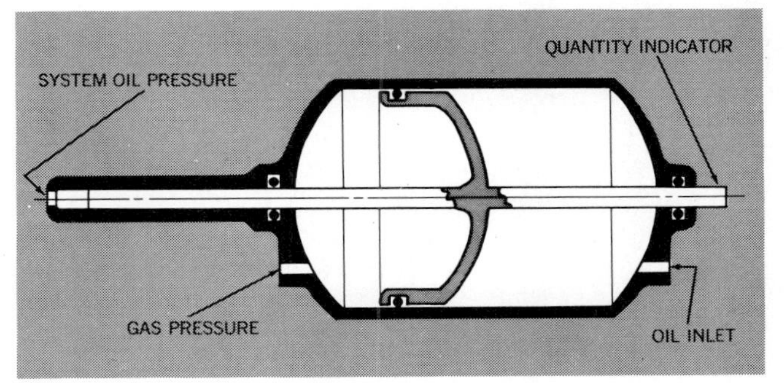

Fig. 6.3. Bootstrap reservoir.

to assure proper system starting characteristics. Once pressure has built up in the system, the pressurized oil is used to maintain reservoir pressurization. This design is known as a "boot-strap" reservoir.

Because an airless reservoir is usually sharply limited in size by various weight or installation requirements, careful calculation of system volume changes is mandatory. The effect on system pressure of insufficient accommodation for volume change can be quite startling. As an example, consider a Type II ($-65°$F to $275°$F) system which holds 20 gallons of MIL-H-5606 oil at $-65°$ F. From Table 2-2 the coefficient of thermal expansion is $4.60 \times 10^{-4}/°$F. At 275°F, the upper operating temperature for MIL-H-5606, the increase in system volume would be $(275-(-65))\ (4.60 \times 10^{-4})\ (2) = 3.13$ gallons. An increase this great would cause severe pressure rises unless proper provisions were made. To determine the degree of pressure rise, equation 2.10 should be employed. From Fig. 2.6 the value of secant bulk modulus of MIL-H-5606 at 0 psig and 275°F can be estimated at 120,000 psi. Inserting this and the initial volume of 20 gallons and the differential volume of 3.13 gallons in equation 2.10,

$$P = (120{,}000)\left[\frac{3.13}{20}\right] = 18780\ psi$$

Of course, the *actual* pressure rise would be lower than that calculated because the tubing and other metal parts would expand as the pressure rose. This expansion decreases the *effective* bulk modulus of elasticity to a value well below that of the oil alone.

One way of minimizing the effects of interchange of oil volume between the high pressure and low pressure sides of a system which employs an airless reservoir is the use of a self-displacing accumulator as shown in Fig. 3.15. This is a combination accumulator-reservoir. The interchange of oil volume from one side of the pump to the other merely results in a shift in the piston position. The actual accumulator can then be designed to handle only leakage requirements and thermal volume changes.

6.2. RELIEF VALVES

All hydraulic systems with fixed displacement pumps use a relief valve to limit system pressure. Most systems using variable delivery pumps also use relief valves to limit pressure surges or to compensate for failed pump pressure controls.

A relief valve is a deceptively simple device. In its most elemental form, Fig. 6.4, the relief valve is a spring-loaded poppet. Flow from the fixed delivery pump is converted to pressure at the load. When the load pressure on the effective poppet area, A_p, creates a force greater than the spring setting, the poppet opens. Flow will then bleed past the poppet to return, or to the reservoir, until the working pressure drops to the relief valve setting. The valve will then close and all the flow go to the load.

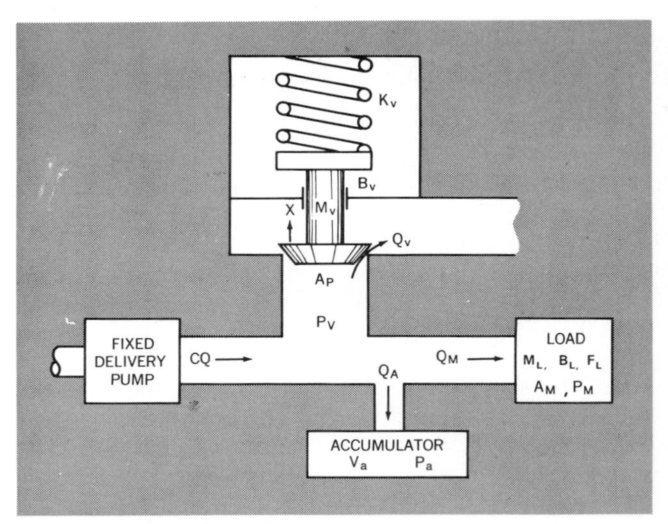

Fig. 6.4. Basic relief valve.

The basic design can be modified to provide damping, or it can be inverted or made in two stages, Fig. 6.5. In each case, however, system pressure is compared with a spring reference. If the error—or difference—between the system pressure and the spring setting is great enough, the valve will open and allow flow from the high pressure region to the low pressure region.

If slow changes in pressure level are considered, a design using only the concepts of Fig. 6.4 would be satisfactory. However, the poppet and spring form a damped mass-spring combination with natural frequency which can be excited by rapid pressure changes. The configurations of Fig. 6.5 are intended to minimize the tendency of the relief valve to chatter. In Fig. 6.5*a*, the poppet must force oil out through an orifice or a small diameter choke as it moves against the spring. The resistance developed by a laminar choke is proportional to the speed of motion of the piston. With a sharp edged orifice, the resistance tends to become proportional to the square of piston velocity. Thus any tendency of the piston to follow high frequency excitation is damped.

In Fig. 6.5*b*, the action of the relief valve is inverted in that the seat moves away from the poppet. Initially, the poppet and seat move downward together against the *combined* spring rate of the two springs. When the poppet contacts the stop, the piston moves away from the poppet against only the low force from the soft lower spring. Thus the initial action is quite stiff, but once the valve opens, it tends to open wide quite quickly.

The two-stage valve, Fig. 6.5*c*, has a similar action. Pressurized oil can flow through the orifice in the top of the main valve piston and operate on the poppet of the pilot stage. When the pressure opens the pilot stage, the pressure of the oil between the main valve and the pilot is suddenly dropped to the outlet level. The large pressure differential across the main valve piston causes the main valve spring to be

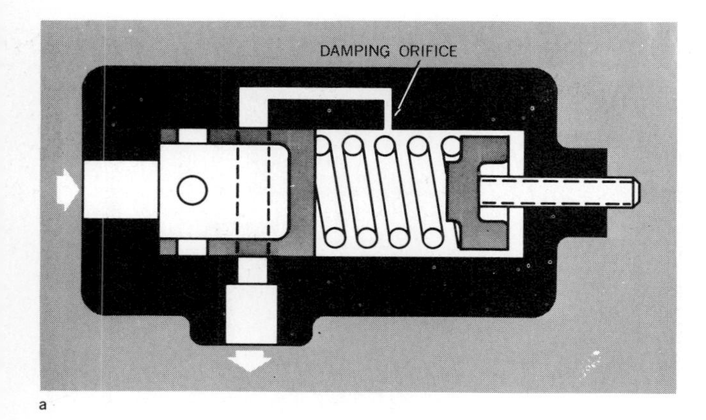

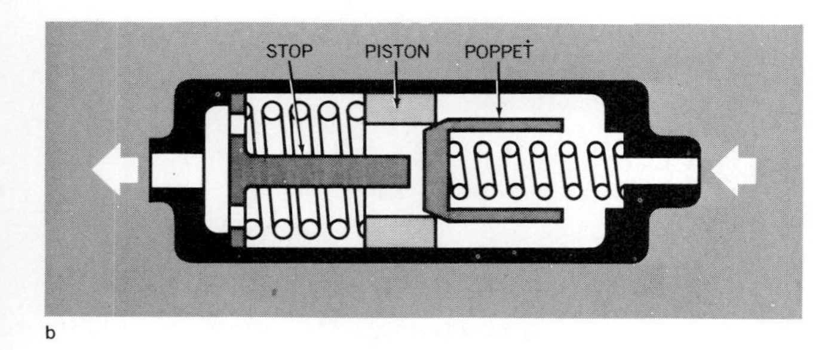

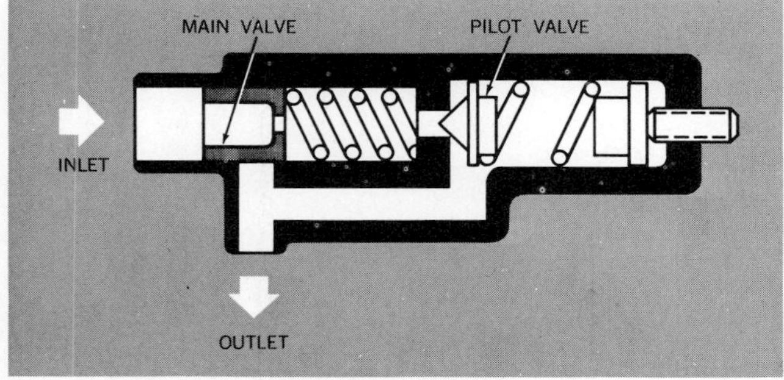

Fig. 6.5. Three relief valve designs.

compressed very quickly and the valve goes wide open.

As with other dynamic elements in a hydraulic system, the relief valve cannot be considered apart from its system. Reethof[1] developed the following analysis of the elements which control the stability of fixed displacement systems where a relief valve controls pressure.

Referring to Fig. 6.4, the summation of forces on the poppet valve is:

$$M_v x s^2 + B_v x s + K_v X = P_v A_p - \delta \qquad 6.1$$

where: A_p = effective poppet area
δ = disturbance force on poppet, lb.
P_v = pressure differential across poppet
M = mass
B = viscous friction coefficient
K = spring rate
X = displacement
s = d/dt

In order that the analysis be linear, each of the values in 6.1 must be considered as small perturbation values around a fixed or quiescent value.

Flow to the actuator consists of three parts: the load velocity flow, the compressibility flow and the leakage flow.

$$Q_m = A_m ys + \frac{V}{\beta} P_v s + LP_m \qquad 6.2$$

where: A_m = actuator effective area
y = actuator displacement
V = volume of compressed fluid
P_m = actuator differential pressure
L = leakage flow coefficient

The forces on the load inertia are:

$$M_l ys^2 + B_l ys^4 = A_m P_m - F_l \qquad 6.3$$

where: F_l = Load force, lb

While strictly true only for small displacements, the flow into or out of the accumulator can be assumed proportional to the rate of change of pressure.

$$Q_a = \frac{V_a}{P_a} (P_v s) \qquad 6.4$$

The flow across the poppet is

$$Q_v = cWX \sqrt{\frac{2P_v}{\rho}} \qquad 6.5$$

where: W = circumference of port

For continuity

$$Q_m + Q_v + Q_a - CQ = 0 \qquad 6.6$$

The above equations can be combined into a fourth order differential equation:

$$(s^4 + a_3 s^3 + a_2 s^2 + a_1 s + a_o)\ P_v = 0 \qquad 6.7$$

For the most critical condition, the coefficients a_0, a_1, a_2 and a_3 are:

$$a_0 = 2\omega_l \omega_v^2 + 2\zeta_l \omega_v c_a \qquad 6.8$$

$$a_1 = 2\zeta_l \omega_l \omega_v^2 + 2\zeta_v \omega_v \omega_l^2 + c_a \qquad 6.9$$

$$a_2 = \omega_l^2 + \omega_v^2 + 4\zeta_l \zeta_v \omega_l \omega_v \qquad 6.10$$

$$a_3 = 2\zeta_l \omega_l + 2\zeta_v \omega_v \qquad 6.11$$

where:

$$c_a = \frac{A_p}{M_v} \left[\frac{P_v}{V_a}\right] Q_v, \qquad \zeta_l = \frac{B_l}{2A_m} \sqrt{\frac{V_a}{P_v M_l}}$$

$$\omega_l^2 = \frac{A_m^2}{M_l} \left[\frac{P_v}{V_a}\right], \qquad \zeta_v = \frac{B_v}{2K_v M_v}$$

$$\omega_v^2 = \frac{K_v}{M_v}$$

The condition for stability is:

$$a_1(a_3a_2 - a_1) > a_3^2a_o \qquad 6.12$$

When

$$a_1(a_3a_2 - a_1) = a_3^2a_o \qquad 6.13$$

the system is marginally stable.

Fortunately the ratio ω_l/ω_v is usually less than 0.10 so that higher powers of ω_l can be neglected. The stability criterion can then be stated.

$$\zeta_v\zeta_l^2\left[\frac{2\omega_v}{c_a}\right]\frac{\omega_l}{\omega_v} > \zeta_v\left[\frac{\omega_2}{\omega_v} - 1\right] + \frac{c_a}{2\omega_v^3} \qquad 6.14$$

Several interesting conclusions can be drawn from 6.14.

1. If the load damping ζ_l is zero, the system stability is determined by the relative magnitude of ζ_v, the valve damping, and $c_a/2\omega_v^3$, the valve natural frequency.

2. Without adequate relief valve damping ζ_v, the system will be unstable.

3. Increasing the accumulator volume will increase the margin of stability, increase the damping ratio ζ_l and decrease the load system natural frequency ω_l.

6.3. PRESSURE REDUCING VALVES

Occasionally, portions of a hydraulic system must be operated at a pressure lower than that generated in the balance of the system. A pressure reducing valve is used to drop the pressure to the desired level. Several types are used:

1. *Pressure reducing valve.* This valve, Fig. 6.6, reduces upstream pressure to a predetermined downstream pressure regardless of upstream pressure variations. In this design the downstream pressure is used to balance the spring setting. The valve piston forms an orifice which drops the pressure to the value established by the spring setting. Because the upstream pressure is balanced within the valve, variations in the upstream pressure have little effect on the downstream pressure.

2. *Back pressure regulators.* These regulate upstream pressure regardless of downstream pressure variations. Most relief valves fall in this classification.

3. *Differential pressure regulators.* This class of valves, Fig. 6.7, maintains a constant pressure difference between upstream and downstream conditions. Here the force balance on the valve spool must consider both upstream and downstream pressure as well as the combined rate of the springs.

Pressure reducing valves can be made to modulate flow or to act as ON-OFF devices. They can be damped or cascaded. No matter how they are used, they react with the system in exactly the same way as relief valves. The same sort of balance between system design and component characteristics must be maintained.

Combinations of these valves and of control valves can be used to handle special requirements. For instance, a flow regulating valve can be made from a combination of a fixed and a variable restrictor, Fig. 6.8. The pressure drop across a fixed orifice is proportional to the square of the flow rate. If this pressure drop can be converted to a force which —when working against a spring—can vary the pressure drop of a properly proportioned variable resistance, the valve can be made so as to provide essentially constant flow despite wide variations in inlet pressure.

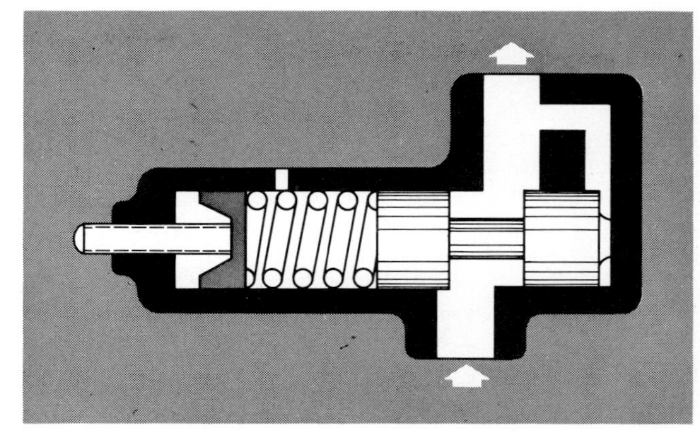

Fig. 6.6. Controlled downstream pressure regulator.

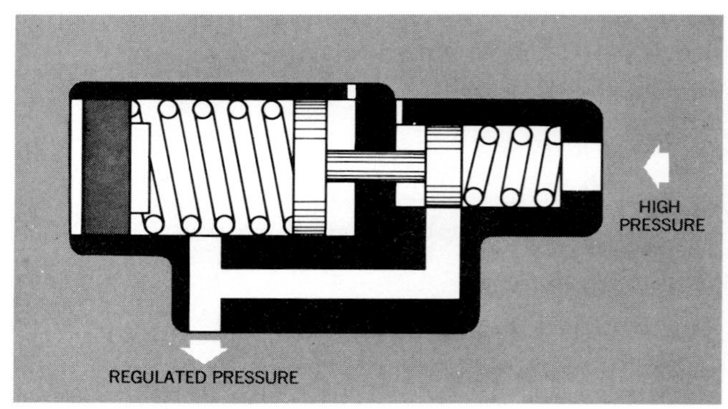

Fig. 6.7. Differential pressure regulator.

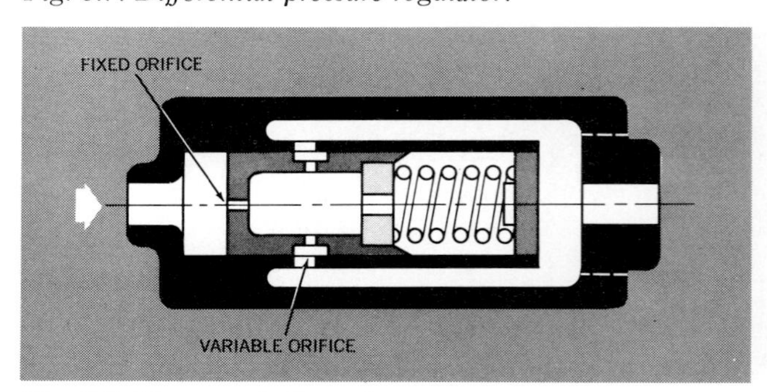

Fig. 6.8. Flow regulating valve.

Relief valves can also be used to establish the priority of one portion of a hydraulic system over another. For example, a branch of a control system might receive its oil supply through a relief valve which is set at some pressure below the system working pressure. So long as the system pressure remains high enough, all parts of the system receive oil. If the system pressure falls below the priority valve setting, the branch on the priority valve will receive no oil. All of the flow will go to branches with a higher priority until the high pressure condition is restored.

In another form of priority scheme the valve which controls flow in one branch is operated by pressure in another branch. Only when the pressure in the high priority branch is sufficiently high may oil be directed to the low priority branch.

6.4 TUBING AND PIPING

Normally, the piping in industrial hydraulic systems need only be sufficiently strong. Only occasionally is weight or space critical. Mobile hydraulic systems in tractors, trucks, diggers, etc., must consider space far more carefully. Shipboard hydraulic systems are usually quite space critical and—especially in submarines—weight critical as well. Weight is the paramount factor in tubing in aircraft and space vehicle hydraulic systems.

Where steel pipe is used, the strength recommendations of the American Petroleum Institute should be followed. Commercial and industrial systems using drawn tubing should be designed in accordance with manufacturer's data. Military systems often use Specification MIL-T-7081 for 6061 aluminum alloy tubing, MIL-T-6845 for ⅛ hard type 302 or 304 corrosion resistant steel tubing and MIL-T-8808 for annealed type 321 or 347 corrosion resistant steel tubing.

The three military specifications require that the tubing shall be capable of withstanding an internal hydrostatic pressure calculated from the following formula:

$$P = \frac{2KtF_{tu}}{D - 1.4t} \qquad 6.15$$

where: $K = 0.6 + 0.4\frac{F_{ty}}{F_{tu}}$

t = tubing wall thickness

D = tubing OD

F_{ty} = yield tensile stress

F_{tu} = ultimate tensile stress

Equation 6.15 may be compared to equation 1.3, the "Thin-wall" formula. Equation 6.15 is the more conservative of the two because it assumes that the pressure acts on a projected width which is greater than the tubing inside diameter. The hoop tension S in equation 1.3 is replaced by the factor $0.6F_{tu} + 0.4F_{ty}$ *in* 6.15.

For thick walled cylindrical members ($t > 0.1r_1$) the formulae must be modified to account for the radial variation in stress. The hoop tension at the inner wall is

$$S_1 = P\frac{r_2^2 + r_1^2}{r_2^2 - r_1^2} \qquad 6.16$$

The meridional stress is

$$S_2 = P\frac{r_1^2}{r_2^2 - r_1^2} \qquad 6.17$$

The stress in a radial direction acting at the inner wall is

$$S_3 = P\frac{r_2^2}{r_2^2 - r_1^2} \qquad 6.18$$

For thick walled tubes the displacement in a radial direction at the inner bore due to pressure is

$$\Delta r_1 = P\frac{r_1}{E}\left[\frac{r_2^2 + r_1^2}{r_2^2 - r_1^2} + \mu\right] \qquad 6.19$$

Where weight is critical, the system designer must look carefully to the effect on the tubing thickness of raising both the ultimate tensile stress and the ratio of the yield stress to the ultimate. An increase in either of these means a possible reduction in wall thickness for the same pressure rating. These considerations have led to the use of ¼ hard type 304 tubing and to the employment of very strong alloys such as AM 350.

The type of tubing connector or "fitting" to be used has a bearing on the choice of material and material condition. In general, tubing joints which require deformation of the tubing are more difficult to make when the tubing is harder and stronger. Such tubing joints include flared connections, swaged connections and "bite" type connections such as the MS flareless joint. Soldered, brazed, welded or cemented connections are not affected, but the thermal effects of brazing and welding may anneal the hard tubing in the region of the joint.

6.5. CONNECTORS

Connectors—more frequently called fittings—are used to join lines and to connect lines to components. No matter what the configuration of the connector used, it performs three tasks: (1) the connector must join to the line in a firm, leakproof manner, (2) the connector must carry any loads imposed on it by the hydraulic system or by the line and (3) the connector must provide a seal between the parts being joined.

All connectors can be grouped into two general categories: separable and permanent. The separable connector contains at least one mechanical joint which can be easily separated and reconnected a number of times. Separable connectors include pipe joints, flared joints such as SAE, AN, MS and MC connectors, flareless bite-type connectors of various configurations and connectors which utilize a removable gasket or other sealing element. Permanent connectors may be brazed, welded, swaged, soldered or adhesively bonded.

The choice as to the type of connector used is usually determined by the nature and intended use of the hydraulic system. In general, the system designer should incorporate as few connectors as possible and permanent joints should be made wherever reconnections do not need to be made. This is because making a connection can generate contamination in a system and each time a system is opened, dirt, oxygen and water can get into the system. Also, any joint tends either to be structurally weaker than the lines it joins or else it weakens the line in the area of the joint. Where weight and size are design considerations, the use of permanent connectors can provide significant reductions.

6.6. SEPARABLE CONNECTORS

Most hydraulic system joints are made with threaded connectors. These can range from simple pipe joints (which may be more akin to permanent than to separable connections) to extremely carefully controlled flared connectors used on space vehicles. The most elemental form of threaded connector is shown in Fig. 6.9. This consists of flanges which must be connected in some way to the lines, a nut to pull the flanges together and some form of sealing element. The sealing element can be an elastomer, a plastic, some form of gasketing material, soft metal or part of the tubing.

Fig. 6.9 is called a series pre-loaded joint. All of the compressive force of the preload induced by the threaded connection is transmitted through the sealing element. Any change in applied load is felt by the seal. The nut or other tension member must have enough compliance to allow the sealing member to flow or relax without the preload being reduced below the level necessary to contain leakage.

In the parallel preloaded joint, Fig. 6.10, the seal only carries the compressive load up to the point where the connector flange halves are drawn together. At this point, sealing efficiency depends more on the design of the sealing element than on the force induced by the threaded connection.

During all operating conditions, the nut must maintain enough compressive load on the seal to prevent leakage. The factors which affect the ability of the nut to provide this minimum load are (1) the hydraulic pressure acting on the seal, (2) the hydraulic pressure acting on the tube ends, (3) bending loads caused by misalignment or by the influence of a vibrating environment (4) dimensional changes due to thermal gradients across the connector and (5) dimensional changes due to temperature changes. The compressive load which the nut can exert to counteract these forces is limited by the yield stress of the nut material, the critical tensile dimension of the nut and the maximum operating temperature of the connector.

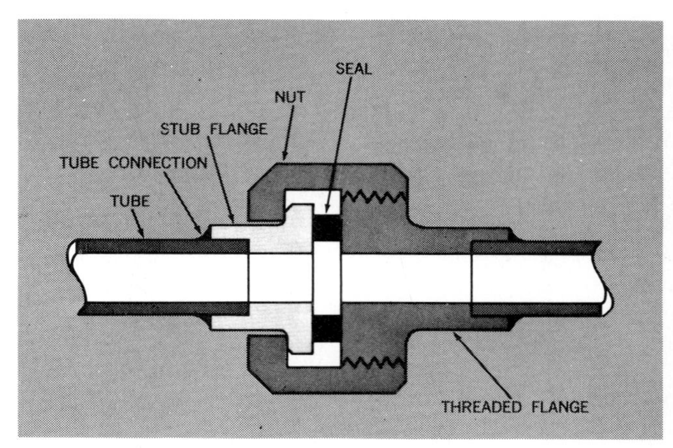

Fig. 6.9. Series pre-loaded joint.

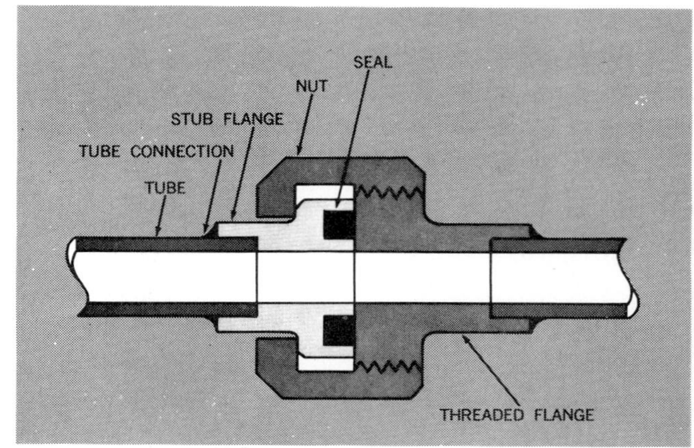

Fig. 6.10. Parallel preloaded tube connector.

Pressure on the seal can increase the sealing effectiveness or can cause the seal ultimately to fail depending upon the design. This consideration will be discussed more fully in Chapter 10.

The pressure end load results directly from system pressure

$$F_e = \frac{\pi}{4} d_e^2 P \qquad 6.20$$

where d_e = effective seal diameter

If the system will be subjected to any pressure pulses or pressure surges, the maximum value of these must be used rather than the working system pressure.

The bending load which can be applied by misaligned tubing on a well supported connector, Fig. 6.11, is determined from the misalignment force times the distance from the fitting. The nut can be considered as a beam, so the stress becomes

$$S = \frac{F\,l}{Z} \qquad 6.21$$

where S = bending stress

Z = section modulus of nut $\left(\frac{I}{c}\right)$

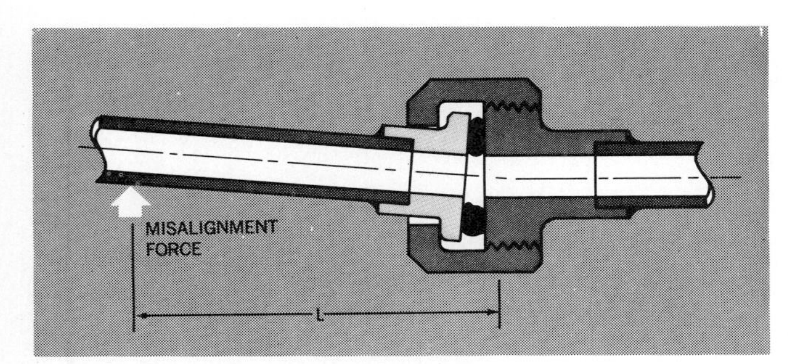

Fig. 6.11. Bending load on connector.

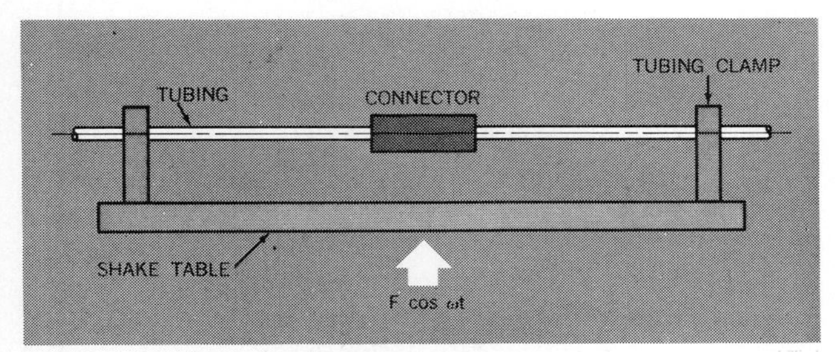

Fig. 6.12. Effect of vibratory environment.

If the misalignment deflection is known rather than the misalignment force, the force can be computed from the deflection equation for cantilever beams of cylindrical cross section

$$F = \frac{3\pi E y (D^4 - d^4)}{64\ l^3} \qquad 6.22$$

where: y = amount of deflection
E = modulus of elasticity

Where the structural condition of the piping and connector is different from that shown in Fig. 6.11 the appropriate beam formulae should be applied.

The effect of a vibratory environment on the connector can be assessed by assuming a tubing section with the connector in the middle, Fig. 6.12. The tubing is firmly supported at its ends. In a vibratory environment, the end supports will impose on the tubing a driving force which can be assumed to be periodic and describable by

$$F(t) = F \cos \omega t \qquad 6.23$$

For a conservative estimate, assume the mass of the tubing and connector to be concentrated at the connector. The tubing will have a spring rate K. The force balance with equation 6.23 is

$$F(t) = F \cos \omega t = M\ddot{y} + Ky \qquad 6.24$$

This can also be written

$$\frac{F}{M} \cos \omega t = \ddot{y} + \frac{K}{M} y \qquad 6.25$$

Assume that the deflection y has a periodic variation with time

$$y(t) = A \cos \omega t + B \sin \omega t \qquad 6.26$$

where A, B = constants

The second time derivative of 6.26 is

$$\ddot{y} = -A\omega^2 \cos \omega t - B\omega^2 \sin \omega t \qquad 6.27$$

Substituting 6.26 and 6.27 into 6.25

$$\frac{F}{M} \cos \omega t = A \cos \omega t \left[\frac{K}{M} - \omega^2\right]$$

$$+ B \sin \omega t \left[\frac{K}{M} - \omega^2\right] \qquad 6.28$$

From the relationship of 6.26

$$y(t) = \frac{F/M \cos \omega t}{K/M - \omega^2} \qquad 6.29$$

Under the worst conditions of deflection, $\cos \omega t = 1$

$$F_{max} = y_{max} (K - M\omega^2) \qquad 6.30$$

If the deflection under vibration can be measured and the frequency (in radians per second) determined, the force causing the vibration can then be computed. This force and the tubing length can be used in equation 6.21 to evaluate the stress due to vibration.

Equation 6.30 provides some other information of value to the system designer. For any driving force F, the flexure will be reduced if the tube stiffness is increased. An increase in tube diameter or wall thickness will increase the stiffness. Note, however, that an increase in mass may—if $M\omega^2$ is smaller than K—increase the amount of flexure. Therefore, an increase in tubing diameter is better than merely increasing the wall thickness. When the vibration frequency is at the section natural frequency, a theoretically infinite deflection could result from a zero driving force. Fortunately, damping internal to the tube and connector will prevent this, but tube sections that have natural frequencies which match the driving vibration frequencies should be carefully avoided.

When hot oil flows through a connector, the internal compressively-loaded parts will tend to expand more than the external nut. This results in a tighter fit and a greater load on the sealing member. Of course, if the expansion is too great, the compressively-loaded members may overstress the nut and cause it to fail directly or as the result of thermal cycling. When a cold oil flows, the effect is opposite, that is, the load on the sealing member is reduced. The preload must be great enough to assure that this reduction does not allow leakage.

Elastomeric materials which might be used as sealants have high coefficients of thermal expansion. Care must be taken that such seal materials do not contract away from the preloading members when subjected to very low temperatures.

One way of keeping track of the several loads and load reactions is to construct a load-deflection diagram, Fig. 6.13. The curve is the load-deflection characteristics of the tensile member. This curve passes through the plot origin. As the load is increased the deflection increases uniformly. A minimum load, as shown, is necessary in order to contain low pressure. The preload force on cold compression members must be greater than the minimum sealing load. The load-deflection curve for the compression members starts from the load point on the tensile curve and has a slope which is negative with respect to the tensile curve. The total deflection of the compression members is given by the dimension between the ordinates.

When the system pressure is applied to the connector, the tensile load is increased. This additional tensile load would also relieve the compressive load due to the preload. This is not a satisfactory condition as the low compressive load would allow leakage. Therefore, the pressure must be dropped, and a new preload chosen as shown, in order to handle the pressure load. Similarly, any loading due to flexure, or misalignment would require a higher preload.

When the system temperature is raised the compression members expand thermally and a new group of plots is developed. The maximum strength of the nut must be great enough to handle the hot preload forces plus the pressure load plus any flexure or misalignment loads.

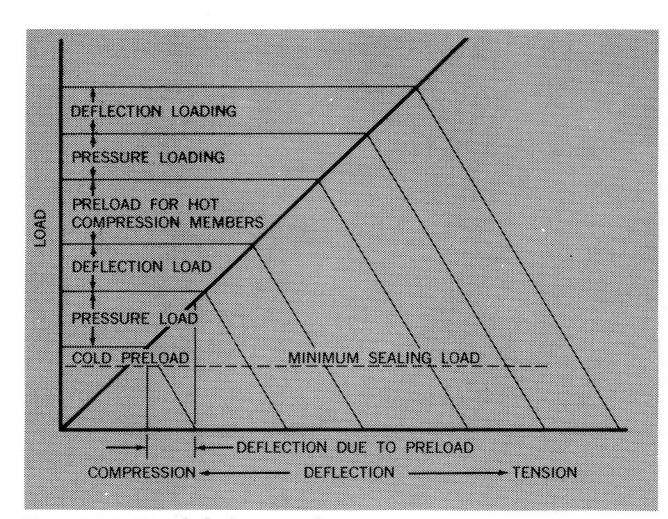

Fig. 6.13. Load-deflection diagram.

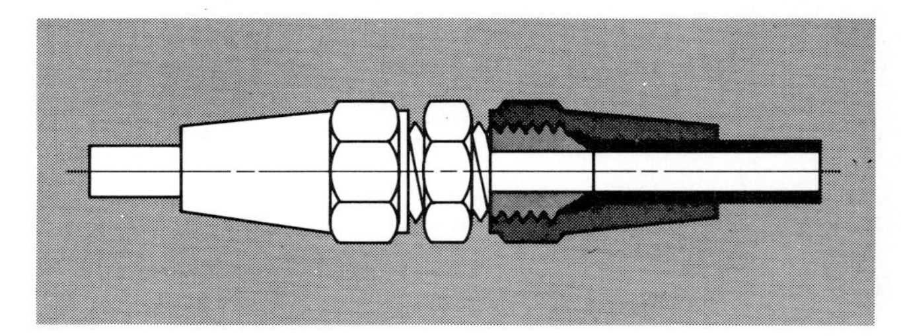

Fig. 6.14. Simple 45° flared connector.

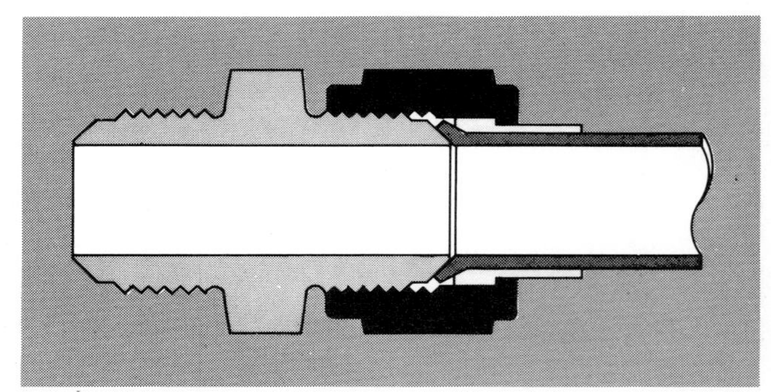

Fig. 6.15. 37° flared connector.

6.7. TYPES OF SEPARABLE CONNECTORS

Perhaps the connector most widely used in power hydraulic systems is the flared tube joint. Fig. 6.14 shows the simplest form in which the screwed assembly of a nut and fitting traps the flared end of the tubing under compression. This simple form of connection usually uses a 45° flare. Fig. 6.15 is a more advanced form of flared connection in which a 37° flare is trapped between the fitting and a sleeve. In this design, no relative rotation exists between the tubing and any contacting members. This design is typical of AN, MS Flared, MC (NASA), JIC and SAE flared connector systems.

When properly made, a flared tube connection is an extremely good joint. One drawback to flared connectors lies in the method of manufacture. Because of the competitive nature of sales of tube connectors, manufacturers are forced to use low cost fabrication methods. Among these methods is "plunge cutting" of the body contour. In plunge cutting, a formed cutting tool shapes the body contour in a single rotation of the work piece. Because of variations in cutting efficiency in advancing and retracting the cutting tool, the perimeter of a cross section of the nose cone of the fitting usually—when sufficiently magnified—has a shape similar to that shown in Fig. 6.16. The use of such a fitting will cause leaks unless the tubing is quite ductile or the connection is very firmly made.

Most tube end flares are made in hand or power tools in which the tube end is swaged out over a split die. Flares made in this manner tend to have a ridge of metal extruded from the outer surface at the parting line of the die. This surface non-uniformity can aid in providing a leakage path. Several companies make small truncated cones of ductile metal which can be placed over the nose cone of flared connectors. When the joint is tightened up, the ductile cone tends to fill in the non-uniformities in the fitting nose and the tubing flare. These conical seals work quite well, particularly when the tubing is of intractible material such as stainless steel.

In order to produce a sound, leak-free joint with the flared system without resorting to use of conical washers, NASA has developed the MC fitting. This connector looks exactly like the AN and MS flared fittings and is interchangeable with them. The MC

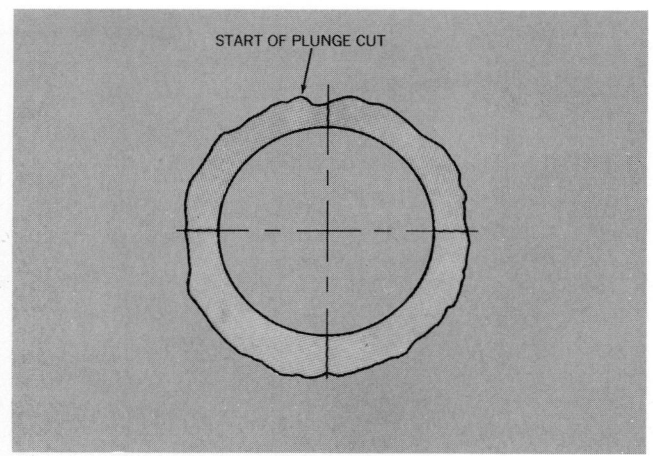

Fig. 6.16. Magnified perimeter of section of nose cone.

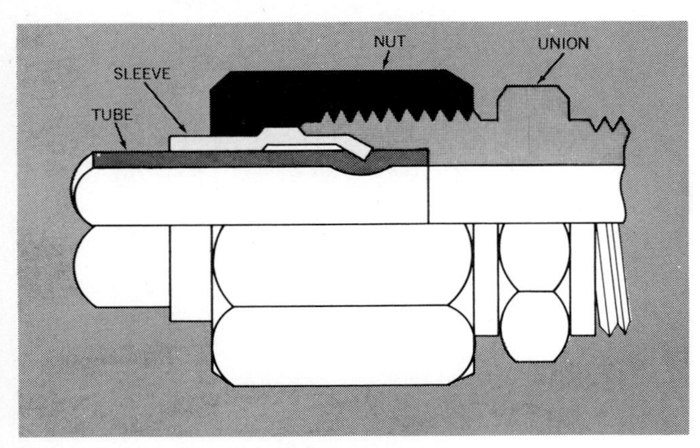

Fig. 6.17. Bite-type connector.

fitting, however, calls for extremely tight tolerances on the roundness of the nose cone and the straightness of elements of the nose cone. In manufacturing the MC fitting nose, a single point cutter is used which generates the equivalent of a very fine screw thread. The threads are usually removed by hand polishing subsequent to the machining operation.

Instead of relying on a tube flare, some connector systems use a sleeve which cuts or embosses a groove on the outside of the tube either during connector assembly or as a result of a presetting operation. Fig. 6.17 shows the MS flareless connection system which is typical of these designs.

A number of proprietary separable connector systems are on the market. Most of them use combinations of flares and biting ferrules to perform the function of gripping the tubing.

6.8. PERMANENT CONNECTORS

Any line connection which doesn't require later separation can use a permanent form of joint. Sometimes, as shown in Fig. 6.18, a separable connector is attached to its tubing by means of swaging, brazing or welding. Occasionally soldering or adhesive bonding is used but these have not found favor in high pressure systems.

When compared to familiar separable connectors, permanent connectors are leak-free, reliable, small, low in weight and they usually develop low pressure loss. The drawbacks of permanent connectors are large capital investment, difficulty of inspection, lack of an easy method of making repairs and the necessity of making all joints in a well equipped shop.

Silver brazed connectors have been used on submarines for some time. The connectors—usually of cast bronze—have a groove which contains a ring of brazing filler metal, Fig. 6.19. In making the joint, the pipe-fitter heats the pipe and fitting with a torch until the filler metal melts and flows around the clearance by capillary action. The presence of a fillet all around the joint is used as proof of the goodness of the joint. In order to improve the quality of the joint and to reinforce the inspection procedure, more recent practice has included heating the joint inductively by electricity. The actual wetting of the faying surfaces is measured ultrasonically.

Portions of the hydraulic systems of several military airplanes have been both silver brazed and gold brazed. Because the smaller airplane lines cannot use the ultrasonic inspection technique, acceptance of the joint often relies on the use of X-ray photographs of the brazing pattern.

Brazing is similar to soldering in that parts are joined by wetting by a separate filler metal. However, brazing requires that the operation be con-

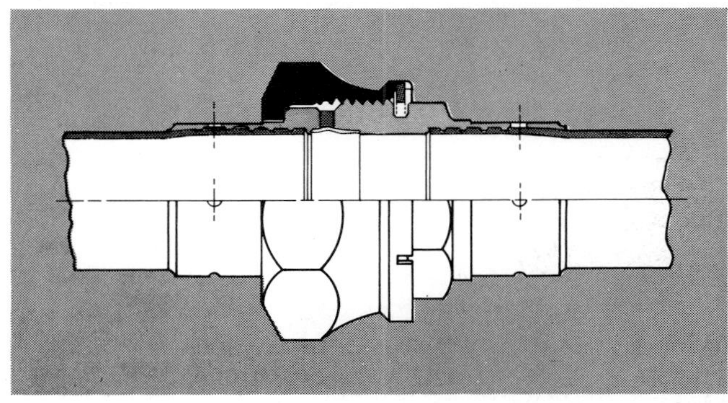

Courtesy The Deutsch Company.

Fig. 6.18. Tubing attached to fitting by swaging.

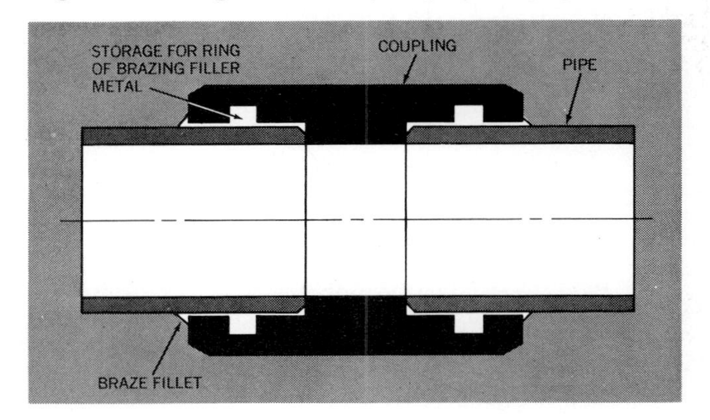

Fig. 6.19. Submarine type brazed hydraulic connection.

ducted at a temperature in excess of 800°F. Any oily material or surface oxides will prevent the filler metal from wetting the faying surfaces. In most brazing applications, oily materials are removed mechanically or by washing. Brazing flux is used to absorb surface oxides and to prevent the formation of oxides during the brazing operation.

Most brazing fluxes are glassy borate compounds. Where, as in high performance servo systems, the presence of glassy contaminants in the hydraulic system is undesirable, fluxless brazing must be used. In fluxless brazing, both the inside and outside of the joint must be bathed in an extremely inert gas during the brazing operation. Especially dried argon is the usual inerting gas. Often the brazing filler metal for fluxless brazing will contain a small amount of lithium. The lithium will attack any small amount of oxides which might inadvertantly be formed during the brazing cycle.

Because brazing depends upon the capillary flow of the molten filler metal, the faying surfaces must be round and concentric. For maximum strength, the gap between the faying surfaces should be between 0.0005 and 0.003 inches. This means that the tubing OD should be held to a total tolerance of 0.003 inches for all diameters. Ovality must be strictly controlled. For best results, the tube ends and the fitting should be clamped during the brazing operation to avoid axial and angular misalignment.

Welding has been proposed to eliminate the brazing problems. Where heavy wall piping is used, hand welding works quite well. Light walled tubing, however, requires aligning jigs and gas inerting. The usual technique is TIG (Tungsten Inert Gas) welding in which a non-consumable electrode is driven around the tubing in an argon atmosphere. One or two passes may be employed, Fig. 6.20. The configuration may vary from butt welding, with or without a filler ring, to sleeve welding.

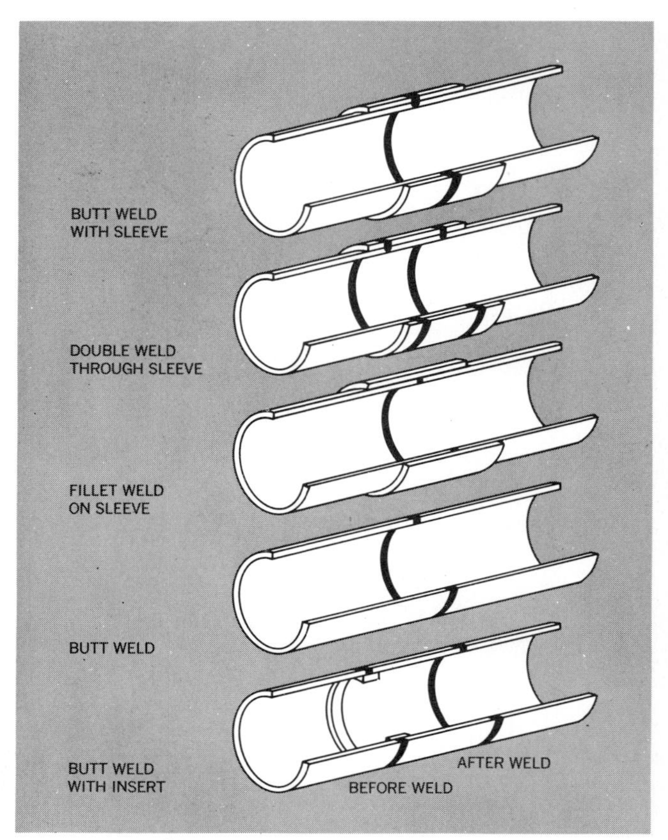

Fig. 6.20. Weld joints.

6.9. FLEXIBLE CONNECTIONS

Wherever relative motion exists between two things which must be connected hydraulically, a flexible connection must be made. Typical of such requirements are the flexing arms of power shovels or the rotation of a trunnion-mounted hydraulic cylinder. Most such flexible connections are made with hose. Hydraulic hose may be strengthened elastomeric tubing fitted with metal end connections or it may be a metal bellows strengthened by external wire braid. The hose may be lined with PTFE tubing for low pressure drop and low contamination generation.

The installation of hose is easy and its use reliable if the manufacturers recommendations are followed. Some users prefer to install hose under some compression. Subsequent pressurization will extend the hose to a neutral length. Operating life is claimed to be better with this type of installation than when the hose is at neutral length when unpressurized.

When compared to piping or to rigid tubing, hose expands considerably when pressurized. This relative softness makes the hose act as if it were on accumulators. A system using hose acts if it had an oil of low bulk modulus of elasticity. Most hose manufacturers do not measure bulk modulus directly. However, tests to military specification MIL-H-38360 and to SAE document ARP 604 do include a test of volumetric expansion. The data from these tests can be extrapolated to yield the combined bulk modulus of the hose and the contained oil. Because the hose is far softer than the oil column, these data can be safely used for all common hydraulic oils. Table 6-1 gives values of bulk modulus (β) which was developed based on data supplied by the hose manufacturers.

As these data are based on highest quality hose, other hose might be expected to yield values of effective bulk modulus somewhat lower than those shown.

6.10. COILED TUBING

Cooke and Stouffer[2] developed the theory of the use of coiled tubing for flexible connectors under a government contract. Their findings were subsequently augmented and published in SAE document ARP 584. The following discussion is based on their studies.

The best combination of dimensions for coiled tub-

TABLE 6-1
EFFECTIVE BULK MODULUS OF HOSE AND OIL

Hose Size	β
Extrapolated from data supplied by Aeroquip, Jackson, Mich., on high pressure (677 series) hose	
-4	31,500 psi
-6	40,000
Extrapolated from data supplied by Resistoflex Corp., Roseland, New Jersey on PTFE lined hose	
Medium Pressure Hose	
-4	46,870
-6	37,180
-8	47,470
-10	53,500
-12	38,200
High Pressure Hose	
-4	28,800
-6	52,900
-8	47,500
-10	48,300
-12	49,500
-16	44,900

TABLE 6-2

Tube Size	Wall Thickness	Pitch	Radius
-4	.028	.375	1.000
-5	.035	.438	1.125
-6	.042	.500	1.375
-8	.058	.625	1.750
-10	.078	.750	2.250
-12	.083	.875	2.625
-16	.120	1.125	3.500

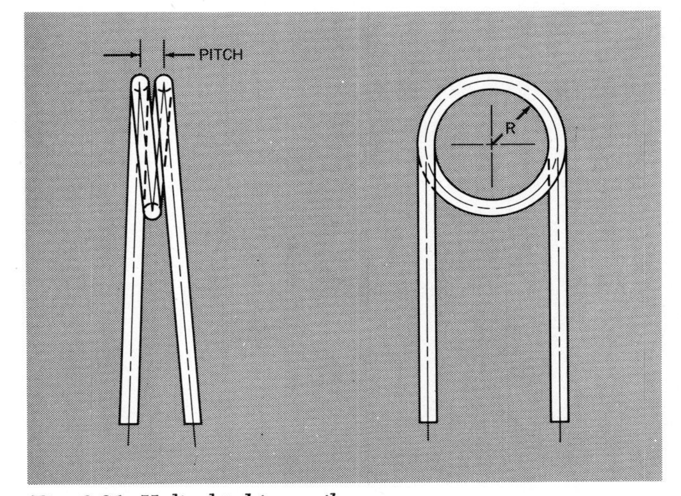

Fig. 6.21. Helical tubing coils.

ing yields a maximum deflection of 7 degrees per coil. Table 6-2 gives a list of dimensions for some sizes of Type 304 stainless steel tubing. These relate to a coil of the configuration of Fig. 6.21. They are based on an adverse combination of 200,000 cycles of mechanical flexure and 200,000 cycles of pressure pulsation.

A similar set of dimensions refer to torsional tubing shown in Fig. 6.22 and Table 6-3. For systems which operate at temperatures no higher than 160°F, a maximum twist of 13 degrees is recommended. Lower limits apply to torsion tubes in systems operating at higher temperatures.

One manufacturer offers tubing pre-coiled into a tight spiral for direct replacement of hose. This spiral tubing can be formed into simple and compound curves for special applications so that bending stresses induced by the relative motion can be minimized.

Coiled tubing which is subject to pressure cycling eventually fails at the neutral axis. This is because the tubing is distorted in cross section during bending. Originally round, during bending it is forced into a rounded triangle with one apex on the inside of the bend. Subsequent pressure cycling tends to return the tubing to its original circular cross section. This action puts the tubing wall near the neutral axis into bending. Such a flexure at the neutral axis occurs

TABLE 6-3

Tube Size	Wall Thickness	Radius	L	A
-4	.028	1.000	11.125′	4″
-5	.035	1.125	14.000′	5″
-6	.042	1.375	16.750′	6″
-8	.058	1.750	22.250′	8″
-10	.078	2.250	27.875′	10″
-12	.083	2.625	33.375′	12″
-16	.120	3.500	44.500′	16″

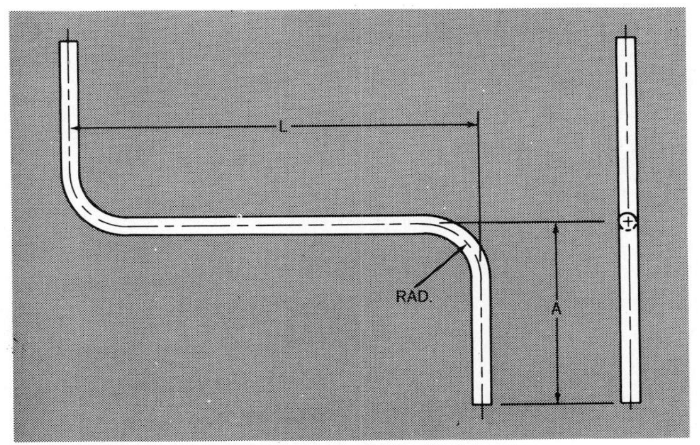

Fig. 6.22. Torsional tubing bends.

each time the internal pressure changes. Fatigue failures can occur as a result. For best service life, tube flattening should be limited to a maximum of 3% reduction of the tubing outside diameter.

A stack of tubing coils is subject to oscillations excited by a vibratory environment. If such oscillations or if service flexing causes the adjacent coils to contact each other, galling and wear can occur. The coils can be interleaved with damping material or wrapped with cord to damp out oscillations. Tying the stack of coils together often results in early failure due to the development of local stress concentrations.

6.11. SWIVEL FITTINGS

In some installations, rigid tubing terminating in a swivel fitting provide an excellent solution to the problem of making hydraulic connections between parts having relative motion. Both cylindrical and spherical swivels are used. Nearly all cylindrical swivels are variations of the basic design, Fig. 6.23. A cylindrical outer body is free to rotate relative to the inner body. Oil is taken axially into the inner body and directed radially through one or more ports to the outer body. Suitable retainers prevent relative axial movement between the body parts.

Spherical swivels are based on the general design, Fig. 6.24. The spherical swivel allows for some axial angularity as well as for relative rotation.

6.12. EXTENSION FITTINGS

Crissey has described an ingenious fitting which combines swivel connections with a trombone action. This is the pressure balanced and volume balanced extension connector, Fig. 6.25. These connectors have been tested successfully for 2 million operating cycles at a minimum rate of 25 cycles per minute with simultaneous impulse loading at 35 cycles per minute. The manufacturer claims that the O-ring seals operate 1 million cycles between changes.

REFERENCES

1. Reethof, G. "On the Dynamics of Pressure Controlled Systems" Paper #54-SA-7, ASME Mfg, Pittsburgh, June 1954
2. Cooke, C. H. & Stauffer, R. D. "Coiled Tubing for Aircraft Service" WADC Technical Report 55-121, FEB. 1955

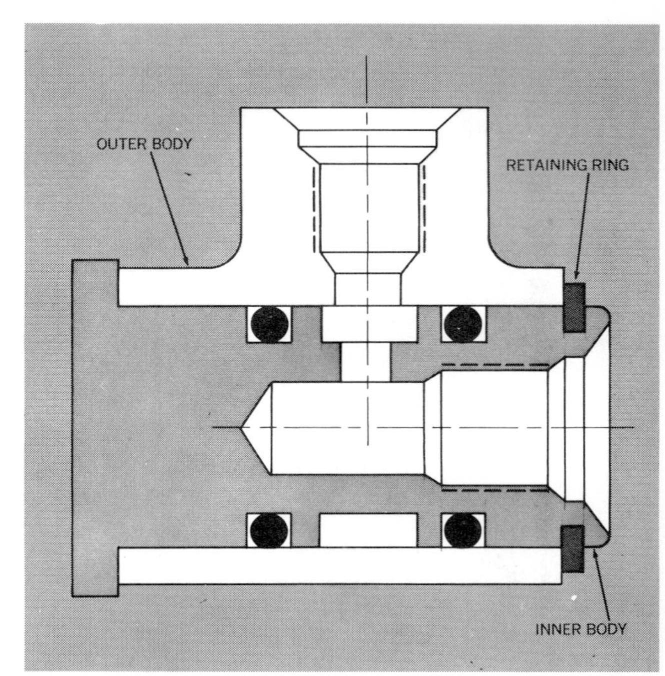

Fig. 6.23. Cylindrical swivel.

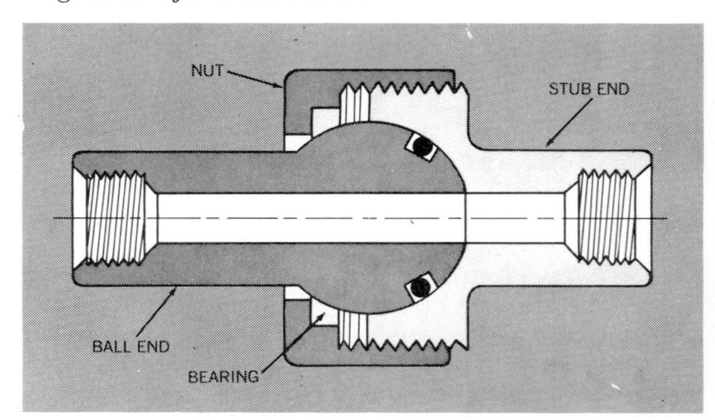

Fig. 6.24. Spherical swivel.

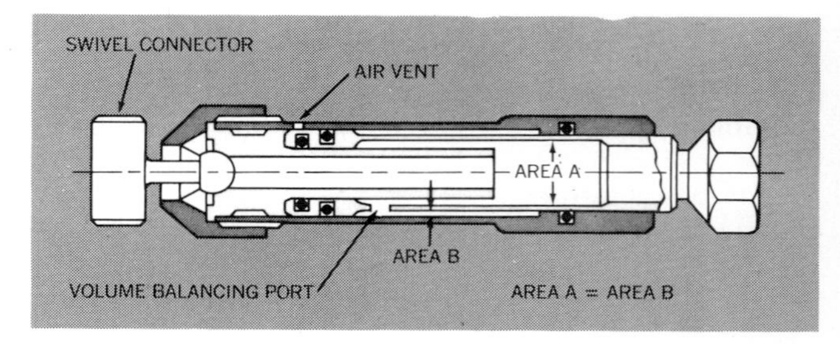

Fig. 6.25. Extension fitting.

CHAPTER SEVEN

Contamination and Filtration

7.1. CLEAN HYDRAULIC SYSTEMS

When electro-hydraulic servo valves were first used extensively on control mechanisms, the metering edges wore rapidly and the valves had a short useful life. Occasionally, a stuck valve was reported. At that time—the early to mid 1950's—there was no adequate method for evaluating hydraulic filters and no means for determining fluid cleanliness. The solution was to place many filters in the system and hope for the best.

At that time, filters were available only in phenolic impregnated paper for moderate temperatures and in sintered bronze balls for higher operating temperatures. Filters were, and often still are, specified by "nominal" ratings in which the filter was supposed to remove 98% of all particles whose two greatest dimensions were larger than the rating size. Unfortunately, even though several filter manufacturers were qualified by the military against this requirement, absolutely no means for proving conformance to the requirement existed.

With the advent of high quality membrane filters, the picture changed rapidly. The membrane filter allowed accurate microscopic counting of contaminant particles and accurate evaluation of available filters. The 10 micron nominal filter, which was the industry standard, was shown often to pass 100 and 150 micron particles. Nominal 5 micron and nominal 2 micron filters were shown sometimes to be poorer filters than the nominal 10 micron filters.

System cleanliness requirements have been related to servo valve clearances and wear rates. Contamination and its sources have been identified and described. Vastly improved filters have been developed. Military filtration specifications which rated filters on their absolute capability to screen out particles larger than the rating have been written. Filters have been qualified to these criteria. Even the inadequate "nominal" rating has been given meaning and significance by trying to test with closely controlled artificial contaminant. The nominal rating has become a measure of effective pore size and could be termed as EPS rating. Because clean systems and a high degree of filtration are now fashionable in hydraulic circuits, many systems may be cleaner than they really need to be. This, however, from an operational reliability standpoint is far superior to being unable to get any system as clean as it should be.

TABLE I

SIZE RANGE IN MICRONS (MICROMETERS)

CLASS	5-10	10-25	25-50	50-100	100 PLUS
0	2,700	670	43	16	1
1	4,600	1,340	210	28	3
2	9,700	2,680	380	56	5
3	24,000	5,360	780	110	11
4	32,000	10,700	1,510	225	21
5	87,000	21,400	3,130	430	41
6	128,000	42,000	6,500	1,000	92

7.2. CONTAMINATION

In a hydraulic system, contamination can be defined as any amount of material which is a chemically reactive hazard or a mechanical impairment to the proper function of the system or any of its component parts. Such contamination can be particulate matter, gums or sludges, water or gasses.

Particulate matter found in hydraulic systems consists of rust, paint chips, metal flakes, glass, fly ash, various fibers, lint, plaster, rubber, sand, weld slag, screws, nails, small tools, etc. Gums and sludges result from combustion products, oxidative breakdown of the fluid and reaction with water or oxygen.

Contamination can result from sources external to the system, can be built into the system or can be generated by the system. The oil, in the process of being formulated, collects dust and ash from the air, water from various sources and process residues from the equipment used to refine and formulate the oil. Steel drums used to ship oil conceal paint, dirt and metal chips in their crevices. Handling shocks and shipping vibration send these into the fluid. When the fluid is put into the working system it is further contaminated by machining lubricants, cutting chips, rubber particles and left-over dirt. Once the system is activated, metal surfaces in moving contact will shear off metal asperities and put them in the system as contaminant.

It is clear that the oil, the components and the system can be dirty at the start. How dirty can they be and how can they be cleaned up?

To provide uniform standards for discussion, a number of technical organizations have established grade levels of contamination.

Each of these rating systems size the contaminant in microns. A micron or micrometer is one millionth of a meter. 1 micron equal 0.0000394 inches. As graphic examples of micron sizes; the human eye can resolve particles of about 40 microns diameter, an average human hair is about 75 microns in diameter and a red blood corpuscle is 7.5 microns in diameter.

Table I shows a classification system that has received tentative agreement from SAE and ASTM. Table I defines the number of particles in each size range allowable in a 100 ml sample.

In the classification of Table I, Class 1 represents the condition of hydraulic fluid MIL-H5606B as formulated. Class 2 is intended to be the contamination level in a good missile system. Critical systems in general might conform to Classes 3 and 4. Class 5 and Class 6 are representative of typical industrial systems.

Huggett[1] reported that Douglas Aircraft measured new MIL-H-5606B oil and found it generally in Class O of Table I. Douglass also measured oil samples from the reservoirs in missile field launchers and found that most of the systems fell in Classes 2 to 4 of Table I. Piccone[2] stated that missiles manufactured at Martin-Denver worked well with hydraulic systems of Class 3-4 of Table I. Some evidence[3] shows that actual system contamination cuts across the classifications shown.

For examples of the size of contaminant particles, designers should consider the typical clearances in electro hydraulic servo-valves, Table II.

TABLE II
CRITICAL CLEARANCES IN
ELECTRO-HYDRAULIC SERVO VALVES

AREA	CLEARANCE
Spool to Sleeve	1-10 Microns Diametrical
Flapper-nozzle	25-38 Microns
Metering Orifice	25-38 Microns
Nozzle Clearance	254-1016 Microns
Bleed Orifice	305-660 Microns

7.3. MINIMIZING CONTAMINATION

The fluid must be clean and shipped and stored in containers which do not allow contamination to enter. Components must be designed and handled to minimize retention of contaminants. System materials must be chosen with potential system contamination as a criterion. Parts must be packaged in contamination-free containers. The assembled system must have adequate filtration.

Achieving cleanliness starts with knowledge and mental attitude. Workers, conscious of the need for cleanliness, must be taught to avoid cutting O-rings on assembly, leaving metal chips inside components and handling clean parts with dirty fingers.

Parts should be designed to be easy to clean. The interior cavities of all hydraulic system components should be smooth and continuous. Pockets, dead ends, labyrinths and cavities should be avoided. The part should be easy to disassemble for cleaning. Feather edges and sharp corners create hanging burrs which later come loose. Threaded joints and screwed fasteners give trouble.

Parts should be cleaned as they are made. Welding, cutting, and grinding should not be done on finished or assembled items. Protective closures must be placed on all parts whenever they are not being used. All threads should be carefully deburred. Whenever possible, assembly should be done in a clean room. Functional testing should be performed on clean test benches using clean oil. When not in use, the parts should have caps or plugs on all ports and openings. The completed and tested unit should be packaged in a contamination-free durable container or bag.

System design concepts and handling procedures are fully as important in preventing contamination as are component design and fabrication techniques. To start off, the system designer should try to determine the amount and nature of the contaminant which the system can tolerate. He should then assure that no chemical incompatibilities exist at any point in the operating temperature band. Of particular importance are the chemical relationships among the fluid, the seals and the metals used in the pumps.

Preventing contamination is easier and simpler than cleaning out dirt. Care in assembly and installation, good work practices and common sense can minimize contamination. Avoid screw-loaded packing glands or be very careful of the type of packing used in them. Use guides in installing O-rings. Minimize tubing lengths and the use of complex fittings. Design the system mounting to reduce the effects of vibration and shock. Avoid opening a joint in a clean system. Be careful about the handling and cleanliness of uncoupled disconnects. Deburr all tube ends. Flush out the system with *clean* oil prior to putting it into service.

7.4. FILTERS AND FILTRATION

After components are assembled in the system and the system flushed out, two problems of contamination control still remain. First, the residual contamination in the components or that generated by assembly techniques must be cleaned out. Second, any contamination generated by the operation of the system must be removed as it is generated.

Contamination removal is the job of filters. Filters are mechanical screens which remove particles from the fluid stream by forcing the fluid to pass through a multiplicity of holes or paths which are so small that they block the passage of particles. Filters used in hydraulic systems are either of the depth type or of the surface type.

In a depth type filter the medium has appreciable thickness in the direction of the flow path. Contaminants can be trapped anywhere along the flow path through the filter medium. Among types of depth filters are: wound wire mesh, sintered porous metal, etched stacked discs, pressed paper, matted fibers, fired ceramic and bags of diatomaceous earth. Depth filters do not have controlled openings. The size of the particles which can pass through the filter is a function of the depth of the filter, the physical nature of the filtration medium and the velocity of the oil through the medium. Depth filters can trap and hold large quantities of contamination. They are often used as bypass equipment to clean up a main system prior to the system's being put in use. By making repeated passes through a large depth filter, a hydraulic oil can become very clean. Many test stands and flow benches use depth filters on a bypass loop constantly filtering the oil in the reservoir.

Depth type filters use many different mechanisms to trap contamination: inertial impingement, absorption, brownian diffusion, interception and gravitation attraction. The usual method of analyzing depth filters assumes that the filter medium consists of a series of surface-type filters.

Assume, as an example, that a hydraulic system contains a volume V of fluid with initial contamination W_o. The fluid is circulated at rate Q through a filter having a total efficiency E. Various mechanisms in the system add contamination at a weight rate A.

At time t, the change in the total contamination in the system dW equals the contamination added Adt minus the amount retained on the filter $(Q/V)\ W\ Edt$, or

$$dW = Adt - W\ Q/V\ Edt \qquad 7.1$$

$$\frac{dW}{dt} + \frac{Q}{V} E\ W = A \qquad 7.2$$

If the initial contaminant is considered and the removal assumed to approximate an exponential decay, the general solution to the amount of contaminant present at any time, as determined by Tao,[4] is

$$W = W_o e^{-\int \frac{QE}{V} dt} + Ae^{-\int \frac{QE}{V} dt} \int e^{\int \frac{QE}{V} dt} dt \qquad 7.3$$

The first term on the right side indicates that the initial contamination is removed at a rate dependent on the flow rate and on the total system volume. It is also removed at a rate which is dependent on the total efficiency E. E changes with time and loading.

The second term suggests that a certain amount of the added contaminant is passing through the filter and is retained in the system.

The key to understanding the relationship of equation 7.3 is the time dependent nature of the filter efficiency. In general, as contaminant is picked up by a filter, the number of openings decreases and the average size of the openings grows smaller. Therefore, in terms of contaminant removal ability, the efficiency of a filter improves with use. Of course, the price paid for this efficiency increase is greater pressure drop. The ultimate in efficiency is a completely clogged filter which passes no contamination—or flow!

At any stage in the useful life of a depth filter, the overall efficiency E is equal to the efficiency e of each layer times n the number of layers. This relationship is used by manufacturers of depth type filters to justify the superiority of depth filters over surface types. Actually *new* depth filters tend to be less efficient and to pass more large particles than new surface filters. The greater capacity of the depth filter usually enables them to be used longer before being clogged. Depth filters also tend to remove more small particles than do surface type filters.

Surface type filters are very short in the direction of flow. They function as fine screens and must catch all the dirt on the first surface presented to the fluid. Contaminant which is not caught on the first surface will pass right on through the filter medium. Surface filters frequently found on hydraulic systems use woven wire mesh or sintered wire mesh media. Porous membranes are used to evaluate other filters or to provide super-fine cleaning. Sometimes several layers of surface type media will be used to provide depth filtration.

7.5. FILTER THEORY

Analysis of surface filter size and efficiency is based on the concept of laminar flow through a multiplicity of straight, constant-diameter capillary passages. Henry D'Arcy developed the relationship:

$$v = -\frac{K}{\mu}\frac{dP}{dl} \qquad 7.4$$

Where: v = Velocity of fluid in capillary passage
P = Pressure
l = Length of capillary
K = Permeability constant

Equation 7.4 can be changed to provide an expression for total flow through a porous medium.

$$Q = \frac{K\ N\ \pi\ d^2\ \Delta P}{4\ \mu\ T} \qquad 7.5$$

Where: d = Capillary diameter
T = Thickness of porous medium (length of capillary)
N = Number of flow passages in porous medium

The ratio of void volume to bulk volume in a porous medium is porosity ϕ.

$$\phi = \frac{N\ \pi\ d^2}{4A} \qquad 7.6$$

Where: A = Surface area of porous medium

Equation 7.6 can be substituted into 7.5:

$$Q = \frac{K\ \phi\ A\ \Delta P}{\mu\ T} \qquad 7.7$$

Equation 7.7 can also be substituted into the Hagan-Poiseuille law:

$$Q = \frac{d^2}{32}\ \frac{\phi\ A\ \Delta P}{\mu\ T} \qquad 7.8$$

Equations 7.7 and 7.8 are the same when

$$K = \frac{d^2}{32} \qquad 7.9$$

This discussion has assumed straight pores. Tortuous flow paths add to the length of the capillary. A tortuosity factor τ (tau) relates the average length of the tortuous path to the length of the ideal straight path. Equation 7.6 becomes

$$\phi = \frac{N\ \pi\ d^2\ \tau}{4\ A} \qquad 7.10$$

and the permeability factor is

$$K = \frac{d^2}{32\ \tau^2} \qquad 7.11$$

A unit called the darcy exists as a measure of porosity. A porous medium has a permeability of one darcy when a single phase (liquid only or gas only) fluid of 1 centipoise viscosity which completely fills the voids of the medium will flow through it under conditions of viscous (laminar) flow at the rate of 1 cc/sec per cm² of cross sectional area under a pressure gradient of 1 atmosphere per centimeter.

The equation defining a darcy (k) is

$$Q = \frac{k\ A}{\mu}\ \frac{dP}{dl} \qquad 7.12$$

where each of the units has values as defined. To use English units, equation 7.12 must be changed

$$k = 2.4\ \frac{Q\ \mu\ \ dl}{A\ \ dP} \qquad 7.13$$

Where: Q = cubic inches per second
μ = microreyns
l = inches
A = square inches
P = psi

As an example of the magnitude of the darcy, consider a 15 micron absolute filter which is 0.06 inches thick in the direction of flow. MIL-H-5606 at 100 F has an absolute viscosity of about 1.67 microreyns (lb-sec/in^2 x 10^6). A typical pressure drop for a 15 micron filter is 7.5 psi per square inch per gpm. Under these conditions the porosity is

$$k = \frac{(2.4)\ (231/60)\ \ (1.67)\ \ (0.060)}{(1)\ \ \ (7.8)}$$

$$= 0.1185 \text{ darcy}$$

7.6. FILTER RATING

Rating systems try to do two things:

1. Define the shape, maximum pore size and pore distribution of the filter matrix to obtain the desired filtration rating.

2. Define the number of pores which must be present in the filter to provide adequate service life.

Specifying these two parameters in meaningful terms becomes tricky because the nature (sand, metal chips, gums, sludges) of the contaminant varies widely from system to system. Even with a single type of contaminant—for example, metal chips—the specifying problem isn't much easier. A depth type filter might do an excellent job of removing long curly chips but freely pass globular chunks of metal. A surface filter, in the same system, might screen out the globular contaminant but pass long stringy particles which have a small cross-section.

In the "nominal" rating, the filter is supposed to catch 98% of all incident particles greater than the rating. The filter is evaluated by subjecting it to a known amount of AC test dust (also known as Arizona Road Dust) of a standard set of size ranges. The effluent is trapped and examined microscopically to see how many particles of what size were passed through the filter. Until membrane filters became available, evaluating to the nominal rating was impossible. When membranes became available, tests on paper-phenolic filters, which were made to specification MIL-F-5504, were conducted using the membranes to trap effluent. Two difficulties: First, the size and percentage grading of the AC dust was not easily checked prior to the test. Second, even if the filter did meet the specification requirement (which most did not), the "nominal" rating put no upper limit on the size of particle which could be passed. The first limitation has been partly overcome by using 10-20 micron glass beads in addition to the AC test dust as a test contaminant in MIL-F-5504.

SAE Committee A-6 sponsored specification MIL-F-8815 which first established an "absolute" rating system. In this system, the filter may not pass *any* spherical particles larger than its rating. Because glass beads are available in very accurately sized lots, they have been chosen as the test contaminant. The pressure drop at a rated flow across the filter element was picked as the best for dirt holding capacity.

Wire cloth can be woven with a twilled Dutch weave, Figure 7.1, which gives accurately spaced openings in the weave. The flow path through these openings is not straight. Therefore, long stringy (dendritic) particles can be caught. The uniformity and accuracy of weave of the wire cloth is very good.

Other methods of building filters to give an absolute rating acceptable for modern hydraulic control systems are wound wire filters, sintered wire mesh

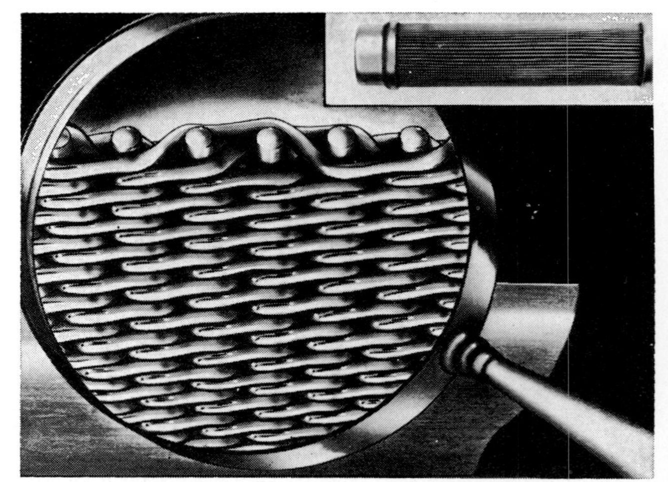

Fig. 7.1. Wire cloth element. Courtesy PurOlator Products

Fig. 7.2. Metal edge element. Courtesy PurOlator Products

filters, membrane and metal edge filters, Figure 7.2.

Surface filters have a limited dirt holding capacity. Because all contaminant must be held on the surface, increased dirt holding capacity comes only with increased surface area. Surface filter elements are usually deeply pleated or folded in order to get more filter area into a small housing. Recently, several filter manufacturers have supplied carefully engineered combination filters which provide, in one assembly, both depth and surface filtration, Figure 7.3. Figure 7.4 shows a photomicrograph of part of a filter element in which a metallic depth filter is bonded to a woven wire mesh filter element.

Figure 7.5. shows a typical plot of the cleanliness of high quality oil before and after filtration. With recirculation, the filter will continue to clean the oil. Contamination particles tend to agglomerate, making big particles out of little ones. As the particles agglomerate, the filter can catch them more easily. Also, a dirty filter catches more fine particles than a clean one because the entrapped particles make the openings smaller.

The total amount of dirt which can be retained by a filter can be measured by choosing an arbitrary value of increased pressure drop as the measure of dirtiness. Figure 7.6. shows the amount in grams of AC Fine dust which can be caught on Dutch weave filters of various micron ratings. The amount is given as specific contamination, milligrams of contaminant per square inch of filter surface, at a pressure differential increase over the original value of 40 psi. Assuming that AC Fine dust represents system contamination, the allowable level of contamination in the system can be computed from Figure 7.6. by multiplying the shown specific contamination value by the design filter area. Conversely, the size of filter can be determined from expected contamination levels.

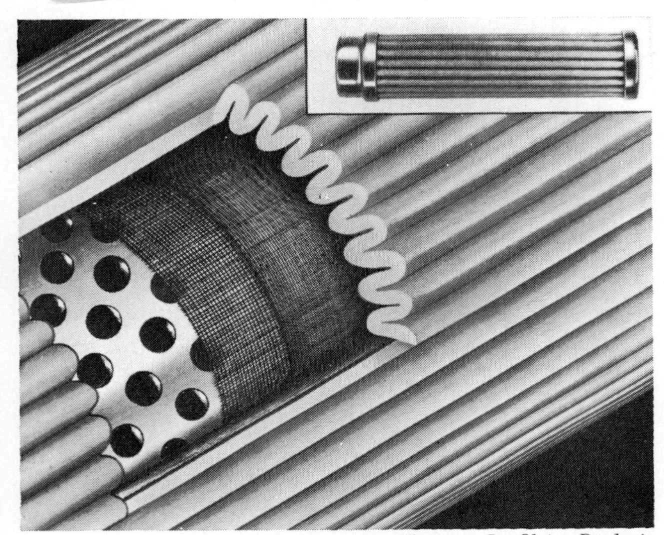

Courtesy PurOlator Products

Fig. 7.3. Depth and surface filtration.

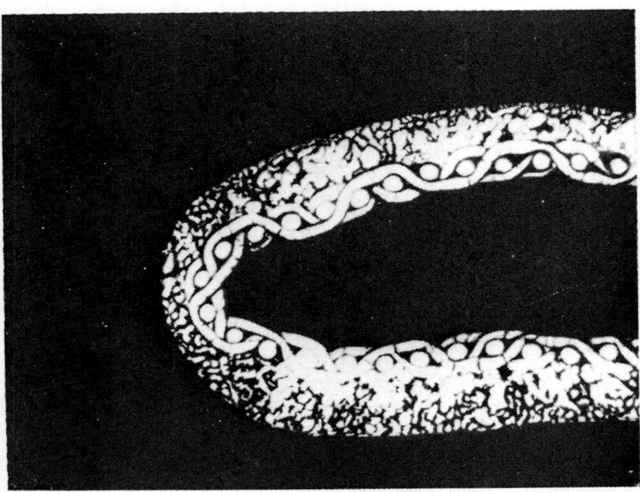

Courtesy Aircraft Porous Media

Fig. 7.4. Metallic depth filter bonded to woven wire mesh element.

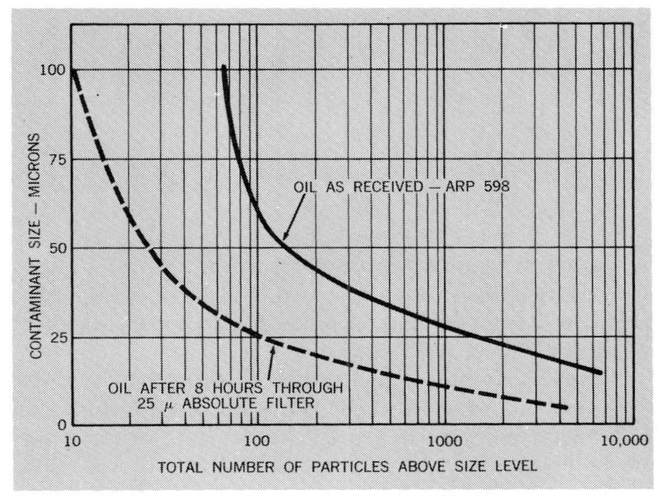

Fig. 7.5. Cleanliness before and after filtration.

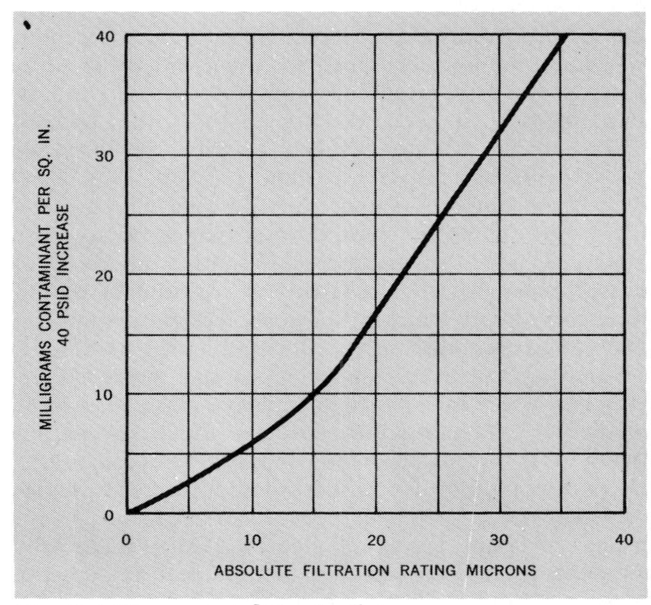

Fig. 7.6. Measure of dirt retention.

7.7. FILTER AND SYSTEM EVALUATION

Much effort has been made to develop accurate means for counting, grading and/or weighing contaminant.

ARP 598 "Procedure for the Determination of Particulate Contamination of Hydraulic Fluids by the Particle Count Method" is used to determine contaminant 5 microns (1 micron = 0.001 mm = 3.94 x 10^5 inches) and larger. In this method, a carefully obtained 100 ml sample of oil is drawn through a membrane filter (HA Millipore or equivalent). The filter must be a 47 mm diameter disc in a holder which exposes 35 mm diameter to the fluid. The disc must contain grid lines on 3.1 mm centers. Thus, each grid square has an area 1/100 that of the exposed area.

By means of a mechanical stage on a microscope, the number and size of particles on ten grids chosen at random are counted and measured. These values are then multiplied by 10. Because no attempt at the methodical selection of areas to be counted is made and because the particle distribution is seldom uniform, the repeatability of this technique is not good.

Detweiler[5] has proposed using a mask which exposes 10 squares on the membrane. These ten squares are so located that every 120° segment has either three or four squares. The pattern also spirals out from the center in such a way that rotation of the mask would not result in duplication of any grid. To allow such choice of grids without using a mask, one manufacturer supplies membranes with an identifying code printed in each grid square.

Several companies have offered automatic particle counters in order to make the task of counting and measuring more easy. One machine requires that the particles be trapped on a membrane as in ARP 598. The membrane is made transparent by means of immersion oil. An intense light beam is passed through a lens system similar to that of a conventional microscope. The beam scans the membrane and impinges on a cathode ray tube. The spot intensity on the cathode ray tube is amplified, counted and monitored. Particle sizing is performed by pulse width selection. This machine works well with spherical particles but gives erratic readings with dendritic dirt.

Another machine counts and sizes particles by utilizing the change in solvent resistance which occurs when a particle suspended in a conducting medium passes between two electrodes. In use, the hydraulic oil is mixed with isopropanol and dichloroethane. The particles, suspended in the fluid, are drawn through a small aperture having an immersed electrode on either side. As each particle traverses the aperture, it momentarily changes the resistance of the liquid between the electrodes. The changing resistance produces a voltage pulse proportional to the size of the particles.

A third counter uses the shadow of a particle as a measure of its size. The fluid sample is forced past a small window. A light beam passes through the window and impinges on a photo tube. Any particle, in passing by the window, causes a change in the output signal from the phototube. The signal changes are amplified and converted into counts of particles of various sizes.

The several automatic counters can only be evaluated against the ARP 598 procedure. This procedure is known to give imprecise results. The counters correlate with ARP 598 results within ±20% at each particle size level. The automatic counters might properly be used as comparison instruments in which the absolute count may not be as important as the count relative to that of fluid of known cleanliness.

Several other evaluation tests are in common use. Degree of filtration tests are specified in MIL-F-8815 and MIL-F-27656. In each of these tests, a known weight of contaminant is added to the test fluid. After the fluid is forced through the filter, the effluent is strained through a membrane. The increase in weight of the membrane is the weight of contaminant which passed through the filter. MIL-F-8815 specifies the use of APM-F-9 glass beads as the contaminant. These beads have a median diameter of 12.5 microns and range from 2 to 80 microns. MIL-F-27656 requires a mix of equal weights of APM-F-9 beads and AC Fine dust or carbonyl iron E. AC Fine dust has a median diameter of 7.5 microns and ranges from 0 to 80 microns. Carbonyl iron E has a median diameter of 4.2 microns. It ranges from 1 to 8 microns in diameter.

The Bubble Point Test is used as a quick and easy method of determining the largest pore of a filter medium. Figure 7.7 shows how results of the Bubble Point Test can be correlated to the size of the largest bead passed for any specific filter medium. At present, Bubble Point data on one filter are not comparable to data on a different manufacturers filter. In the Bubble Point Test the filter element is immersed in a suitable fluid, usually isopropyl alcohol, Solox 190 alcohol or MIL-H-5606 fluid. The top of the filter is located ½ inch below the surface of the liquid. The element is slowly rotated. Air pressure is raised until a bubble of air appears on the outside of the element. The air pressure at this junction is the Bubble Point value.

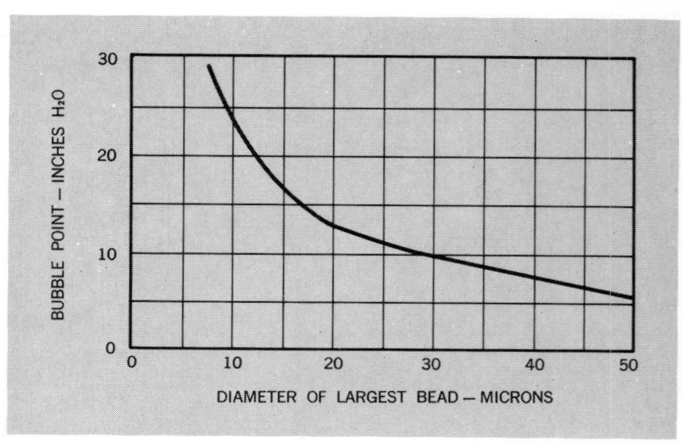

Fig. 7.7. Bubble point test.

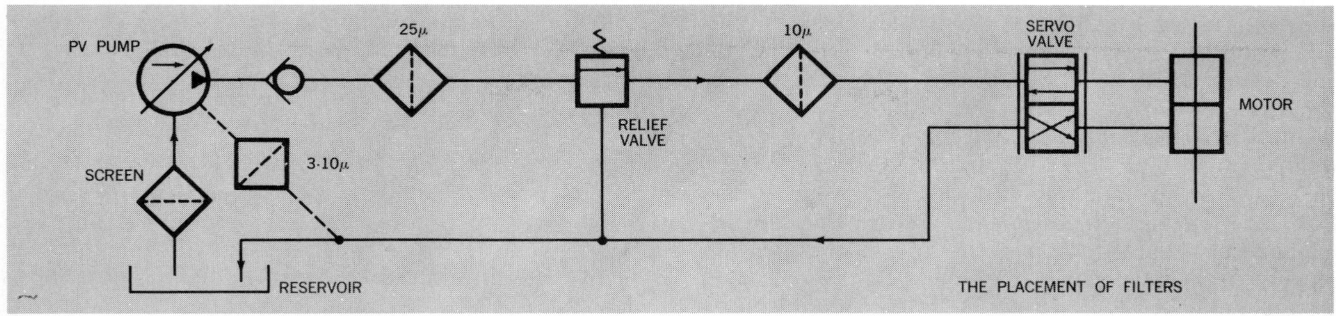

Fig. 7.8. Circuit location of hydraulic filters.

Theoretically, the relationship between Bubble Point and circular pore size is determined by

$$\text{Pore Diameter} = \frac{4\ \gamma\ \cos\ \theta}{P} \qquad 7.14$$

Where: γ = surface tension of fluid
θ = angle of contact between fluid and filter

In actuality, the following equation is used:

$$D = \frac{K_1}{P_B} \qquad 7.15$$

D = pore diameter, microns
P_B = bubble point pressure in inches of H_2O
K_1 = empirical constant

The value of K_1 ranges from 200 to 360 depending on the fluid, the fluid temperature and the type of filter medium.

The Bubble Point Test can be extended by continuing to raise the air pressure until air boils freely out of the whole filter element. At this point greatly increased air flow can be obtained with but very small increases in pressure. By determining another constant K_2 for equation 7.15 the Boiling Point can be related to the mean pore size of the filter element. A value of $K_2 = 320$ provides a means of determining mean pore size which has some experimental validity.

7.8. PLACEMENT OF FILTERS IN HYDRAULIC SYSTEMS

Specifying of filter size and fineness and location of filters in a system require the relative evaluation of such factors as size and weight of the filter bodies, accessibility of the installation location, the number and nature of the critical elements in the system, the dimensions of the clearances to be protected, the nature and amount of expected contamination and the degree of system maintenance to be expected. Some of these considerations, such as the amount of contamination, cannot easily be determined or even estimated.

In general, filters are specified and located by some variation of the following procedure.

1. Determine the areas critical to contamination. These usually are the servo control valves, metering orifices and pumps. Other critical areas may occur in some systems.
2. Plot the size of the critical dimensions in the critical areas. For instance, the diameter of metering orifices and the working clearances in valves and pumps should be noted.
3. Determine size of filters to be placed immediately in front of critical elements. Because they will often be fine filters, they usually cannot be filters of great capacity if their size is to be reasonable.
4. Choose coarser filters of greater capacity to clean up the system oil. Place these in places, such as the pump outlet, where they will catch most of the system contaminant. Thus the fine filters can be used only for fine filtration.

In typical systems, filters are at the pump pressure outlet, on the pump case drain line, at the entrance to servo control valves and in the return line up stream of the reservoir. The areas being protected are 1) the pump, and 2) servo control valves. Piston-to-bore and spool-to-sleeve radial clearances run from 1-6 microns. Metering orifice and flapper-to-nozzle clearance are 25-35 microns.

One familiar design technique is to place a large capacity, relatively coarse (25 micron absolute) filter at the pump outlet and fine filters either in the servo valve package or immediately upstream of the valve. Fine filters are never placed in the pump suction line because of the possibility of a large pressure loss inducing cavitation at the pump inlet. Instead, filters are often placed in the return line ahead of the reser-

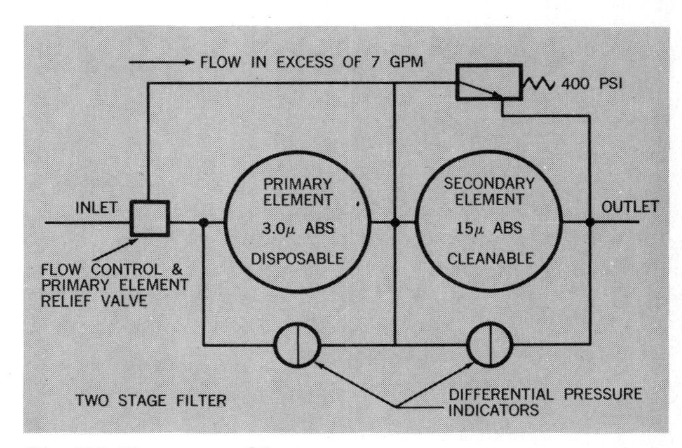

Fig. 7.9. Two stage filter.

voir and in the pump case drain line between the pump and the reservoir. Figure 7.8 shows a hypothetical system with filtration located with this design technique.

Another approach to the placement of filters is described by Meredith.[6] The hydraulic systems involved have two pumps drawing from a common reservoir. A 10 micron nominal filter in the return line had an internal relief valve to bypass the filter during periods of very high flow demand. When a failure occurred in one pump, contamination could bypass the filter and initiate failure in the other pump.

The solution developed was to install a two stage filter in the return line in place of the 10 micron nominal filter. As shown in Figure 7.9 the two stage filter consists of a 3 micron absolute first stage and a 15 micron absolute second stage. The first stage filter is a throw-away element. The second stage filter is cleanable. Differential pressure indicators on each stage show when maintenance is necessary.

When flow rates are lower than 7 gpm, all of the oil passes through both filter elements. During periods when the demand is greater than 7 gpm, the flow in excess of this amount bypasses the first stage and goes through the second stage only. If the pressure drop across the second stage were to exceed the very high level of 400 psi, a relief valve would allow bypassing of both stages.

The filters are placed so that both return and case drain oil must pass through the filter before entering the resevoir. The system filling port is so placed that supply oil must also pass through the filter. With the primary system filtration in the return line, only a small capacity 25 micron absolute filter is used in the pressure line. With this installation a much greater pump life is claimed.

7.9. GASSES IN OIL

Gasses can be present in hydraulic oil in three ways. Gas can be in a free pocket in a high point of the system, gas can be entrained as bubbles, or gas can be dissolved in the oil. Free gas is usually air which either was in the system when it was filled or which collected as a result of air bubbles coming out of entrainment. Entrained gasses appear when the oil sweeps air out of a free pocket and carries it along the stream. Entrained gasses also can appear wherever high local oil velocities are generated. This condition can occur at the inlet to a pumping piston or at a metering land in a valve or across an orifice. The local conversion of pressure to velocity can reduce the oil pressure below the vapor pressure. Bubbles of oil vapor can then appear in the fluid stream. Dissolved air is present in nearly all oils.

Dissolved air generates no problems in hydraulic systems so long as it stays dissolved. When it comes out of solution it becomes an entrained gas. The amount of air which can be in solution in the oil is proportional to the pressure and inversely related to the temperature. Therefore, air will come out of solution in systems which rise in temperature during operation. Air will also come out of solution at the same points of locally low pressure mentioned above.

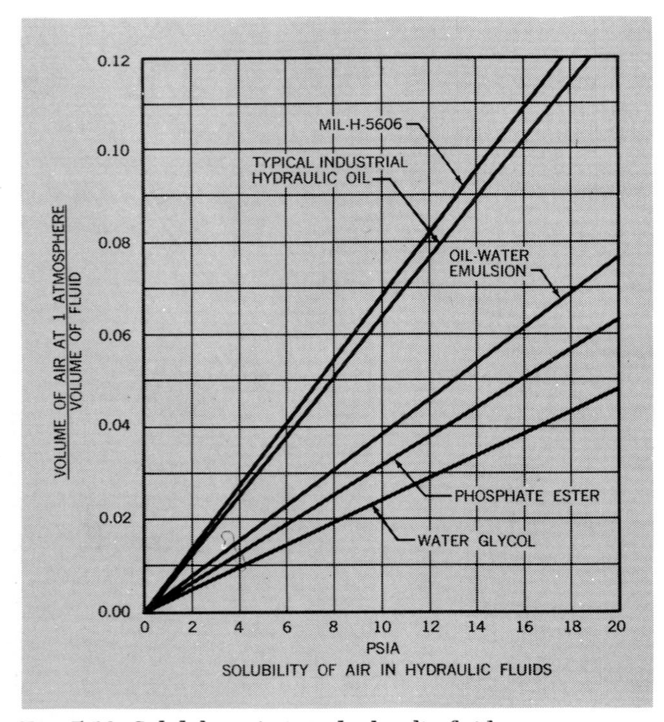

Fig. 7.10. *Solubility of air in hydraulic fluids.*

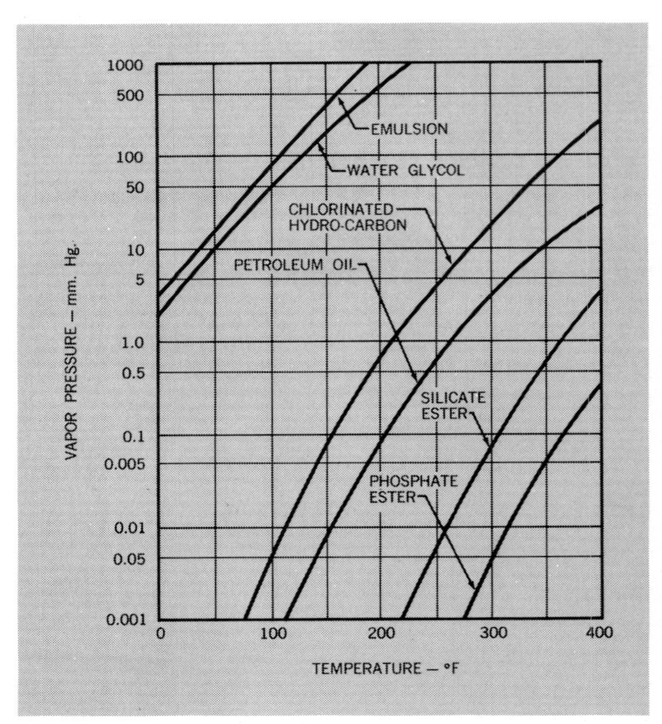

Fig. 7.11. *Variation of vapor pressure with temperature for hydraulic oils.*

Air or vapor which comes out of oil will re-dissolve when the pressure is raised. However, the rate of absorption decreases as bubble size increases. Hence, any bubbles which grow in size through coalescing with other bubbles will tend to stay as entrained gas rather than re-dissolving.

The act of absorption of bubbles created at metering lands or orifices can be very harmful. The bubbles can re-enter solution or re-enter the liquid state at a very high rate. The rapid collapse of a bubble can generate extremely high local oil velocities. If these high velocities occur near metal parts, the high velocities can be converted to impact pressures. This is the phenomenon known as cavitation. Cavitation can cause rapid wear of pump pistons and slide value metering lands.

Common hydraulic oils can hold a surprisingly large amount of air in solution. Typically, oils as received contain dissolved gasses equivalent to 6.5% by volume of entrained gas. After pumping, this may rise to as high as 10% by volume. Figure 7.10 shows the room temperature solubility of air in various hydraulic oils as a function of pressure. Because vapor bubble generation depends on vapor pressure, Figure 7.11 shows vapor pressure plotted against temperature for typical hydraulic oils.

Entrained or free air is harmful to hydraulic systems primarily because its presence sharply reduces the stiffness of the actuator oil column. The compressibility of the air acts as a soft spring in series with the stiff spring of the oil column. The spring rate of a fluid column is

$$K = \frac{A\ \beta}{l} \qquad 7.16$$

where: K = spring rate lb/in
A = area of fluid column in^2
l = length of fluid column—in
β = bulk modulus of elasticity lb/in^2

The importance of a high bulk modulus and, by corollary, a high spring rate appears when an analysis is made of the natural frequency of the actuator-load combination. The natural frequency is determined by the square root of the ratio of the effective spring rate and the effective mass.

$$\omega_n = \sqrt{\frac{K_e}{M_e}} \qquad 7.17$$

ω_n = natural frequency
K_e = effective spring rate
M_e = effective mass

The effective mass is the sum of the load mass, the mass of the piston and the mass of the oil in the actuator. The presence of a few percent of air in the oil would have little effect on the total effective mass. The effective spring rate is determined by considering the load spring, the oil spring and the springiness of the air in series, or, if the air and oil are combined into an effective bulk modulus β_e, then

$$\frac{1}{K_e} = \frac{1}{K_{load}} + \frac{l}{A\ \beta_e} \qquad 7.18$$

Thus the effective spring rate will always be *lower* than spring rate of the softest element. Because a high natural frequency of the actuator is desired in order to withstand flutter loads or to avoid limiting servo capability, the presence of air in the oil column sharply limits the usefulness of the actuator.

7.10. GAS REMOVAL

Gasses mixed with oils are usually removed by some method which 1) coalesces small bubbles into large ones, 2) separates gas and oil by means based on differences in density in intensified acceleration fields and 3) physically removes the gas from the oil container.

The forces acting on a bubble in oil are 1) buoyancy and 2) viscous drag. These balance to determine the velocity with which the bubble moves through the liquid. The buoyancy force is given by

$$F_b = \frac{4}{3}\ \pi\ r^3\ \rho g \qquad 7.19$$

Where: r = radius of bubble
ρ = mass density of oil

Properly, this force should be reduced by the weight of the gas in the bubble but this is so small compared to the weight of the displaced oil that it can be ignored.

The viscous drag force is given by

$$F_d = C_d\ \frac{\rho}{2}\ \pi\ r^2\ v^2 \qquad 7.20$$

Where: C_d = drag coefficient
v = bubble velocity

C_d is dependent upon Reynold's Number. The relationship is, for value of N_R less than 1000.

$$\log(10\ C_d) = -\ 0.28\ \log N_R \qquad 7.21$$

For most cases of interest to hydraulics engineers, a value of C_d = 1.4 may safely be used.

If the buoyancy and viscous forces are equated

$$\frac{4}{3}\ \pi\ r^3\ \rho g = \frac{1.4}{2}\ \rho\ \pi\ r^2\ v^2 \qquad 7.22$$

or

$$v = 1.38\ \sqrt{rg} \qquad 7.23$$

Assuming that the ease of removal of entrained gasses from oil is related to velocity of movement of the bubble through the oil, then this analysis suggests the following techniques for aiding gas removal:

1. Increase Reynolds number so as to decrease the drag. This can be done most easily by warming the oil to reduce viscosity.
2. Increase the size of the bubble. This is also accomplished by heating the oil. Other methods of agglomerating or coalescing small bubbles into large ones should be helpful.
3. Increasing the gravitational field will increase the velocity of travel. This is normally done by swirling or centrifuging.

This discussion does not touch on the technique of pulling a vacuum above the oil. If the foregoing analysis is valid, the vacuum would not aid in removing entrained gasses. However, a vacuum could result in reducing the amount of gas in solution. Henry's Law states, that the amount of gas in solution is directly proportional to the absolute pressure. If dissolved gas can be released by pressure reduction and removed to the surface by methods discussed above, the total of the gas in the oil can be reduced below the level where entrained gasses can exist in working portions of the system.

Occasionally, gas removal systems utilize a screen or filter to separate bubbles of gas from the oil. This can be effective if the pressure drop across the screen or filter can be kept very low. The low pressure drop is necessary because the only mechanism which prevents the bubble from flowing through the screen is the surface tension of the oil.

Surface tension around a gas bubble in oil tends to reduce the size of the bubble. If any effect, such as trying to force the bubble through a filter, tends to alter the shape of the bubble so as to stretch or bend the gas-oil interface, the stretching and bending will be resisted by the molecular cohesion of the oil molecules. However, these are fairly low level forces.

As examples, the surface tension γ (gamma) of typical oils is

MIL-H-5606	70°F	28 dynes/cm
Petroleum oils	70°F	35-38 dynes/cm
MIL-L-7808	70°F	26 dynes/cm

The pressure necessary to develop curvature in a gas bubble is given by 7.14. When $\cos\theta$ equals unity

$$P_T = \frac{4\gamma}{D} \qquad 7.24$$

Where: P_T = pressure due to surface tension
D = bubble diameter

If an oil which contained gas bubbles were to be strained through a 25 micron absolute filter, the maximum bubble which could pass would be 25 microns or 2.5×10^{-3} cm in diameter. The pressure due to surface tension if the fluid were MIL-H-5606 and the filter pore were circular would be

$$P_T = \frac{(4)\ (28)}{2.5 \times 10^{-3}} = 4.48 \times 10^4 \ dynes/cm^2$$

$$= 0.652 \ psi$$

At any differential pressure across the filter greater than this, the bubble will pass through.

REFERENCES

1. Huggett, H. L. "Servo Valve Internal Leakage as Affected by Contamination" *Proceedings* SAE Aerospace Fluid Power Conference, May 1965, pps 238-244
2. Piccone, M. "Control of Contamination in Rocket Booster Hydraulic Systems" Ibid, pps. 245-254.
3. Anon "Contamination in Hydraulic Systems" *Machine Design,* Oct. 26, 1967, pps. 200-202.
4. Tao, Ting C. "Media and Filtration Mechanics" Fram Aerospace, Pawtucket, R.I., March 1966
5. Detweiler, J. I. Weaknesses in Particle Counting Techniques" *Journal* of the American Association for Contamination Control, May 1964
6. Meridith, D. B. "Dual Filters Increase Aircraft Reliability" *Hydraulics and Pneumatics,* Feb. 1964

CHAPTER 7

PROBLEMS

1. What is the porosity in darcies of a 25 micron absolute filter which is 0.05-inch thick in the direction of flow? Assume round pores. (Hint: assume MIL-H-5606 oil at 100°F. Use Figures 7-8, 2-4 and 2-7 to obtain necessary data)

2. What is the natural frequency of a column of aircraft phosphate ester at 150°F which is 2 inches in diameter and 20 inches long?

3. Calculate the natural frequencies of the same size column of MIL-H-5606 oil at 120°F and 3000 psi which contains 1%, 2% and 5% of entrained air by volume.

4. A bubble of air 0.04-inch in diameter is entrained in SAE 10 oil at 160°F. What is the buoyancy force? What is the velocity of rise of the bubble ? What is the force due to viscous drag?

5. A 15 micron filter having circular pores is in a bath of SAE 10 petroleum oil at 70°F. What pressure differential across the filter is necessary before gas bubbles can be driven through it?

6. A reservoir contains MIL-H-5606 at 70°F. Because the oil contains air bubbles, a screen must be installed between inlet and outlet to remove the air. The pressure drop due to flow through this screen mav not be greater than 0.5 psi. What size circular holes should be specified in the screen? If the screen is 0.030-inch thick and has a porosity of .005 darcy, what must the area of the screen be if the flow rate is 10 gpm?

CHAPTER EIGHT

Control Valves

8.1. FUNCTIONS

Control valves perform two primary functions: they direct the flow of oil from the power generating devices and transmission system to the power transducers and they aid in establishing the system pressure conditions by restricting the flow of oil. These valves fall into two general categories, on-off (shutoff) valves and modulating valves.

Shutoff valves function so as to be either wide open (zero resistance to flow) or completely shut (infinite resistance to flow). In order to make the shutoff as

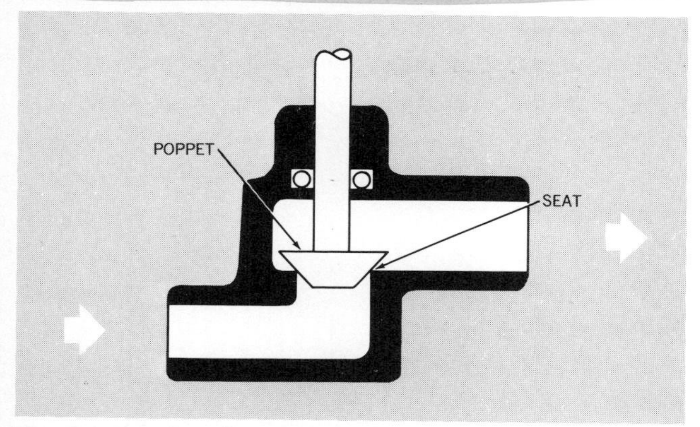

Fig. 8.1 Poppet valve.

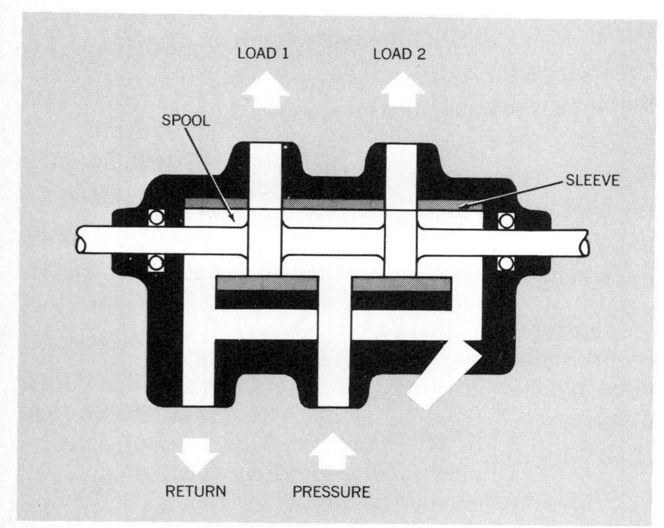

Fig. 8.2 Spool and sleeve valve.

complete as possible, these valves are usually poppet-type as shown in Figure 8.1. However, they may use a ball, blade, gate or other mechanism. Occasionally a spool and sleeve such as is shown in Figure 8.2 is used in a shutoff valve.

Modulating valves are controllable to any degree of opening in their design range. This design range may, or may not, include full shutoff. These valves are often of the spool and sleeve construction. A few plate valves and modulating poppet valves have been tried but they have not been successful. The first stage of electrohydraulic servo valves is usually either a flapper-nozzle or a jet pipe.

8.2. SHUTOFF VALVES

These valves are used to control actuator devices in which the actuator is either stopped or is travelling at its maximum rate. With the exception of the transition time between zero and maximum rate, no in-between speeds are possible. Valves of this type can vary from simple manual or solenoid operated on-off valves to complex sequencing and directional controls. They are also used to control a specific type of servomechanism known as a "bang-bang" servo.

Because maximum flow rate through shutoff type valves is desired, they are usually designed for minimum pressure loss through the valve. When the valve is shut it is completely shut and when it is open it is wide open. These valves are also often required to operate very quickly. Fluid compressibility and metal elasticity are the factors which slow down the effective operating rate of shutoff valves.

Once its task is known, the specifying of a shutoff valve is relatively easy. The nature and magnitude of the input and the limitations on available power will determine the choice between manual or powered operation. The task to be performed will control the complexity of the valve. Other considerations are the size, weight, shape and allowable pressure loss.

The pressure loss in the valve will determine the available combinations of actuator rate and load carrying capacity. The pressure loss—or the flow resistance characteristic—will also determine the maximum power which the valve-actuator combination can deliver to a load.

Power is proportional to flow times pressure. The relationship is developed in the following way. The pressure head in feet per psi for any fluid is given by

$$H = \frac{144}{w}$$

Where: H = pressure head in feet per psi
w = weight density of fluid lb/ft^3

and the weight per gallon is

$$W = \frac{231}{1728} w$$

One gallon per minute times 1 psi is

$$\frac{(144)\ (231)\ w}{1728\ w} = 19.25 \text{ ft-lb/min}$$

Because

$$1 \ HP = 33{,}000 \text{ ft-lb/min.}$$

$$\frac{HP}{(GPM)\ (PSI)} = \frac{33{,}000}{19.25} = 1714.281$$

Therefore

$$HP = \frac{(GPM)\ (PSI)}{1714}$$

For ease in analysis flow control valves are usually assumed to act as if they were orifices. Then the flow through the valve can be described by a modification of the orifice equation.

$$Q = Y\sqrt{\Delta P} \qquad 8.1$$

Where: Y = Valve conductance

For shutoff valves in the open condition, Y is a constant. Because ΔP equals $P_S - P_L$, 8.1 can be rewritten

$$Q = Y\sqrt{P_S - P_L} \qquad 8.2$$

Because the horsepower to the load is given by

$$HP\ LOAD = \frac{P_L Q}{1714} \qquad 8.3$$

then

$$HP\ LOAD = \frac{Y}{1714} P_L \sqrt{P_S - P_L} \qquad 8.4$$

Differentiating 8.4 with respect to P_L

$$\frac{d\ (HP)}{d\ P_L} = -\frac{Y\ P_L}{(2)\ (1714)\ \sqrt{P_S - P_L}} + \frac{Y}{1714}\sqrt{P_S - P_L} \qquad 8.5$$

To determine when the power is maximum, 8.5 can be equated to zero

$$-\frac{Y\ P_L}{(2)\ (1714)\ \sqrt{P_S - P_L}} + \frac{Y}{1714}\sqrt{P_S - P_L} = 0 \qquad 8.6$$

Simplifying

$$-\frac{P_L}{2} + P_S - P_L = 0$$

And

$$P_L = \frac{2\ P_S}{3} \qquad 8.7$$

Equation 8.7 shows that, when the control valve is considered to act as an orifice, the maximum power is delivered to the load when 1/3 of the pressure drop occurs across the valve. This rule is very valuable as a first approximation to the task of matching valve characteristics to any known requirement.

The actual maximum horsepower which can be delivered to the load can be determined by substituting 8.7 into 8.4.

$$HP_{(MAX)} = \frac{2\ Y\ P_S}{(3)\ (1714)}\sqrt{\frac{P_S}{3}} = \frac{\sqrt{3}\ Y\ P_S^{3/2}}{7713} \qquad 8.8$$

Because of the usefulness of the relationship of 8.7, many manufacturers rate their valves at the horsepower which they would deliver to a load which absorbs 2/3 of the total input pressure. This horsepower rating can be determined from the basic power equation. Assume that the total pressure drop occurs across the valve.

$$HP_{VALVE} = \frac{P_S\ Q}{1714}$$

Using the relationship of 8.2

$$Q = Y\sqrt{P_S}$$

and

$$HP_{VALVE} = \frac{Y\ P_S^{3/2}}{1714} \qquad 8.9$$

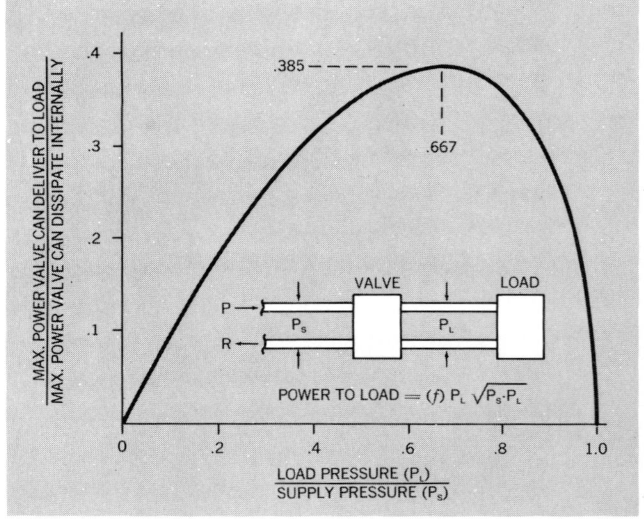

Fig. 8.3 Determining available power to the load.

Dividing 8.8 by 8.9

$$\frac{7713}{\sqrt{3}\ Y\ P_S^{3/2}} \cdot \frac{1714}{Y\ P_S^{3/2}} = 0.385$$

Thus the maximum horsepower which can be delivered to a load can be determined by computing the horsepower dissipated when full pressure drop occurs across the valve. This amount, when multiplied by .385 gives the maximum horsepower which a valve can deliver to a load. These relationships are shown in Figure 8.3. When the load pressure is zero, no power is being delivered to the load. When the load pressure equals the supply pressure, the load is stalled and no flow occurs. This is also a condition of zero power. The maximum power is delivered when $P_L = 2/3\ P_S$. This maximum power to the load is .385 times the maximum power dissipated when all of the pressure drop occurs across the valve.

8.3. MODULATING VALVES

Many of the relationships developed for shutoff valves are valid for modulating valves except that the (Y) term of 8.1 is not constant. The equivalent orifice diameter varies as the valve is opened and closed. To some extent the conductance coefficient Y is a function of Reynolds Number. Also, depending on the type of modulating valve used, the shape of the opening affects the value of Reynolds number.

8.4. CONTROLLING THE VALVE

While manually controlled on-off, on-off-on, on-on and other forms of valves are frequently used, modern hydraulic control systems often require that the valve function at a remote location or that it function in sequence with other operations. This usually means that some sort of remote control or powered control is used to command the valve. The control methods available are electrical, fluidic, hydraulic or pneumatic. Each type has its own advantages.

Analogs of electrical systems describe modulating valves graphically. Figure 8.4 shows the relationship

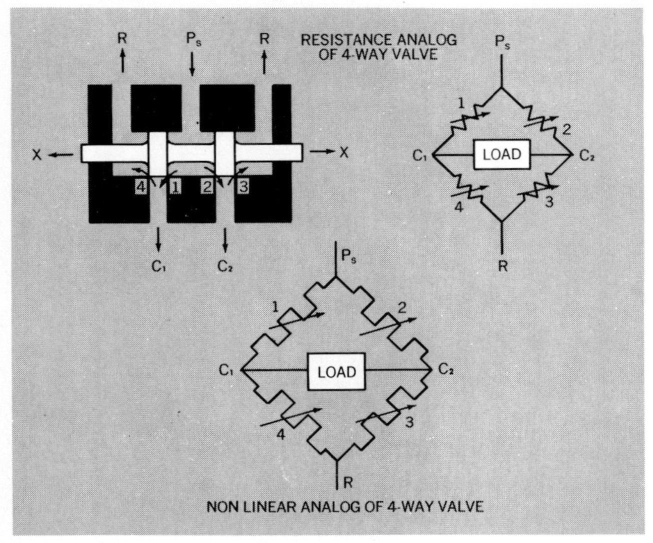

Fig. 8.4 Electrical analogs.

between a Wheatstone bridge and a four-way valve. Each of the metering ports of the valve bears a correlation—of a sort—with a comparable resistor in the Wheatstone bridge. Because all of the modulating valves used in hydraulic systems use metering ports which act as if they were orifices, the analogy with a resistance is not a good one. The orifices compare more closely with a nonlinear varistor. They will, therefore, be shown by a square wave rather than by a sawtooth—the conventional symbol for a resistance. Therefore, the nonlinear analog bridge of Figure 8.4 is the more proper presentation of the four-way spool valve.

8.5. SPOOL AND SLEEVE VALVES

The most commonly used modulating valve configuration is the spool and sleeve shown in Figure 8.2. These valves can be displaced manually, by a torque motor, by a tractive solenoid working against a nonlinear spring or by a pair of tractive solenoids which displace the valve on a time-sharing basis. The spool valve may be connected directly to the working load or it may be used to drive a larger valve.

Many spool valves which are used to control servo systems use square ports and straight, sharp-edged metering lands. Flow through the metering port of this configuration is constricted in a manner similar to flow through an orifice. The axial force on the spool equals the axial component of the net change of momentum.

$$F_{(AXIAL)} = Qv\rho \cos\theta \qquad 8.10$$

but

$$Q = cA\sqrt{\frac{2\,\Delta P}{\rho}} \quad and \quad v = \sqrt{\frac{2\,\Delta P}{\rho}}$$

so

$$F_{(AXIAL)} = \left[cA\sqrt{\frac{2\,\Delta P}{\rho}}\right]\left[\sqrt{\frac{2\,\Delta P}{\rho}}\right]\rho \cos\theta$$

Simplifying and defining the area A to equal w, the peripheral width of the orifice times x, the axial length of the orifice, then

$$F_{(AXIAL)} = 2cwx\,\Delta P \cos\theta \qquad 8.11$$

Since θ is always less than 90° for such a valve configuration, the axial force is always positive and tends to close the valve. Figure 8.5 shows how θ varies with radial clearance. For large valve clearances, θ tends to approach 69° when the lands are sharp edged.

If the land edges are rounded, the jet angle changes. The angle is 45° for the case where both lands have the same radius and the clearance equals the axial

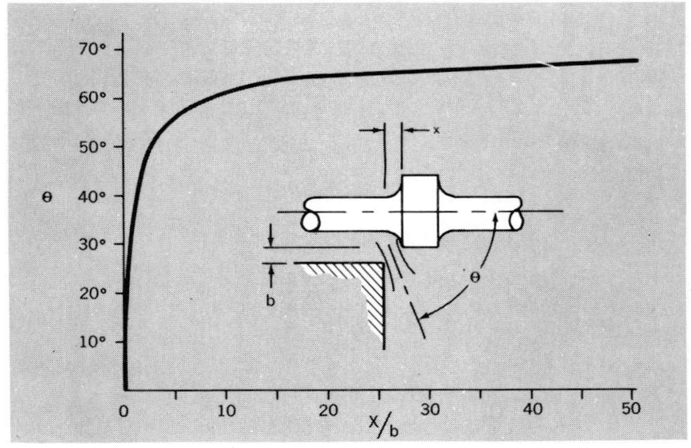

Fig. 8.5 Effect of valve opening on flow angle.

displacement. For the condition of zero clearance, positive displacement and radiused edges, θ is greater than 45° and less than 69°.

In an actual 4-way valve such as is shown in Figure 8.2, two orifices are in series. Therefore, the reaction

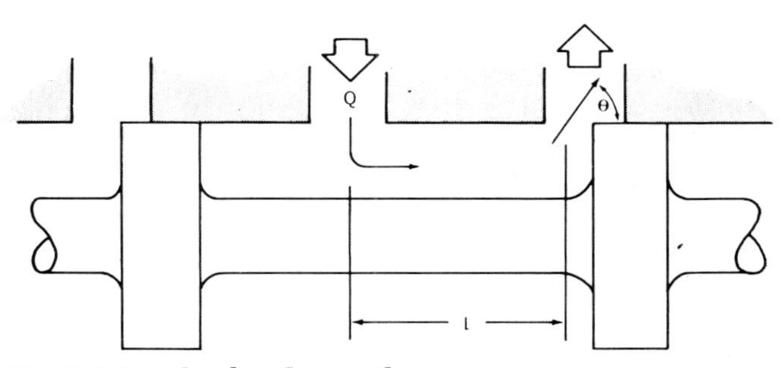

Fig. 8.6 Spool valve flow path.

force is twice that of equation 8.11.

This discussion assumes that the flow is steady. If the valve spool moves, then the flow rate will change and the oil in the valve must be accelerated. If the oil is flowing as shown in Figure 8.6 then the pressure differential necessary to produce the acceleration would act on the valve faces so as to close the valve. If, however, the flow direction were to be reversed, the pressure differential would tend to open the valve. This would destabilize the valve if the force due to this acceleration were greater than the force determined by equation 8.11. The acceleration force is given by

$$F = Ma = \rho l \frac{dQ}{dt} \tag{8.12}$$

and

$$\frac{dQ}{dt} = cw \sqrt{\frac{2\ \Delta P}{\rho}} \frac{dx}{dt}$$

substituting

$$F = \rho l c w \sqrt{\frac{2\ \Delta P}{\rho}} \frac{dx}{dt}$$

or

$$F = cwl\sqrt{2\rho\ \Delta P} \frac{dx}{dt} \tag{8.13}$$

The total expression for the flow forces is then

$$F_{(AXIAL)} = 2\ cwx\ \Delta P\ \cos\theta + cwl\sqrt{2\rho\ \Delta P} \frac{dx}{dt} \tag{8.14}$$

This expression takes the form of a spring which is linear with x and a damping term which is linear with the time derivative of x. However, the damping term can be affected by the sign and magnitude of l. As shown in Figure 8.6, l is the axial distance between the centers of the incoming and outgoing flows. With oil flowing *outwards* from the metering port, l is positive, the damping is positive and the valve element is stable because the flow forces oppose the direction of displacement from the no-flow condition. When the oil flows inwards, l is negative and the stability of the valve element depends upon the relative magnitude of the two elements of equation 8.14.

In an actual 4-way valve, one flow is outward and one is inward. The spacing of the ports should be such that the positive damping effects are greater than the negative effects.

Because of these flow forces, and the way they change with valve displacement, the centering, or flow reaction, forces on straight edged, square ported valves tend to appear as in Figure 8.7a. The drop in force at large valve displacements is usually an undesirable effect. Careful valve port shaping can often change the reaction force curves to those shown in Figure 8.7b.

Spool valves, as shown by Figure 8.8, can be made

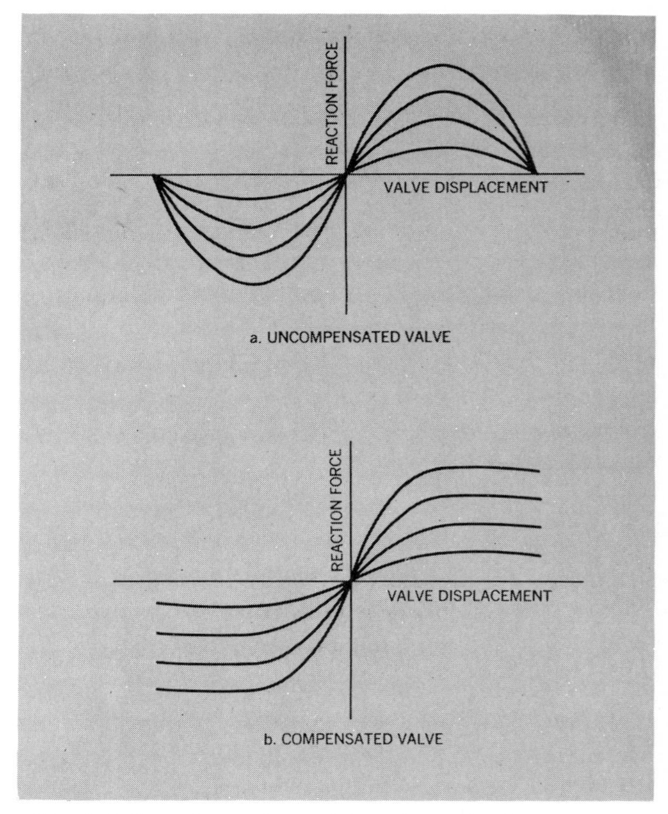

Fig. 8.7 Flow forces in modulated spool valves.

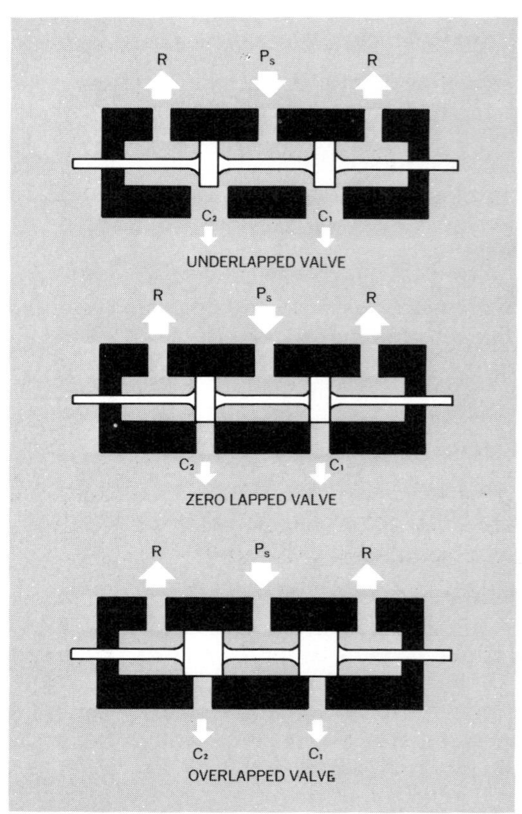

Fig. 8.8 Spool land widths.

under lapped, over lapped, or with zero lap. The zero lap condition is also sometimes called line-to-line or critically lapped.

The under lapped valve, or open center valve, is so designed that the valve lands are narrower than the ports. When the valve is centered, this arrangement permits a constant flow of oil from the high pressure side to the low pressure side. The cylinder ports are then at about 50 per cent of the supply pressure. Usually the design valve movement is far greater than the underlap, but occasionally a valve is so designed that the stroke lies entirely within the under lap region.

Overlap is rarely created intentionally. When the valve lands are wider than the ports, the dead zone so created makes the valve unresponsive to small signals.

8.6. VALVE CHARACTERISTICS

Assuming that the valve is either zero lapped or else is operating in a linear portion of its control motion, the three factors which describe the valve characteristics are flow gain, pressure gain and the flow/pressure coefficient. These are

Flow Gain

$$K_Q = \frac{\partial Q_L}{\partial x_v} \qquad 8.15$$

Q_L = load flow
x_v = valve displacement

Pressure Gain

$$K_P = \frac{\partial P_L}{\partial x_v} \qquad 8.16$$

P_L = load pressure

Flow/Pressure Coefficient

$$K_c = -\frac{K_Q}{K_P} = -\frac{\partial Q_L}{\partial P_L} \qquad 8.17$$

It follows that

$$\Delta Q_L = K_Q\,\Delta x_v - K_c\,\Delta P_L \qquad 8.18$$

Pressure gain and flow gain are directly affected by lap condition. Figure 8.9 shows flow gain curves for the three lap conditions. Note the difference in flow gain for the underlapped valve between operating in the underlap region and in normal operation.

Pressure gain is affected by lap as shown in Figure 8.10. Full system pressure to the load cannot be obtained with an underlapped valve until the metering land which opens the port to the load has completely blocked the escape of system pressure to the return. In an overlapped valve, once the overlap is passed, pressure gain is higher than with an underlapped valve because only a slight displacement creates full system pressure in one load port while opening the other to the return line. Line-to-line valves have the same pressure gain as overlapped valves but

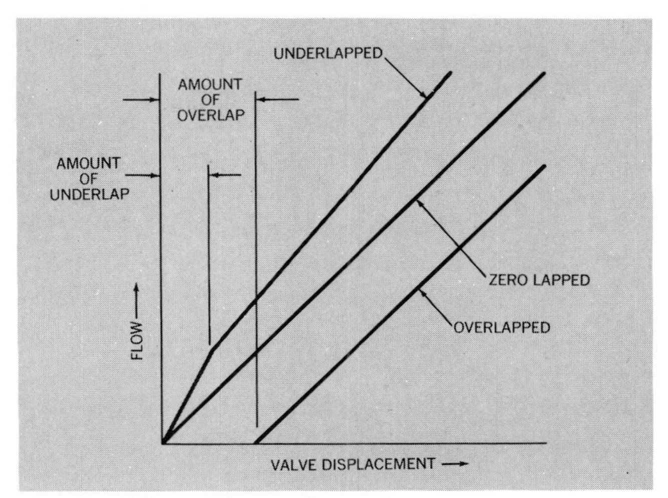

Fig. 8.9 Effect of lap on flow gain.

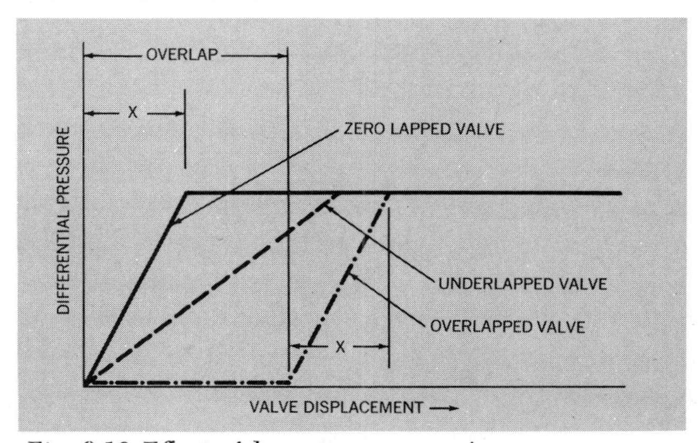

Fig. 8.10 Effect of lap on pressure gain.

no lost motion of the spool exists.

The expressions for the valve factors can be derived by considering a symmetrical line-to-line valve as shown in Figure 8.4. Pressures and flows will be identified to the metering edge numbered in the Figure. Because of symmetry

$$Q_2 = Q_4 \quad so \quad \Delta P_2 = \Delta P_4$$

$$Q_1 = Q_3 \quad so \quad \Delta P_1 = \Delta P_3$$

$$P_S = \Delta P_1 + \Delta P_4$$

$$\Delta P_1 = \Delta P_2 + \Delta P_L$$

$$\Delta P_1 = \frac{P_S + P_L}{2} \qquad 8.19$$

$$\Delta P_2 = \frac{P_S - P_L}{2} \qquad 8.20$$

When x_v is positive (spool displaced to the right of neutral), $Q_3 \approx O$ *and* $Q_L = Q_2$

$$Q_L = cA\sqrt{\frac{2\,\Delta P_2}{\rho}} = cwx_v\sqrt{\frac{P_S - P_L}{\rho}} \qquad 8.21$$

$$\frac{\partial Q_L}{\partial x_v} = cw\sqrt{\frac{P_S - P_L}{\rho}} = K_Q \qquad 8.22$$

When x_v is negative

$$\frac{\partial Q_L}{\partial x_v} = cw\sqrt{\frac{P_S + P_L}{\rho}} = -K_Q \qquad 8.23$$

Solving 8.21 for P_L

$$P_L = P_S - \frac{\rho}{x_v^2}\left[\frac{Q_L}{cw}\right]^2 \qquad 8.24$$

then

$$\frac{\partial P_L}{\partial x_v} = \frac{2\ P_S - 2P_L}{x_v} = K_P \qquad 8.25$$

and

$$K_c = \frac{K_Q}{K_P} = \frac{c\ w\ x_v}{2\sqrt{\rho}\ (P_S - P_L)} \qquad 8.26$$

For the special case of the valve spool being in the null position Q_L, P_L and $x_v = 0$,

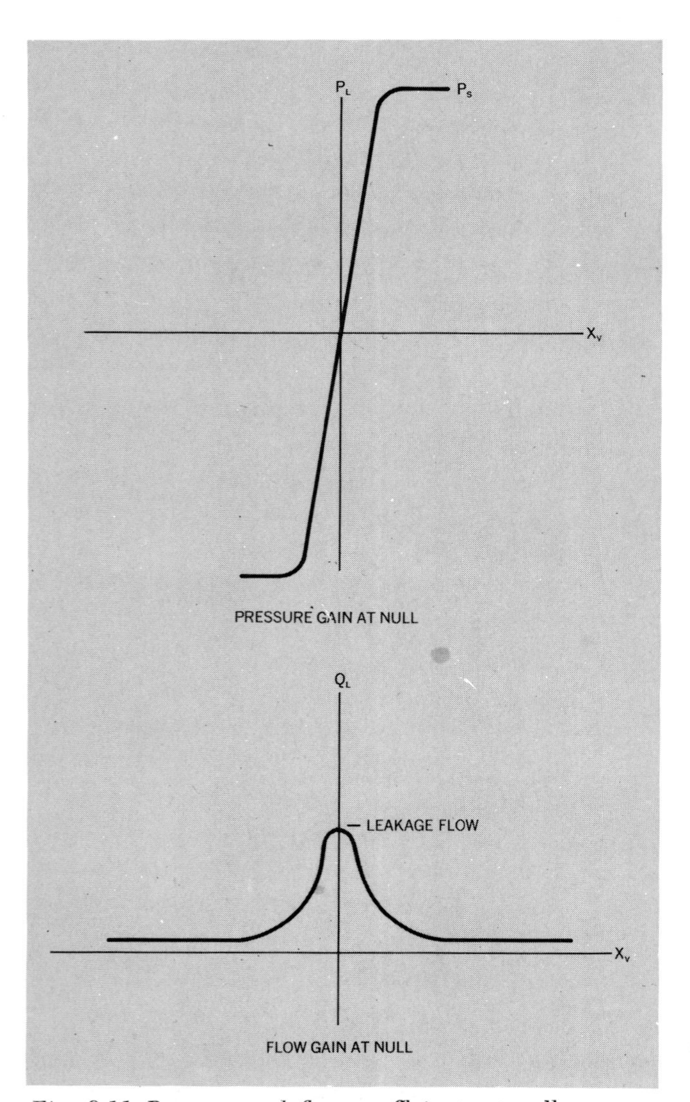

Fig. 8.11 Pressure and flow coefficients at null.

so

$$K_{Q_o} = cw\sqrt{\frac{P_S}{\rho}},\ K_P = \infty \quad and \quad K_c = 0$$

Leakage modifies these so that the actual values of the coefficients in and around null are as shown in Figure 8.11.

If the valve at null is assumed to have a rectangular orifice and the leakage flow is assumed to be laminar, this relationship will hold

$$Q = \frac{\pi\ b^3\ w\ \Delta P}{32\ \mu}$$

At null $\Delta P = P_S$ and w can be considered the "area gradient (in²/in)" as the valve opens.

$$\frac{Q_o}{P_S} \approx \frac{\partial Q_L}{\partial P_L} = K_{co} = \frac{\pi\ b^3\ w}{32\ \mu} \qquad 8.27$$

The pressure gain at null can be re-evaluated by dividing 8.27 by the derived value of K_Q at null.

$$K_{P_o} = \frac{K_{co}}{K_{Q_o}} = \frac{32\ c\ \mu}{\pi\ b^3}\sqrt{\frac{P_S}{\rho}} \qquad 8.28$$

8.7. VALVE-LOAD CONFIGURATIONS

Spool valves can be connected to their loads in a variety of ways. The form of connection can also vary depending upon whether the source of hydraulic power is an essentially constant pressure (CP) supply such as a pressure compensated variable delivery pump or a constant flow (CQ) supply such as a fixed delivery pump.

The valve configuration of Figure 8.2 is often used with a CP source. The circuit analogy of Figure 8.4 shows that these valves are full bridges with four variable arms. The circuit analogy holds true no matter what the degree of lap. Figure 8.12 shows the flow-pressure coefficients (K_c) for a zero lapped 4-way valve. The curves form a family of parabolas based on the amount of valve opening or valve conductance. For a wide-open valve, $Y = 1$. If the valve were to have a modest amount of underlap, the curves would be modified as shown in Figure 8.13. Here while the basic relationship still tends to be parabolic, variations in pressure will give variations in flow even when the valve is in a neutral position. When the valve is so designed that it operates wholly in the underlap range, the curves appear as in Figure 8.14.

Another way to use a spool valve with a CP source is the 3-way configuration of Figure 8.15. Here the differential areas of the load motor are balanced against the source pressure and the controlled pressure. The load motor or piston usually is designed with a 2-1 area ratio. The controlled pressure is then varied up and down from $P_S/2$. As can be seen from the analogy, this configuration is a half bridge with two arms variable. The K_c curves for 3-way valves are

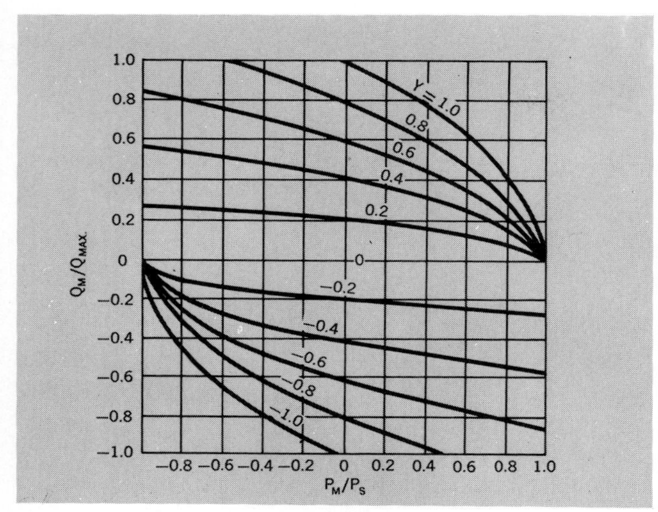

Fig. 8.12. Flow-pressure coefficients for a zero lapped 4-way valve.

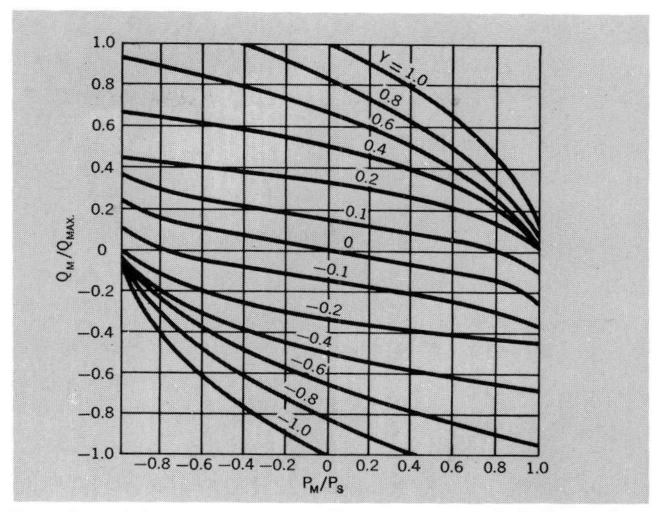

Fig. 8.13 Flow-pressure coefficients for a slightly underlapped 4-way valve.

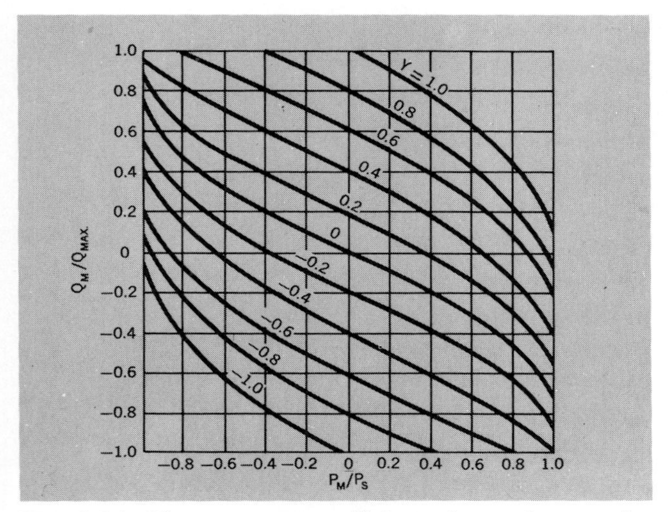

Fig. 8.14. Flow-pressure coefficients for a 4-way valve operating entirely in the underlap range.

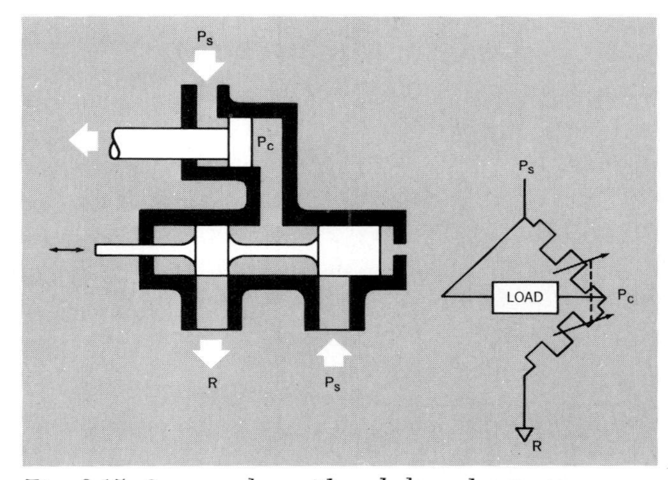

Fig. 8.15. 3-way valve with unbalanced actuator.

identical with those of 4-way valves except that the range of P_m/P_s goes only from −0.5 to 0.5.

An under lapped valve is employed with a pure CQ source. If the valve were not underlapped, the pressure would build up, under the no-load condition, until some part of the system ruptured. Therefore, when a zero lapped valve is used with a CQ source, a relief valve is incorporated to limit the pressure. Then, when all or part of the flow is passing through the relief valve, the source is really CP. Only when the load absorbs all of the flow is this system truly CQ.

Figure 8.16 shows a method of control of CQ power using a spring to balance the controlled pressure. From a hydraulic control standpoint, this is a quarter bridge with only one arm variable.

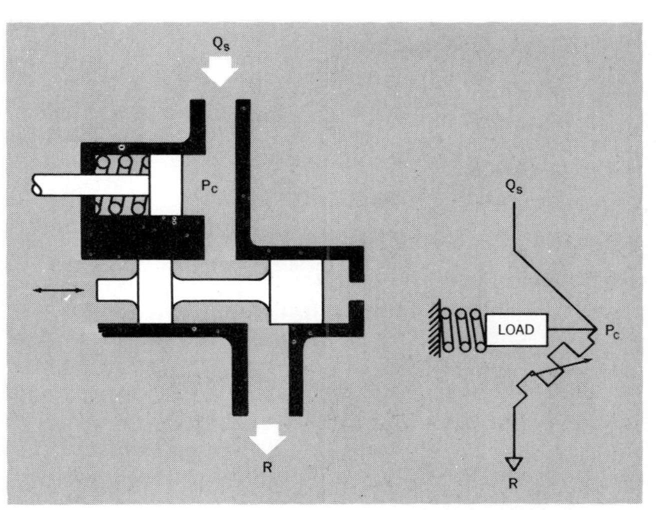

Fig. 8.16. 2-way valve with a constant flow cq source.

8.8. SINGLE NOZZLE—FLAPPER VALVE

A control valve configuration which has been used with success for many years is the single nozzle–flapper valve shown in Figure 8.17. This valve is a half bridge with one arm variable. It is used as the first stage in electrohydraulic servo valves. The K_c curves are as shown in Figure 8.18.

The flow equation for the single nozzle–flapper valve is

$$Q = \pi \; d_n \; (L + x) \; C_D \sqrt{P_c} \qquad 8.29$$

Where: C_D is the orifice coefficient *not* the nozzle constant.

The limit of control is reached when the nozzle and gap areas are approximately equal.

$$L + x = \frac{d_n}{4} \qquad 8.30$$

L is often selected so that $L = d_n/8$. L is usually between .001″ and .002″. In order for the maximum flow output to be equal in both directions

$$\frac{d_n}{d_o} = 1.5 \qquad d_o = 5.33L \qquad 8.31$$

Then the pressure output as a function of flapper position is

$$\frac{P_c}{P_s} = \frac{1}{1.28 \left[\frac{x}{L}\right]^2 + 2.54 \left[\frac{x}{L}\right] + 2.28} \qquad 8.32$$

This relationship is quite linear in the range

$$0.3 < \frac{P_c}{P_s} < 0.7$$

The load flow relationship is

$$\frac{Q_L}{Q_{L\,(MAX)}} = \sqrt{1 - \frac{P_c}{P_s}} - 1.09 \; (1 + \frac{x}{L}) \sqrt{\frac{P_c}{P_s}} \qquad 8.33$$

This is fairly linear when

$$-0.6 < \frac{Q_L}{Q_{L\,(MAX)}} < 0.6$$

For further analysis, reading Morse[1] is recommended.

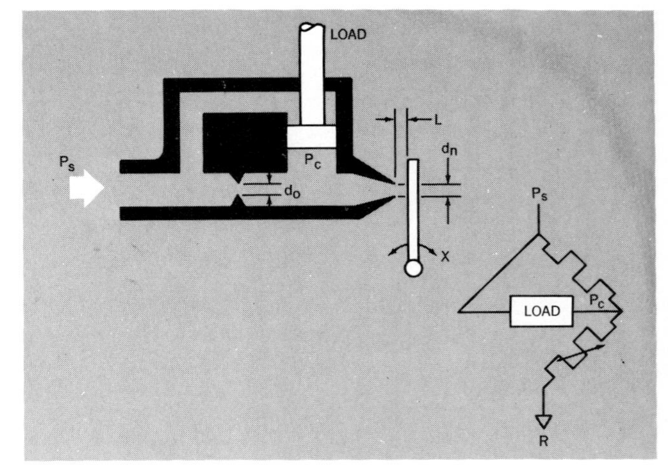

Fig. 8.17. Single nozzle-flapper valve.

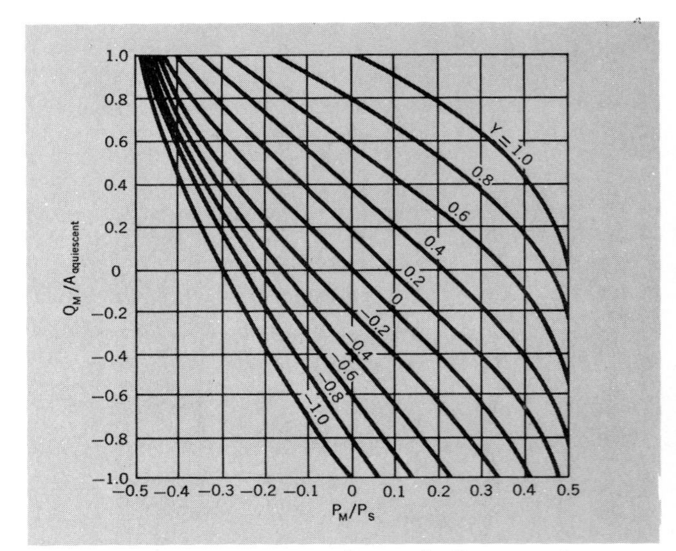

Fig. 8.18. K_c curves for single nozzle-flapper valve.

8.9. DOUBLE NOZZLE—FLAPPER VALVES

Perhaps more electrohydraulic servovalves use a double nozzle–flapper amplifier or first stage than all other configurations combined. As shown in Figure 8.19, this valve is a full bridge with two arms variable. The use of two nozzles makes the design relatively immune to changes in its null or zero differential pressure position with changes in supply pressure and/or temperature. The use of force feedback from the load to the flapper also aids in maintaining linear operation over a wide range of condi-

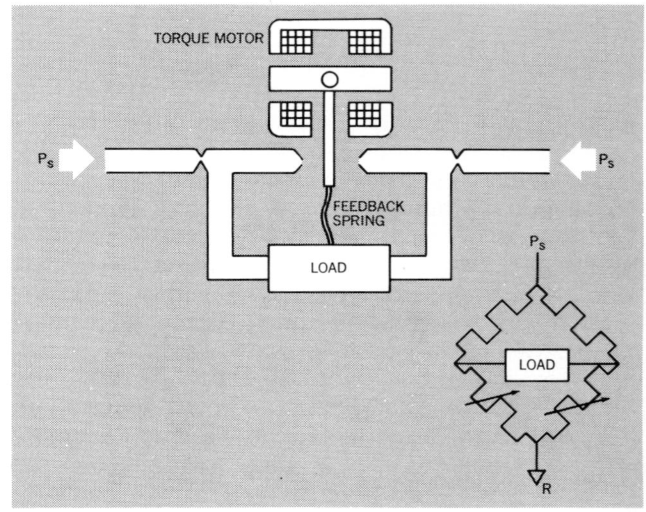

Fig. 8.19. Double nozzle-flapper valve.

tions. The K_c curves as shown in Figure 8.20 are quite similar to those of an underlapped 4-way valve. When the flapper displacement is toward the left, x is considered positive. Then

$$\frac{d_o}{4}\sqrt{P_S - P_1} = (L + x)\sqrt{P_1} \tag{8.34}$$

In the other direction and for the other nozzle

$$\frac{d_o}{4}\sqrt{P_S - P_2} = (L - x)\sqrt{P_2} \tag{8.35}$$

A usual design technique is to make the orifice and nozzle diameters equal. The clearance L is always less than $d_n/4$ and is usually about $d_o/8$. Using these relationships

$$P_1 = \frac{P_S}{\frac{(x/L)^2}{4} + \frac{x/L}{2} + \frac{5}{4}} \tag{8.36}$$

x positive

$$P_2 = \frac{P_S}{\frac{(x/L)^2}{4} - \frac{x/L}{2} + \frac{5}{4}} \tag{8.37}$$

x negative

$$P_C = \frac{-16(x/L)\,P_S}{\left[\frac{x}{L}\right]^4 + 6\left[\frac{x}{L}\right]^2 + 25} \tag{8.38}$$

and

$$\frac{Q_L}{Q_{L\,(MAX)}} = \sqrt{1 + \frac{P_1}{P_S}} - 1.09\left(1 + \frac{x}{L}\right)\sqrt{\frac{P_1}{P_S}} \tag{8.39}$$

x positive

$$\frac{Q_L}{Q_{L\,(MAX)}} = 1.09\left(1 - \frac{x}{L}\right)\sqrt{\frac{P_2}{P_S}} - \sqrt{1 - \frac{P_2}{P_S}} \tag{8.40}$$

x negative

8.10. JET-PIPE VALVES

The valve configuration shown in Figure 8.21 is used as the first stage in some servo valves. Because of its relatively large clearances, the jet pipe valve may be less sensitive to the effects of contamination than are flapper-nozzle valves. The jet-pipe valve is a full bridge, four arm variable configuration similar to the underlapped 4-way slide valve.

The jet pipe valve is operationally similar to an underlapped, 4-way valve. The jet pipe is normally rotated by a torque motor. Oil under high pressure flows out of the jet pipe and impinges on a receiver. Two small diameter holes located side-by-side on the receiver are connected to either end of the load. With the jet pipe centered over the two holes, equal pressures are developed on each side of the load. When the torque motor causes the jet pipe to rotate off center, the jet impinges more on one hole and less on the other. This creates a pressure unbalance across the load.

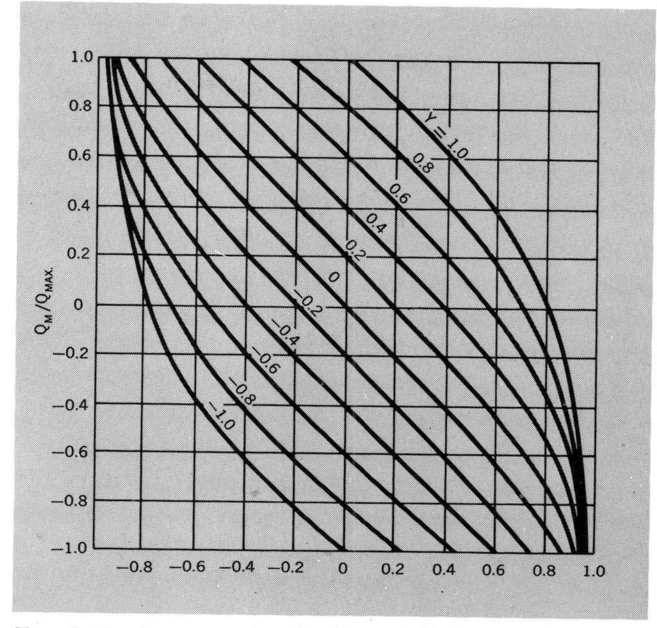

Fig. 8.20. K_c curves for double nozzle-flapper valve.

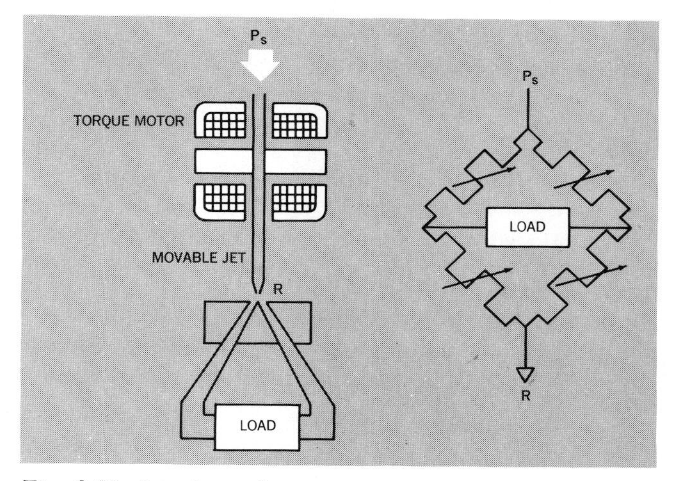

Fig. 8.21. Jet pipe valve.

Concerning the design of jet-pipe valves, Morse[1] says:

The exact shape and size of the parts of a jet-pipe valve cannot be computed in the way that nozzle–flapper valves are analyzed. However, in general, the following guides hold true.

1. L should be at least 2 d_n.
2. The two receiver holes should be as close together as it is possible to make them.
3. The nozzle and receiver diameters should be approximately equal.
4. The nozzle and receiver diameters should be as large as quiescent flow allowances will permit.
5. If the above are followed and if the internal diameter of the nozzle is constant for several diameters up the nozzle then, when the nozzle is centered, P_1 will equal P_2 will approximately equal $P_s/2$.

In a variation of the jet-pipe valve the nozzle is fixed in position and a deflector directs the stream of oil to the load ports. Forces can be applied to the torque motor through load springs to provide for dual mechanical-electrical inputs or to provide mechanical feed-back to the torque motor from the load. The valve, of course, remains a full bridge with four arms variable.

8.11. SERVO VALVES

Any valve which is used as the controlling portion of a servo system is, by definition, a servo valve. This can include on-off-on shutoff valves, poppet valves, sliding plate valves and a wide variety of spool and sleeve valves. Shutoff valves are used in "bang-bang" servos in which the motor can drive in both directions but at maximum velocity only. Modulating valves of various sorts are used in proportional control servomechanisms.

Modulating servo valves usually have sharp edged metering lands and square ports so that the flow vs. displacement curve is very linear.

8.12. ELECTROHYDRAULIC SERVO VALVES

These valves form a unique group of control devices of great importance to advanced hydraulic control systems. Many control systems require a transducer to convert low level electrical signals to hydraulic power. The power amplifiying capability of these valves can be prodigious. A servo system in which an electrical current of 8 ma driving into a 1000 ohm load is used to control a hydraulic flow of 20 gpm at a pressure of 3000 psi is not at all unusual. Using I^2R as the input power and using the relationship of equation 8.3 to determine the output power, the power amplification developed is

$$\frac{(.008)^2\ (1000)}{\dfrac{(.385)\ (20)\ (3000)}{1714}} = 210{,}000$$

Most electrohydraulic servo valves are two stage devices. The first stage is usually a single or double nozzle—flapper or a jet pipe. The flapper or jet pipe is moved by a torque motor. This first stage controls a spool and sleeve second stage, which, in turn, controls the motor or actuator.

Many variations of these basic designs have been used. Spool first stages have been driven by torque motors, by single tractive solenoids working against non-linear springs, by double tractive solenoids working against linear springs and by double tractive solenoids which shuttle the spool back and forth. In many designs the motion of the second stage is used to reposition the first stage. In some designs the second stage motion is converted to force by being transmitted through a spring before being applied to the first stage. Certain configurations of servosystems feed back the output of the actuator to reposition the servo valve first stage. Occasionally, for high flow requirements, a third stage valve is added between the second stage and the load.

A different set of data and nomenclature is used to describe electrohydraulic servo valves than was utilized in past discussions. Output flow and pressure are usually defined in terms of input current or differential current. Load flow curves are plotted as percent of full flow versus differential load pressure with input current as a variable parameter.

Because of their use in servo systems, e-h servo valves are often described by measures of their dynamic performance. These data can be given as plots of amplitude and phase as a function of frequency. With this method, the input current is varied sinusoidally. With fixed amplitude, the frequency is gradually increased. The output is measured. The output versus input amplitude ratio is usually converted to decibels gain or attenuation according to the following formula.

$$db = 20\ log_{10}\ \frac{Ratio\ at\ frequency\ of\ interest}{Ratio\ at\ base\ frequency}$$

The base frequency is chosen at a low enough point to assure no frequency dependent losses.

As the frequency is increased the output will tend to lag behind the input. This time lag can be measured and expressed in degrees where 1 sinusoidal cycle equals 360 degrees. The phase lag plot and amplitude plot are often shown on the same graph. Shown on Figure 8.22 are plots of actual data taken with such valves in the manner discussed above.

Another frequently used means of describing the dynamical nature of servo valves is the transfer function.

> The transfer function of a dynamical element is obtained by writing the time differential equation relating output to input, replacing the operation d/dt by the complex number s and solving for the ratio of output to input.

A transfer function is an easy tool to use analytically. Meaningful transfer functions can be derived for electrohydraulic servo valves, particularly when limited to relatively low frequencies.[2]

Unfortunately, servo valves have a number of non-linear elements which limit the applicability of direct analytical techniques. Among these elements are: electrical hysteresis of the torque motor, change in torque motor output with displacement, change in orifice fluid impedance with flow and with changes in fluid characteristics, change in orifice coefficient with pressure ratio and coulomb friction of the mechanical parts. Many servovalve parts have shapes which are analytically other than ideal. For example, the fixed orifices used in flapper-nozzle valves have diameters in the range of 0.006 to 0.008 inches. With such a small diameter, the length of the orifice becomes a significant percentage of the diameter. The orifice equations are based on sharp edged conditions which obviously cannot apply here. The orifice length is not

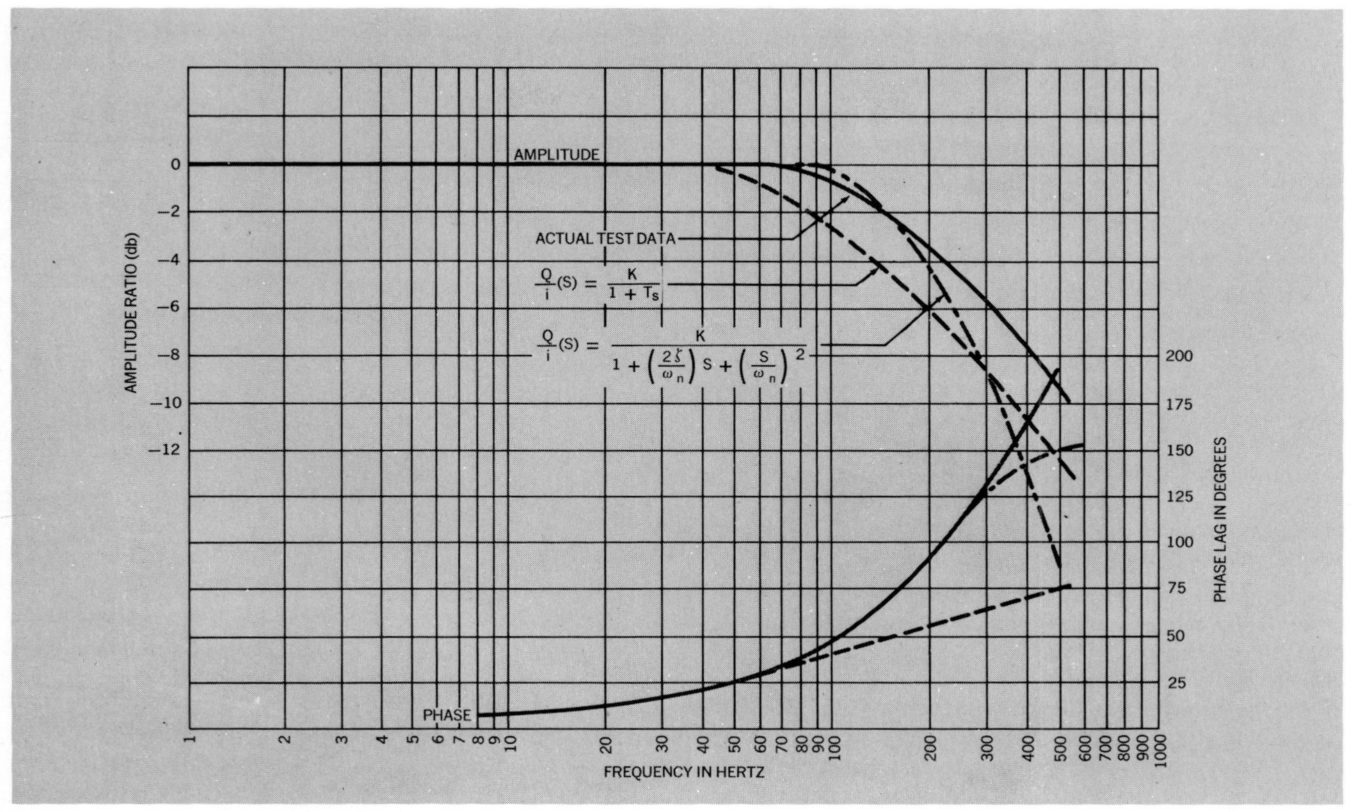

Fig. 8.22. Servo valve frequency response.

great enough to become a laminar flow choke. Therefore, the flow conditions lie somewhere between a first order and a second order relationship between pressure and flow.

Despite these non-linearities, the convenience of transfer functions make their use desirable wherever they are logically applicable. If all the dynamical factors are cranked into the equations, the mathematics become quite complex. However, a number of simplifying assumptions which are theoretically defensible may be made. Among those made for valves having a nozzle-flapper first stage are:

1. An ideal current source (infinite impedance) is used.
2. Negligible load pressure exists.
3. All non-linearities can either be approximated by linear dynamic effects, or can be neglected.
4. The armature-flapper can be represented as a simple lumped-parameter system.
5. Perturbation conditions can be applied to the hydraulic amplifier orifice characerisctics.
6. Fluid compressibility and viscosity effects are negligible.
7. Motions of the flapper are small with respect to spool motion.
8. The forces necessary to move the spool are small with respect to the driving force available.

Comparable assumptions may be made for designs other than double nozzle, single flapper configurations. These assumptions have, as corollaries, the belief that spool mass, friction, flow forces and other spool force effects are negligible. These lead to the block diagram of Figure 8.23 as described by Thayer.[2]

Figure 8.23 describes a system which is third-order in *s*. The system dynamical elements are the armature—flapper mass, spring and damping (a second order system) and the flow integration of the spool (a first order effect). The valve spool corresponds to the actuator piston of a typical positional servosystem.

Of these effects, the rotational mass of the armature/flapper is easy to calculate. The effective stiffness (spring) of the armature/flapper is composed of several effects including the positive spring rate of the flexure tube and the negative spring effect of the permanent magnet flux. The damping force is also complex but can be approximated with a damping ratio ζ of 0.4. On the basis of perturbation flow the orifice bridge is reduced to a simple gain term. This gain is the differential flow unbalance per increment of flapper motion.

The internal loop gain of the servovalve is

$$K_v = \frac{K_2 \; K_w}{K_f \; A_s} \qquad 8.41$$

The hydraulic amplifier flow gain K_2 is related to the nozzle configuration in the following way

$$K_2 = C_o \; \pi \; d_n \sqrt{\Delta \; P_n} \qquad 8.42$$

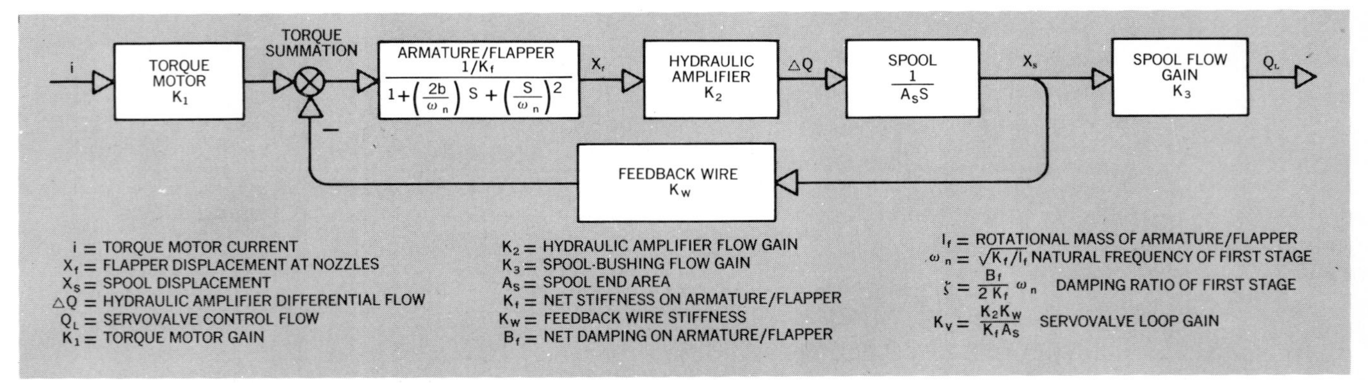

Fig. 8.23. Servo valve block diagram.

Where: C_o = nozzle orifice coefficient
d_n = nozzle diameter
ΔP_n = nozzle pressure drop

Any of the loop gain parameters can be altered to change the response of the servovalve. For example, loop gain can be increased by reducing the spool diameter, by increasing the nozzle diameter, by increasing the nozzle pressure drop or by using a higher torque motor charge level.

Even though the block diagram of Figure 8.23 leads to a cubic equation, a first order transfer function will provide an analytical description of flow control servovalves which is usable up to a frequency of about 50 Hz. A second order transfer function will have validity to about 270 Hz.

8.13. FLOW CONTROL SERVO VALVES

Figure 8.12 shows typical flow vs. pressure curves for flow control servovalves. The valves can be compensated by means of feed-back to give curves as shown in Figure 8.24.

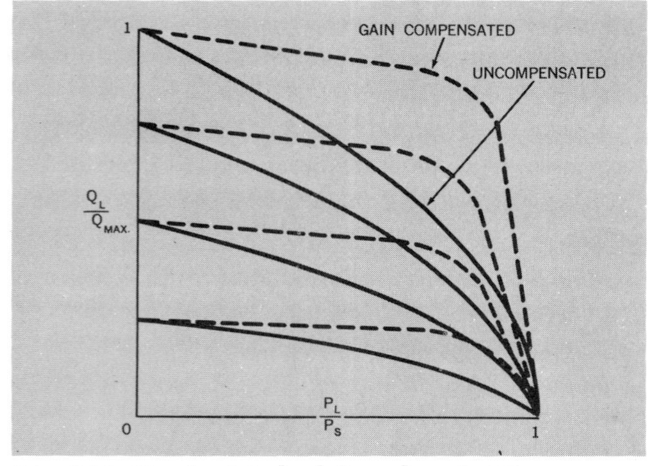

Fig. 8.24. Steady state load flow characteristics.

The first order transfer function which is valid to about 50 Hz[2] is

$$\frac{Q}{I}(s) = K\frac{1}{1 + Ts} \qquad 8.43$$

Where: Q = output flow
I = input current
K = servo valve static flow gain at zero load pressure drop $\frac{\text{cis}}{\text{ma}}$
T = time constant

If a second order transfer function is needed, the following can be used.

$$\frac{Q}{I}(s) = K\left[\frac{1}{1 + \left[\frac{2\delta}{\omega_n}\right]s + \left[\frac{s}{\omega_n}\right]^2}\right] \qquad 8.44$$

Where: ω_n = natural frequency in radians per second
ζ = damping ratio

The several parameters of these transfer functions can be obtained from manufacturers data.

Plots of the transfer functions of equation 8.43 and 8.44 are shown in Figure 8.22 for comparison with actual measured data.

8.14. PRESSURE CONTROL SERVO VALVES

Occasionally, a system is so designed that the differential pressure across the load must be controlled rather than the flow to the load. Pressure control servo valve operation can be described by the transfer function

$$\frac{P}{I}(s) = K_t\left[\frac{1}{1 + \left[\frac{2\delta}{\omega n}\right]s + \left[\frac{s}{\omega_n}\right]^2}\right] \qquad 8.45$$

Where: K_t = Static pressure gain psi/ma

8.15. ACCELERATION SWITCHING (AS) SERVO VALVES

In the AS servo valve, the proportional input current to the first stage is replaced by an alternating switching action. Control of the relative on and off time durations of the positive and negative input currents produces a corresponding rate of change of flow from the servo valve. In transfer function notation, this is

$$\frac{Q}{I_t}(s) = \frac{K}{s} \qquad 8.46$$

Where: K = acceleration switching gain $\frac{in^3/sec^2}{\text{time unbalance}}$

I_t = time unbalance of current

For those wishing further study in electrohydraulic servovalves, ARP 490A[3] is strongly recommended.

REFERENCES

1. Morse, A. C. "Electrohydraulic Servomechanisms" McGraw-Hill 1963
2. Thayer, W. J. "Transfer Functions for Moog Servo-valves" Technical Bulletin 103, Moog Servocontrols, Inc. January 1965
3. ARP 490A "Electrohydraulic Servovalves" Society of Automotive Engineers, New York, New York

CHAPTER 8

PROBLEMS

1. The supply pressure at a valve inlet port is 2650 psi. The pressure drop from return port to the reservoir is 250 psi. The reservoir is pressurized at 60 psi. The valve is sized to provide maximum power to the load when the load flow is 6 gpm. What is the maximum power?

2. When 3000 psi is imposed across the pressure and return ports of a valve and when the two cylinder ports are connected together (no load), 6.8 gpm will flow through the valve. If the same valve were connected to a load and the same supply pressure were available, what is the maximum power which could be delivered to the load?

3. A hydraulic system has a 3000 psi supply. When the load flow is 3 gpm, maximum power is being delivered to the load. Under these conditions the line loss (laminar) between pump and valve is 80 psi, and the line loss (laminar) between valve and reservoir is 60 psi. The valve acts as an orifice. (a) What load would be an actuator having a working area of 3 square inches support? (b) If the flow rate were changed so that the load velocity were 6 ips, how big a load could be handled?

4. A pressure compensated, variable delivery pump can deliver up to 8 gpm at 3000 psi. At that flow rate the line loss (laminar) between pump and servovalve is 175 psi. The loss (laminar) between valve and reservoir is 125 psi. The reservoir is pressurized at 50 psi. If the system supplies a load which is controlled by a 4-way valve which acts like an orifice, and the valve is sized to deliver maximum load power at the 8 gpm flow, what would this maximum power be? If the actuator has a working area of 3 square inches, how big a load could be moved at a design rate of 8 ips?

5. A 4-way slide valve is 3/8-inch in diameter. The ports cover half of the circumference of the lands. If the pressure drop across each metering land is 500 psi when the valve is displaced 0.050-inch from neutral and $\theta = 69^\circ$, what is the force tending to restore the valve spool to its null position. (Assume $c = 0.6$).

CHAPTER NINE

Power Output Transducers

9.1. DEFINITION

To do work, oil flow and pressure must pass through an output device. Output devices carry the generic name of motors. They may have translatory or oscillatory output rather than a rotary motion. Most hydraulic system motors are cylindrical actuators linked to their loads so that a rotating or arcuate motion path is developed.

Pure rotating motors are seldom found in power hydraulics use. Such motors derive their energy from flow impulse or reaction forces. Water wheels, pelton wheels and turbine engines are examples of these types of motors. Their primary drawback for use in hydraulic systems is their inability to hold a static load in a fixed position.

A cylindrical actuator which rotates a pivoted load by means of a crank arm, and a piston motor which develops rotation through ingenious camming are each an example of a device which is basically translatory in nature but which is used to develop angular motion. In general, however, cylindrical hydraulic actuators which—whether operating singly, in tandem or in parallel—primarily develop straight line motion or which develop very low frequency partial rotation are considered one family of motors. They are usually referred to as actuators.

Rotary motors may be—as with pumps—gear motors, vane motors or piston motors. Again, as with pumps, the motor may have a fixed oil quantity per revolution or may be programmed for variable requirements.

A special family of angular motion motors is the vane motor which is used primarily in an oscillatory fashion. These usually have a total rotation capability of less than one revolution.

9.2. CYLINDRICAL ACTUATORS

The unbalanced design, Fig. 9.1, has only two dynamic seals; the piston seal and the rod seal. This design may be used only with simple loads. Extension requires more oil than retraction. Similarly, a fixed oil flow rate will result in a slower extend rate and a faster return rate. In general, for any practical valve, the valve setting must be changed between extend and retract or differences in load or speed or power must be accepted. Fig. 9.2 shows one way in which an unbalanced actuator can handle a load which moves in other than a straight line.

A balanced actuator, Fig. 9.3, overcomes the effect on actuator control of differential area. With this design, any valve which is reasonably symmetrical will give uniform control characteristics in each direction. The primary objection to this design is the greatly increased space required in the axial direction. The working area of the piston should carry the full load at design speed with a pressure differential across the piston of two-thirds of the pressure available across the valve ports.

To save space and still get the control advantages of the balanced actuator, a partially balanced design, Fig. 9.4, is often used.

Where multiple hydraulic systems are used to provide reliability, a tandem configuration, Fig. 9.5, or a parallel design, Fig. 9.6, can be used. The tandem design has a small cross-section but great length. Its use makes the balanced configuration almost mandatory. The parallel design is shorter and broader. It can use actuators of balanced, unbalanced or partly balanced configuration. Where tandem actuators are used or where parallel actuators are force-summed to a common load, provisions must be made by-passing stuck valves so that a failure of one system will not lock the other system. Fig. 9.7 shows a triple parallel force-summed servo-actuator designed for advanced servo studies.

Actuators used on servo control systems must position heavy loads quickly and accurately. Therefore, they should act as very rigid, very stiff structural

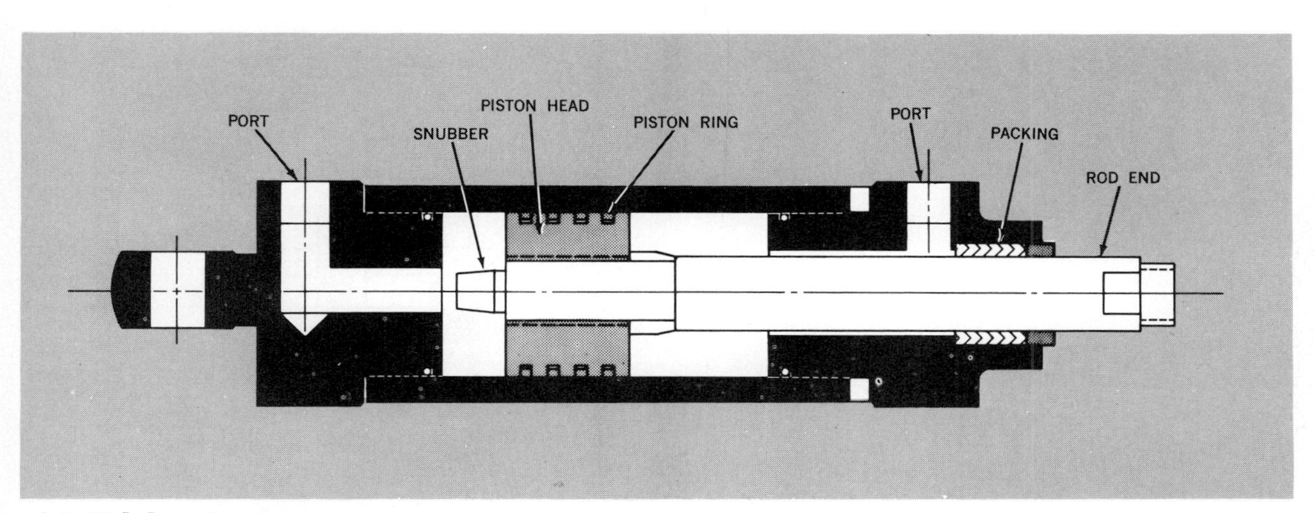

Fig. 9.1. Unbalanced actuator.

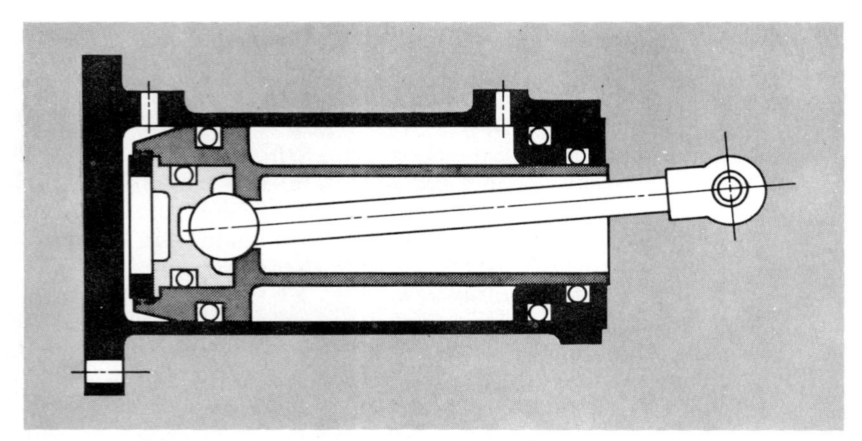

Fig. 9.2. *Unbalanced actuator for non-linear loads.*

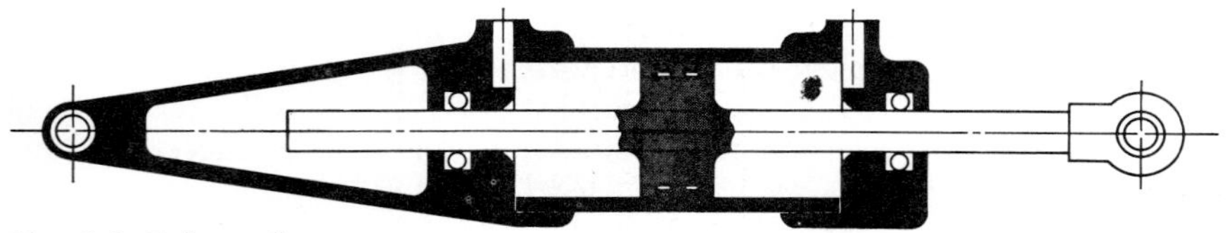

Fig. 9.3. *Balanced actuator.*

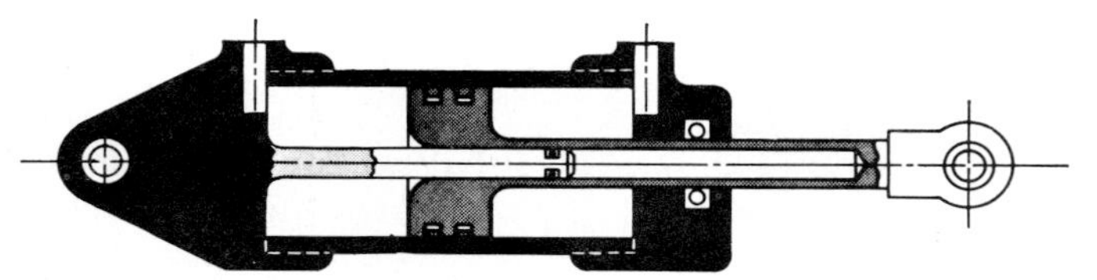

Fig. 9.4. *Partially balanced actuator.*

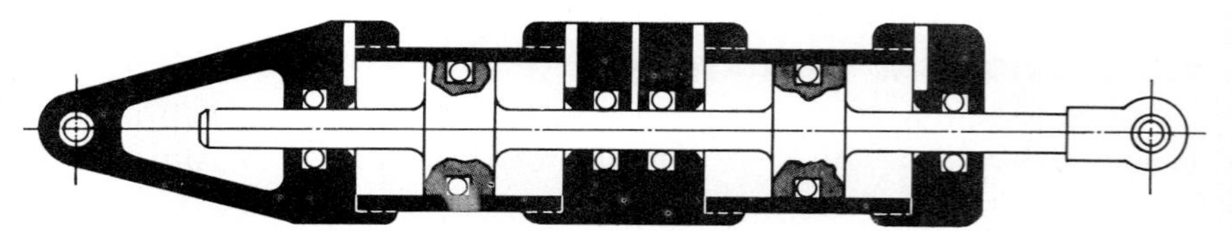

Fig. 9.5. *Tandem balanced cylinder.*

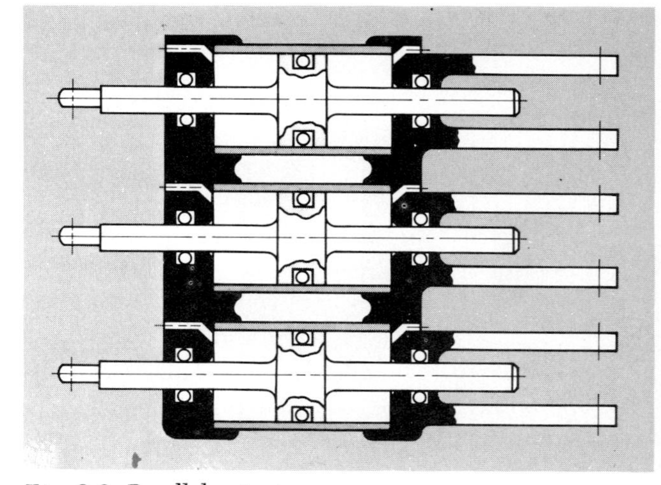

Fig. 9.6. *Parallel actuators.*

members. The fluid in the actuator is, however, compliant. A reasonable assumption can be made that, under servo control or under vibratory loads, the softest portion of the structural link is the fluid in the actuator. As was seen earlier, the spring rate of a fluid column is

$$K = \frac{A\,\beta}{l} \tag{9.1}$$

This relationship is applied in Fig. 9.8. Consider that an external load is applied axially to the rod and that the load is supported only by a change in the compression of fluid. This actuator has two fluid columns, one on each side of the piston. In any practical design, the pressure conditions will be such that the pressure in neither side will go below the oil vapor pressure. The system acts like two springs working in opposition in such a way that initial compression is never exceeded and neither spring ever acts in tension. In this interesting configuration, the springs are really acting in parallel. The spring rate of the system is the *sum* of the individual spring rates.

If the actuator piston is considered to be negligibly thick, the two spring rates are

$$K_1 = \frac{A\,\beta}{c_p\, l} \tag{9.2}$$

and

$$K_2 = \frac{A\,\beta}{(1 - c_p)\, l} \tag{9.3}$$

where: K_1, K_2 = spring rates
l = actuator stroke length
c_p = proportionate length of one side.

The total spring rate is

$$K = \frac{A\,\beta}{c_p\, l} + \frac{A\,\beta}{(1 - c_p)\, l} = \frac{A\,\beta}{(c_p - c_p^2)\, l} \tag{9.4}$$

This spring rate is plotted against piston position. The system spring is softest when the piston is in the mid position. The combined spring rate approaches infinity when the piston nears either end.

For an unbalanced actuator, where $A_1/A_2 = k$, then

$$K_1 = \frac{A_1\,\beta}{c_p\, l} \qquad K_2 = \frac{A_1\,\beta}{k\, l(1 - c_p)}$$

$$K_T = \frac{A_1\,\beta\,(k - kc_p + c_p)}{k\, lc_p\,(1 - c_p)} \tag{9.5}$$

Cylindrical actuator design variations seem endless. Certainly, the requirements of specific applications should dictate the design. Also, the desire for very short fluid columns between the valve and the actuator has led to many designs in which the valve and actuator are integrated physically and operationally.

Fig. 9.9 shows a balanced actuator with an electrical feedback device such as a potentiometer or linear variable differential transformer (LVDT). Figure 9.10 is an actuator with a mechanical positional feed-

Courtesy the Boeing Co.

Fig. 9.7. Three barrel parallel actuator instrumented for testing.

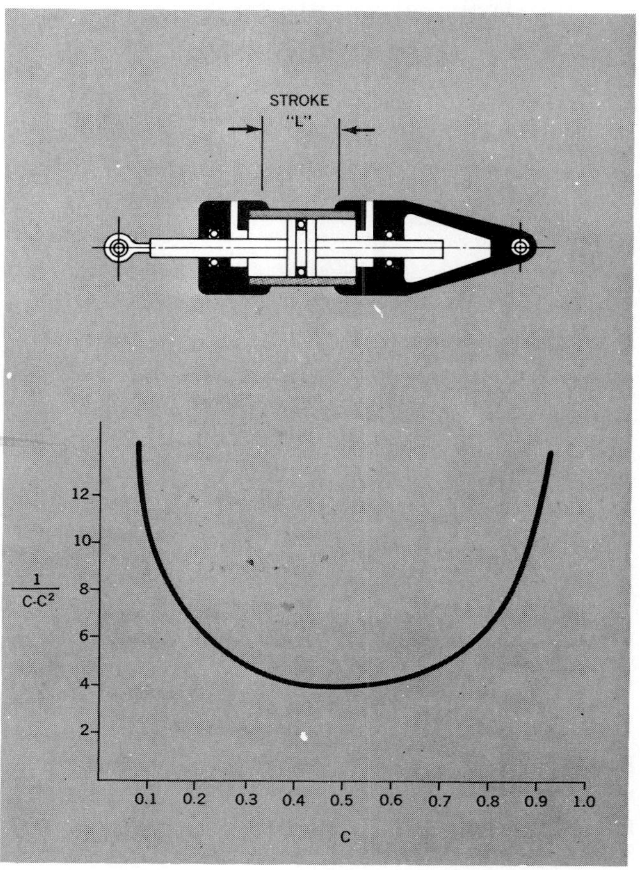

Fig. 9.8. Oil spring as a function of piston position.

back system to an electrohydraulic servovalve. Movement of the actuator is transmitted back to the torque motor armature through a spring. The spring converts actuator motion to a force. This force will increase until it balances the electromotive force developed on the torque motor. Because the forces on the valve flapper are then balanced, the valve will be returned to the centered- or null-position.

The actuator, Fig. 9.11, operates oppositely to the normal. The rod end is rigidly attached to structure and the valve-actuator package moves back and forth. In this design, the mechanical follow-up linkage is identical with the actuator stroke. Oil can be delivered to the system through flexible hoses, swivel joints or other means. This configuration is a favorite with authors of servo texts. Because the actuator will move exactly as much as the valve moves, the servo is said to have a feedback ratio of unity. This makes the computations relatively simple.

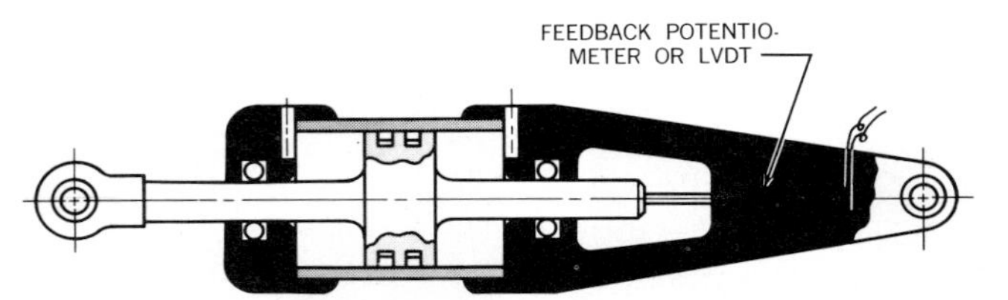

Fig. 9.9. Balanced actuator with electrical feedback.

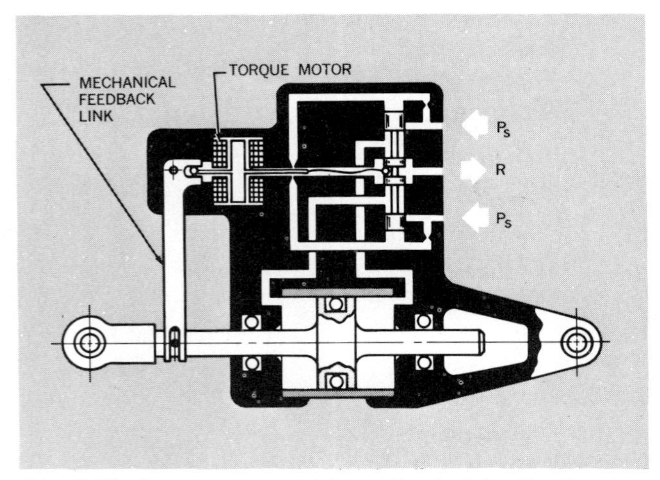

Fig. 9.10. Servoactuator with mechanical feedback.

All the actuators considered so far have been *analog* devices—that is they are capable of an infinite number of positions and move smoothly from one place to another. Actuators can be, however, designed as digital elements. In a digital actuator the allowable positions correspond only to a designed sequence of specific numbers. A digital actuator, Fig. 9.12, conforms to the binary (0,1) numbering system. The actuator is a series of pistons with allowable displacements which vary in accordance with binary values. The number of positions which such an actuator can assume is related to the number of pistons by

$$M = 2^N \qquad 9.6$$

M = number of actuator positions
N = number of pistons

Fig. 9.12 has four binary weighted actuators so that 16 positions are possible. These can be represented by the numbers 0 through 15. A larger number of pistons will provide an increasingly large number of actuator positions.

The binary actuator is usually controlled by one two-position three-way valve per piston. For actuation, the piston is connected to high pressure. Otherwise, it is connected to return line pressure. In comparison with an analog actuator under the control of a modulating servovalve, the digital system is a more complex array of more simple elements. Theoretically, the binary actuator will never deviate from the commanded position by more than ½ the travel of the shortest stroke position no matter how long the total stroke. This means that the analog system would probably be more accurate than a binary servo over short total stroking lengths. When the total stroke is long, as in some machine tool controls, the binary actuator could be considerably the more accurate. Because of this, binary hydraulic actuators have been used to position the control rods in nuclear reactors.

In the conventional binary number system, an error in the operation of the largest piston could result in a position error of 50%. Special binary numbering systems, such as the Grey Code, can be used to avoid such a large dependence on a single valve.

Fig. 9.13 shows two types of diaphragm actuators, one will produce a longer stroke than the other. In each case, a spring is usually employed to provide the retract stroke force. These actuators have no seals and no sliding friction. Contact between the diaphragm and the metal parts is a rolling effect. Another type of no-seal actuator uses an expandable bellows. It expands under pressure and retracts under vacuum.

9.3. ACTUATOR VALVE COMBINATIONS

The most efficient actuator designs incorporate the valve—and often the valve control linkage—into the actuator design. The simplest version is the servoactuator, Fig. 9.11. The output load has a specific position for each position of the input. A motion of the input linkage will displace the valve and allow pressurized oil P_S to flow to one end or the other of the actuator. The porting between the valve and the actuator is such that the actuator will move to close off the valve. Servo engineers call this *negative* feedback. The output motion of this servo and, therefore,

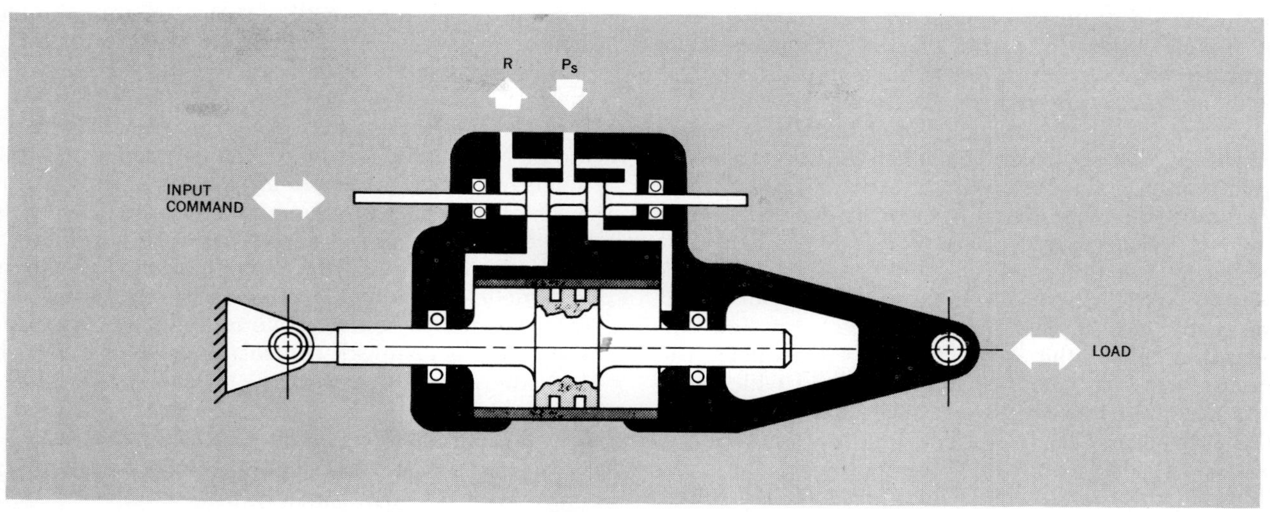

Fig. 9.11. Mechanical input servoactuator.

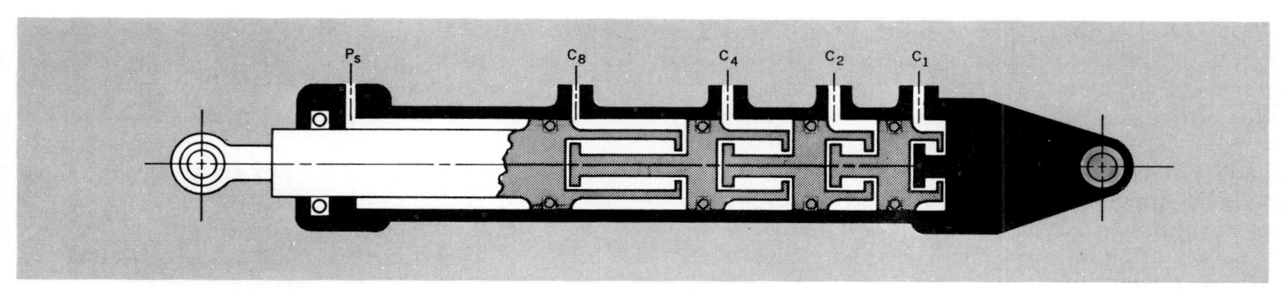

Fig. 9.12. Sixteen position binary actuator.

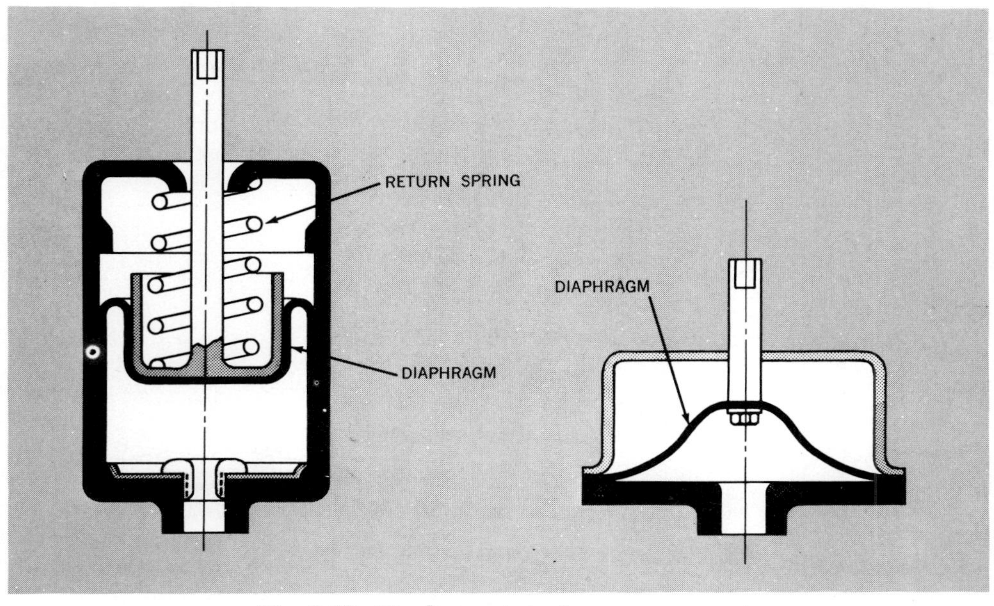

Fig. 9.13. Diaphragm actuators.

the output position, will follow the motion of the input linkage. One limitation of this system is that the input linkage motion must have the same travel length and direction as the actuator.

Another way of achieving stable control with negative feedback is shown in Fig. 9.14. In this system the input motion may be quite different in length and direction from the actuator stroke. To understand how this servo system works, consider the pin joint connecting the differentiating link and the feedback link to be fixed in space. A motion of the input would then result in a smaller motion of the error link and in an opening of the valve. As the valve opens, the actuator will start to move. If the pin joint which connects the input to the differentiating link now be considered fixed and the other end of the differentiating link be considered movable, the motion of the actuator will drag the feedback link and cause the differentiating link to rotate about its upper pivot. If the connections between the actuator and valve are correctly phased, the feedback motion will cause the valve to be closed and the actuator motion to be stopped.

Fig. 9.15 is a more complex servo. In this design an electrical input from a servo valve or a mechanical input from an operator or from a sequencing linkage can control the servo independently or simultaneously. In the electrical control mode, the servovalve controls an actuator (often called a modulating or "Mod" piston) which controls the main valve through a whipple tree connection with the mechanical input. If the two inputs are applied simultaneously, the electrical system can be considered to be in series with the mechanical input and the commands to the control valve will be added algebraically. In this mode, motions of the actuator due to electrical command are not transmitted back to the mechanical input because the resistance to motion of the mechanism back to the input source is usually very much larger than the resistance to motion of the control valve. Therefore, feedback motion will go to the valve rather than to the input linkage. This arrangement is typical of automatic augmentation system connections.

If the lockout immobilizes the modulating piston, the system reverts to the design of Fig. 9.14.

If the mechanical system input were held fixed at the pin joint which connects the differentiating link to the error link and if some sort of electrical stroke measuring device were mounted on the actuator output, the system would be typical of an airplane autopilot connection. In this configuration, the feedback is done electrically and the actuator motion would drag the pilot's control column along with it. In this way the mechanical controls are always in phase with the electrical controls.

9.4. ACTUATOR DESIGN COMPUTATIONS

Design requirements for linear actuating cylinders will vary, of course, with the installation. However, certain design parameters are typical to all problems. Among these are the determination of working area of the piston, the sizing of the rod, and the choosing of the proper inside diameter and wall thickness of the housing.

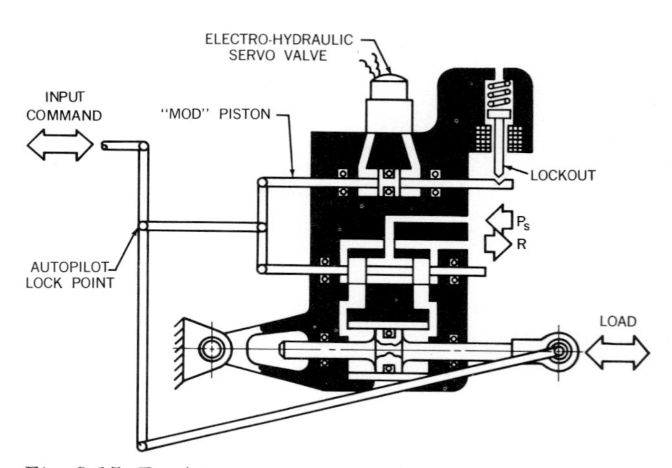

Fig. 9.15. Dual input servo actuator.

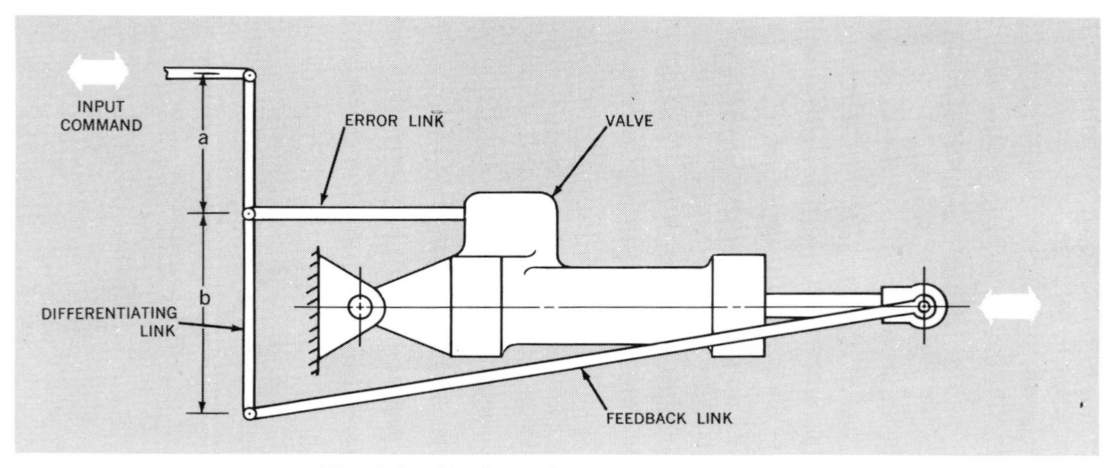

Fig. 9.14. Mechanical input servoactuator.

The working area of an actuator can be set by (1) the stalled, or no-flow, force output, (2) the no-load velocity requirement, (3) a schedule of force-velocity requirements or (4) dynamic load analyses. Where more than one of these is important, the force vs. velocity curves for each condition should be computed. The several curves should be plotted on a common graph as an aid to size selection.

The stalled, or no-flow, force defines a working area of the piston at supply pressure

$$A_p = \frac{F}{P_S} \qquad 9.7$$

Under these conditions the valve will be wide open. If the valve flow-pressure coefficient is known, a parabola relating force to velocity can be drawn which passes through the stalled force point. The simplest description of K_c is the valve conductance

$$Y = \frac{Q}{\sqrt{P_V}} \qquad 9.8$$

The load pressure $P_S - P_V$ and the load velocity is Q/A_p so

$$F = P_S A_p - \frac{\dot{x}^2 A^3_p}{Y^2} \qquad 9.9$$

A typical plot of this relationship is shown in Fig. 9.16.

If a no-load velocity requirement has been established, a parabola based on this point and the valve conductance can also be drawn. Such a plot is shown in Fig. 9.16.

Where a point or points relating required load and velocity are known, a more detailed analysis is possible. The first assumption to be made is that the valve and actuator will be sized to provide the most adverse combination of velocity and force when one-third of the pressure drop is taken through the valve. Because the line losses to and from the actuator will also vary with velocity, they should properly be considered.

If they are laminar losses, then

$$\Delta P = \frac{128 \ \rho \ v \ l \ Q}{\pi \ D^4} \qquad 9.10$$

If the flow is turbulent

$$\Delta P = \frac{2.53 \ \rho \ l \ Q^2}{N_R^{0.25} \ D^5} \qquad 9.11$$

Both because of practical design considerations and ease of computation, laminar flow should be assumed until conditions prove otherwise. The load pressure is

$$P_L = \frac{2}{3} \left[P_S - (\Delta P_P + \Delta P_R)\right] \qquad 9.12$$

If the pressure line loss can be assumed equal to the return line loss

$$P_L = \frac{2}{3} (P_S - 2\Delta P) = \frac{2P_S - 4\Delta P}{3} \qquad 9.13$$

This can be combined with previous relationships to yield

$$\Delta P^2 - \frac{P_S}{2} \Delta P + \frac{96 \ \rho \ v \ l \ \dot{x} \ F}{\pi \ D^4} = 0 \qquad 9.14$$

If the line size is known, a plot of force and velocity can be drawn from this relationship. If a critical combination of force and velocity is known, quadratic solutions of line pressure drop may be made and checked for Reynolds Number. Then the force-velocity curve can be constructed. (When making such calculations, l must include *both* the pressure and return line lengths.) A typical curve based on a critical force-velocity point is plotted on Fig. 9.16.

The fourth measure of valve-actuator sizing assumes that the load must be cycled sinusoidally. Actuator piston velocity is

$$\dot{x} = \frac{1}{A_p} \left(Q - \frac{V}{2\beta} P_L\right) \qquad 9.15$$

The dynamic load relationship is

$$P_L \ A_p = M\dot{x}s + B\dot{x} + \frac{Kx}{s} \qquad 9.16$$

These can be combined to give

$$F = 2 \ A_p \ \beta \dot{x} \left[\frac{Ms^2 + Bs + K}{(2 \ A^2_p\beta + MV + BV) \ s + KV}\right] \qquad 9.17$$

To evaluate this relationship at any frequency s must be replaced with jw, where $j = \sqrt{-1}$. The force-velocity relationship then takes the form of a vector having its tail at the origin. The general equation is

$$F = D \ \dot{x} \ \omega \ (A + jB) \qquad 9.18$$

A typical form for the force-velocity relationship as the frequency is varied over the range of interest is also shown on Fig. 9.16.

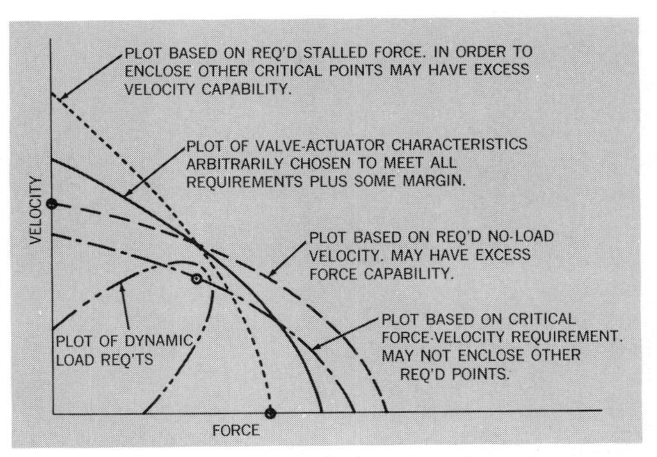

Fig. 9.16. Typical force velocity plots for valve-actuator combinations.

These discussions should emphasize that no one technique for choosing actuator area is usually sufficient. Also, if a range of operation is important, the actuator working area cannot be chosen without consideration of the valve coefficients and the flow losses encountered in the balance of the system. The four criteria discussed gave quite different curves on the force-velocity plot. The final configuration chosen must give an F-V curve which will enclose each of the critical points. As shown in Fig. 9.16, the choice of the best combination of valve coefficient and actuator working area may be a compromise among the plots developed by the criteria studied.

The size of the piston rod must now be determined. The usual practice is to use Euler's Formula

$$F_c = \frac{c_f \pi^2 EA}{\left(\frac{L}{k}\right)^2} \quad 9.19$$

where: F_c = buckling load
E = Young's Modulus;
A = rod area
L = column length
k = radius of gyration
c_f = fixity factor

For a solid rod and a fixity factor of ¼ (one end—the actuator end—fixed and the load end free). Equation 9.19 reduces to

$$F_c = \frac{\pi^2 EI}{4 L^2} \quad 9.20$$

where: I = column moment of inertia
The critical stress is

$$S_{cr} = \frac{\pi^2 E}{64} \left(\frac{D}{L}\right)^2 \quad 9.21$$

If, as is more usual, the rod is hollow

$$F_c = \frac{\pi^2 E I}{4 L^2} = \frac{\pi^3 E (D^4 - d^4)}{256 L^2} \quad 9.22$$

where: D = OD and d = ID

The critical stress is

$$S_{cr} = \frac{\pi^2 E}{64} \left[\frac{D^2 + d^2}{L^2}\right] \quad 9.23$$

If an actuator which was handling a 4500 lb load had a hollow shaft which had a free length of 10″, then

$$4500 = \frac{(9.8)\ (3 \times 10^7)}{256 \times 10^2} (D^4 - d^4)$$

$$D^4 - d^4 \geqq 0.392$$

Solutions to this relationship provide rod dimensions which are barely adequate to support the maximum load. If a safety margin of 1.5 were to be provided, then

$$D^4 - d^4 \geqq 0.586$$

This relationship states that the rod OD *must* be greater than $D^4 = 0.586''$ or $D = 0.875''$. Among acceptable choices are

D	*d max.*	*Max. Stress*
0.9375	0.656	60,000 psi
1.000	0.802	62,400 psi
1.125	1.035	107,000 psi

With the rod and bore sizes chosen, the only remaining task is to choose the actuator wall thickness. This should be determined by the thick wall formula

$$S = P \frac{r_2^2 + r_1^2}{r_2^2 - r_1^2} \quad 9.24$$

where: S = stress, psi
P = pressure, psi
r_2 = external radius;
r_1 = internal radius

For the case of a cylinder with the inside diameter of 2¼ in, a stress limit of 50 kpsi might be desired at the working pressure of 3000 psi.

$$5.0 \times 10^4 = 3 \times 10^3 \frac{r_2^2 + (1.125)^2}{r_2^2 - (1.125)^2}$$

$$r_2 = 1.2''$$

With r_2 chosen, the radial displacement should be checked to assure that the deflection due to internal pressure will not render the piston seals ineffective.

$$\Delta r_1 = P \frac{r_1}{E} \left[\frac{r_2^2 + r_1^2}{r_2^2 - r_1^2} + \mu\right] \quad 9.25$$

where: E = modulus of elasticity
μ = Poisson's ratio

Assuming a barrel of steel where $E = 3 \times 10^7$ psi and Poisson's ratio is 0.3.

$$\Delta r_1 = 3 \times 10^3 \frac{1.125}{3 \times 10^7} \left[\frac{1.44 + 1.26}{1.44 - 1.26} + 0.3\right]$$

$$\Delta r_1 = 0.00172''$$

Because the squeeze on an O-ring is about .015 inches, the calculated radial expansion appears reasonably conservative.

9.5. ACTUATOR DYNAMICS

The steady state velocity of an actuator is

$$V = \frac{Q}{A} \quad 9.26$$

However, the instantaneous velocity under a command change is affected by valve and load characteristics. As such, the actuator velocity can only be described by a non-linear, second order differential equation. Baier[1] has developed an analytical solution to this equation. The following is a modification of this solution.

Assume that each metering port in the valve acts as an orifice

$$\Delta P = KQ^2 \quad 9.27$$

where: K = orifice constant $= \frac{1}{(\text{Conductance})} 2$

With an unbalanced actuator, the flow into one end is

$$Q_1 = A_1\dot{x} \quad 9.28$$

and into the other end

$$Q_2 = A_2\dot{x} \quad 9.29$$

The corresponding pressure drops through the two sets of valve metering orifices are

$$\Delta P_1 = K(A_1\dot{x})^2 \quad 9.30$$

$$\Delta P_2 = K(A_2\dot{x})^2 \quad 9.31$$

For any supply pressure P_S, assuming 0 psig reservoir pressure, the actuator head end pressure is

$$P_H = P_S - \Delta P_1 = P_S - K(A_1\dot{x})^2 \quad 9.32$$

and in the rod end

$$P_R \; \Delta P_2 = K(A_2\dot{x})^2 \quad 9.33$$

The load is assumed to be purely inertial (no spring or friction). If M is the combined mass of the piston and load,

$$M\ddot{x} = [P_S - K(A_1\dot{x})^2] \; A_1 - KA_2(A_2\dot{x})^2 \quad 9.34$$

or

$$\frac{P_S \; A_1}{M} = \ddot{x} + \frac{K}{M} \; (A_1^3 + A_2^3) \; \dot{x}^2 \quad 9.35$$

Equation 9.35 is of the form

$$\ddot{x} + c_1(\dot{x}^2) = c_2 \quad 9.36$$

where: $$c_1 = \left[\frac{K \; A_1^3}{M}\right] \; (1 + k^3) \quad 9.37$$

$$c_2 = \frac{P_S \; A_1}{M} \quad 9.38$$

$$k = A_2/A_1$$

Equation 9.36 can be expressed as

$$\frac{d^2Y}{dT^2} + \left[\frac{dY}{dT}\right]^2 = 1 \quad 9.39$$

Then

$$Y = \frac{x}{c_3} \quad 9.40$$

where: $c_3 = \frac{1}{c_1}$

and

$$T = t \; c_4 \quad 9.41$$

where: $c_4 = \sqrt{c_1 c_2}$

Using classical approaches to the solution of equation 9.39

$$Y = \ln \cosh T \quad 9.42$$

$$\frac{dY}{dT} = \tanh T \quad 9.43$$

Because

$$dx = c_3 \; dY \quad 9.44$$

$$dT = c_4 \; dt \quad 9.45$$

Then

$$\dot{x} = \frac{dx}{dt} = c_3 c_4 \frac{dY}{dT} \quad 9.46$$

$$\dot{x} = \frac{1}{A_1}\sqrt{\frac{P_S}{K(1 + k^3)}} \; \frac{dY}{dT} \quad 9.47$$

Maximum velocity is reached when $\frac{dY}{dT} = 1$

$$\dot{x}_{max.} = \frac{1}{A_1}\sqrt{\frac{P_S}{K(1 + k^3)}} \quad 9.48$$

As an example of these relationships, consider a balanced actuator in which $A_1 = A_2 = 3.0$ sq. in. The piston and load weighs 5000 lbs. The supply pressure is 3000 psi and $\Delta P_1 = 500$ psi at 60 gpm.

$$M = \frac{5000}{(32.2) \; (12)} = 12.9 \; \frac{\text{lb-sec}^2}{\text{in}} \qquad k = 1$$

$$K = \frac{500}{(231)^2} = 9.35 \times 10^{-3}$$

$$c_3 = \frac{M}{KA_1^3 \; (1 + k^3)} = \frac{12.9}{9.35 \times 10^{-3} \; (27) \; (2)} = 25.5$$

$$c_4 = \sqrt{c_1 \; c_2} = \frac{A_1^2}{M} \; [K(1 + k^3) \; P_S]^{1/2} = \frac{9}{12.9}$$

$$[9.35 \times 10^3 \; (2) \; (3000)]^{1/2} = 5.25$$

$$x = c_3 Y = 25.5\ Y \qquad t = \frac{T}{c_4} = \frac{T}{5.25}$$

Assume that $\dot{x}$ and t are desired when $x = 4''$, then

$$Y = \frac{4}{25.5} = 0.157$$

$$T = 0.5, \quad \frac{dY}{dT} = 0.48 \qquad t = \frac{T}{5.25} = \frac{0.5}{5.25}$$

$$= 0.095 \text{ seconds}$$

$$\dot{x} = \frac{1}{A_1}\sqrt{\frac{P_s}{K(1 + k^3)}}\ \frac{dY}{dT} = \frac{1}{3}\sqrt{\frac{3000}{9.35 \times 10^{-3}\ (2)}}$$

$$(0.48) = 22.8 \text{ in/sec.}$$

Gartner[2] and his associates have shown how to use these same relationships to relate piston area to actuator time response. Assume that the displacement x will reach its desired value x_f at time t_f. Then t_f should be minimized by the proper choice of piston effective area. The area is important because, if too small, the piston cannot accelerate an inertial load quickly enough. If the piston is too large, the ultimate velocity achievable may be too low even though acceleration may be better than adequate.

From equations (9.40) and (9.41)

$$x = c_3\ Y = c_3 \ln \cosh T = c_3 \ln \cosh c_4 t \qquad 9.49$$

$$x = \frac{M}{(KA_1^3)\ (1 + k^3)} \ln \cosh \left[\frac{A_1^2}{M}\sqrt{P_s K\ (1 + k^3)}\ t\right] \qquad 9.50$$

Differentiating this with respect to A_1 and setting $dt_f/dA_1 = 0$

$$\frac{3K\ (1 + k^3)}{M} A_1^2 x_f = \frac{2A_1\ t_f \sqrt{P_s K\ (1 + k^3)}}{M}$$

$$\tanh \frac{A_1^2\sqrt{P_s K\ (1 + k^3)}}{M}\ t_f \qquad 9.51$$

By resubstituting this reduces to

$$3Y = 2T \tanh T \qquad 9.52$$

This equation and (9.42) can be solved simultaneously to obtain the unique values of Y and T which satisfy them:

$$Y = 1.145$$

$$T = 1.1812$$

These represent the values of Y and T when $dt_f/dA_1 = 0$. At this condition t_f is either maximum or minimum with respect to the piston area. Inserting the value of Y into the relation of (9.40) yields

$$x_f = 1.145\ c_3 \qquad 9.53$$

$$= \frac{1.145\ M}{(KA_1^3)\ (1 + k^3)}$$

Expressing this in terms of A_1 gives the optimum area for traveling to point x_f in minimum time t_f.

$$A_1 = \left[\frac{1.145\ M}{K\ x_f\ (1 + k^3)}\right]^{1/3} \qquad 9.54$$

The time required to reach x_f is obtained also from the relations of (9.41)

$$t_f = \frac{T}{C_4} = \frac{1.182}{C_4} \qquad 9.55$$

$$= \frac{1.182\ M}{A_1^2\sqrt{KP_s}\ (1 + k^3)}$$

The use of these equations can be illustrated by assuming the same conditions as the previous example except that the optimum area for a 4 in. stroke shall be determined.

The optimum area is

$$A_1 = \left[\frac{(1.145)\ (12.9)}{9.35 \times 10^{-3}\ (4)\ (2)}\right]^{1/3}$$

$$= 5.8 \text{ in}^2$$

$$t_f = \frac{(1.182)\ (12.9)}{(5.56)^2\sqrt{9.35 \times 10^{-3}\ (3000)\ (2)}}$$

$$= 0.0658 \text{ seconds}$$

9.6. ACTUATOR STROKE SNUBBING

With a massive load and sizable velocity, the actuator can be damaged if the piston bottoms against either end. To avoid damage to the load or to the servo system, load alleviation devices are often built into the actuators. Coil springs and belleville springs can be used to soften bottoming forces, Fig. 9.17. A host of design variations use orifice type or laminar type flow restrictors to damp out actuator velocity. Fig. 9.18 shows some of the configurations.

Exact analysis of dashpot flow restrictors is not possible because of the usual lack of knowledge of such factors as friction levels and because of the interdependence of piston velocity and resisting force. However, by making some simplifying assumptions, design parameters having first-cut validity can be developed. Depending upon the nature of the actual requirement, the design parameters will usually have to be updated by actual testing.

As an example, consider Fig. 9.19. The actuator has a dashpot (annular orifice) type of snubber. Balancing the input work against the output work yields

$$E + W_m + W_1 = W_f + W_2 + W_3 \qquad 9.56$$

where: E = kinetic energy $= \frac{Mv^2}{2}$

W_m = mechanical work

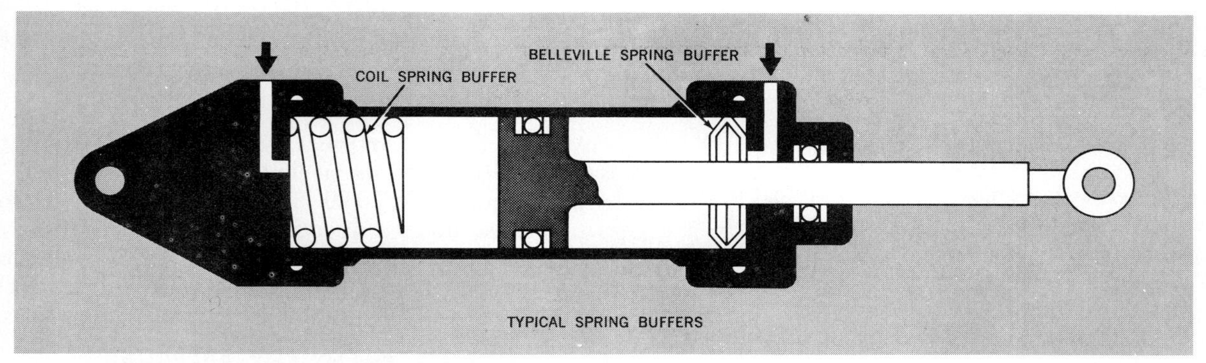

Fig. 9.17. Springs cushion the actuator stroke.

$$W_1 = \int_0^l P_1 A_1 dx$$
$$W_2 = P_2 A_2 l$$
$$W_3 = \int_0^l (\Delta P) A_3 dx$$
$$W_f = \text{friction work}$$

Computation of W_1 is simplified by assuming that the input pressure drops linearly from its initial value to zero. W_1 becomes

$$W_1 = \frac{P_1 A_1 l}{2} \qquad 9.57$$

Friction force must be computed for each design. However, friction forces tend to be small compared to other forces involved. A gross rule of the thumb would be to assign 10 lbs of friction force for each inch of diameter per moving seal. W_2 will vary with actuator velocity. This relationship is hard to handle so that, for trial calculations, assume P_2 to be a constant.

W_3 is evaluated by computing the integral. ΔP is derived from the equation for an annular orifice.

$$\Delta P = \frac{12 \mu Y Q}{\pi d b^3} \qquad 9.58$$

This equation can be modified to consider that the orifice length varies. l can, therefore, be replaced by the variable x. W_3 is then

$$W_3 = \int_0^l \frac{12 \mu Q A_3}{\pi d_3 b^3} x \, dx \qquad 9.59$$

Integrating and evaluating:

$$W_3 = \frac{6 \mu Q A_3 l}{\pi d_3 b^3} \qquad 9.60$$

But

$$A_3 = \frac{\pi d_3^2}{4} \quad \text{and} \quad Q = A_3 v$$

So

$$W_3 = \frac{3 \pi \mu d_3^3 l^2 v_0}{16 b^3} \qquad 9.61$$

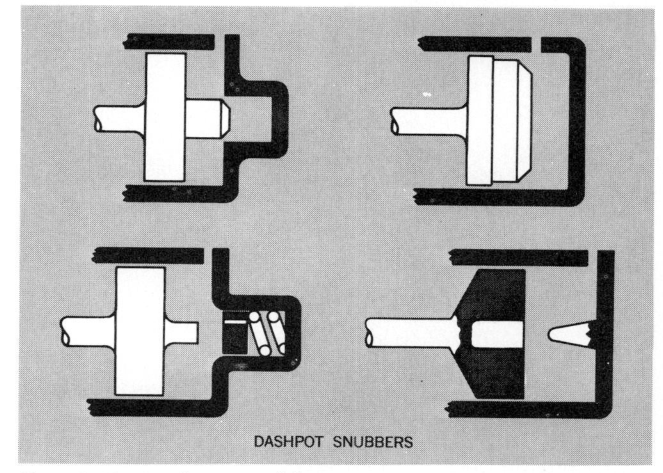

Fig. 9.18. Dashpot snubbers.

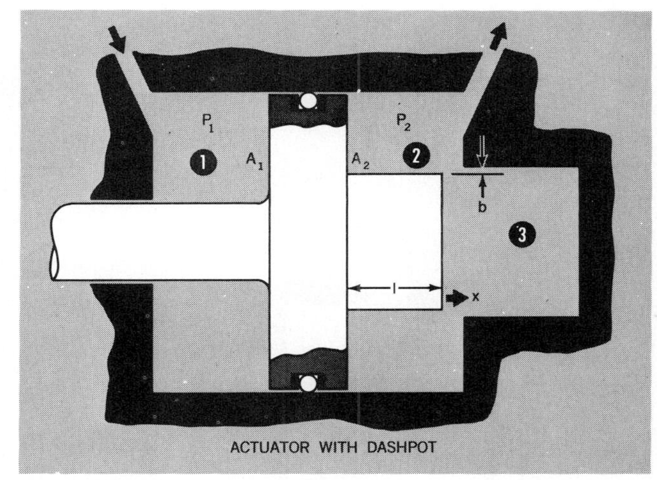

Fig. 9.19. Actuator with dashpot.

Equation (9.56) can now be further evaluated

$$\frac{M v_0^2}{2} + Fl + \frac{P_1 A_1 l}{2}$$

$$= W_f + P_2 A_2 l + \frac{3 \pi \mu d_3^3 l^2 v_0}{16 b^3} \qquad 9.62$$

Equation (9.62) can be solved for any parameter of interest such as dashpot clearance, pressure differential or piston-plus-load mass. Under the assumptions used in this analysis, stopping time is

$$t = \frac{2l}{v_o} \qquad 9.63$$

When applying equation (9.62) to any practical problem, the engineer should realize that, because of these simplifying assumptions, the analytical work must be supported by operational testing to assure proper functioning of the damper.

An equation similar to (9.63) can be developed for a system which uses an orifice rather than an annular choke. The primary change is to W_3. For an orifice, ΔP becomes

$$\Delta P = \frac{\rho}{2} \frac{Q^2}{c^2 A_o^2} \qquad 9.64$$

However,

$$Q = A_3 v = \frac{\pi d_3^2 v_o}{8} \quad \text{and} \quad A_o = \frac{\pi d_o^2}{4}$$

So,

$$\Delta P = \frac{\rho d^4 v_o^2}{4 c^2 d_o^4} \qquad 9.65$$

Then

$$W_s = \frac{\pi \rho d_3^6 v_o^2}{16 c^2 d_o^2} \int_o^l dx \qquad 9.66$$

Integrating and evaluating

$$W_s = \frac{\pi \rho d_3^6 v_o^2 l}{16 c^2 d_o^2} \qquad 9.67$$

The system equation then becomes

$$\frac{M v_o^2}{2} + Fl + \frac{P_1 A_1 l}{2} \qquad 9.68$$

$$= W_f + P_2 A_2 l + \frac{\pi \rho d_3^6 v_o^2 l}{16 c^2 d_o^2}$$

Another form of snubber, or shock absorber, which is frequently used is a flow port consisting of a series of holes, Fig. 9.20. The advancing piston covers up these holes sequentially. The reduced area of the port offers greater flow resistance thereby slowing down the piston. By spacing the port holes, nearly any desired rate of snubbing can be achieved.

9.7. ACTUATOR TRANSIENT PRESSURE

Occasionally, a situation arises where the actuator load is primarily inertial and the load must be rapidly stopped, started or its direction reversed. As in the earlier study of transient pressures in transmission lines, the most severe pressures occur when the actuator and its load are moving at maximum velocity and the control valve is instantaneously closed. The differential equation to be satisfied is

$$O = M\ddot{x} + B\dot{x} + Kx \qquad 9.69$$

$$v(t = o) = v_o$$

For the case where the damping coefficient B may be ignored, this reduces to

$$O = \ddot{x} + \frac{K}{M} x \qquad 9.70$$

$$v(t = o) = v_o$$

Replacing $\frac{K}{M}$ with ω_n^2 and taking the LaPlace transform

$$X(s) = -\frac{v_o}{s^2 + \omega_n^2} \qquad 9.71$$

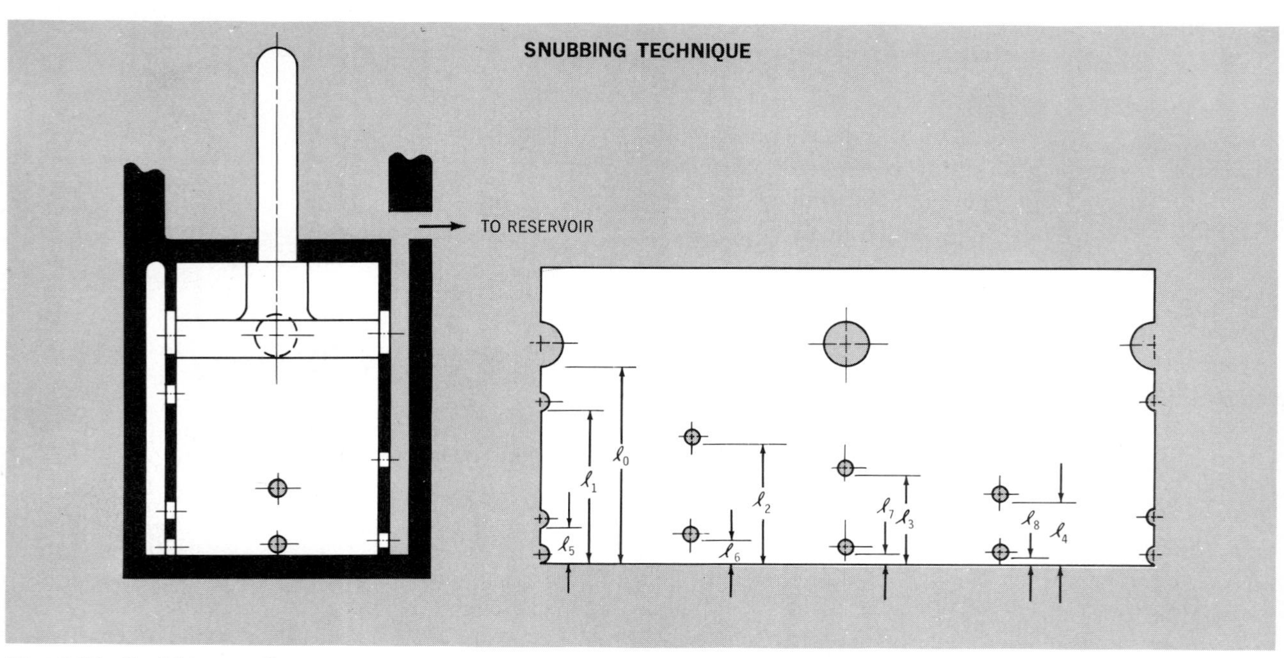

Fig. 9.20. Snubbing technique.

The time solutions are

$$x(t) = -\frac{v_o}{\omega_n} \sin \omega_n t \tag{9.72}$$

$$\dot{x}(t) = -v_o \cos \omega_n t \tag{9.73}$$

$$\Delta P_{max.} = \frac{F_{max.}}{A} = \frac{Kx_{max.}}{A} \tag{9.74}$$

$$= -\frac{K\,v_o}{A\,\omega_n} = -\frac{v_o}{A}\sqrt{K\,M}$$

If piston head leakage and friction can be expressed in units of lbs per inch per second, this value of the damping coefficient B can be inserted and the damped response determined.

The equation now is

$$O = \ddot{x} + \frac{M}{B}\dot{x} + \frac{K}{M}x \tag{9.75}$$

$$v(t = o) = v_o$$

The LaPlace transform is

$$x(s) = -\frac{v_o}{s^2 + \frac{B}{M}s + \frac{K}{M}} \tag{9.76}$$

By completing the square in the denominator and taking the inverse LaPlace transform

$$x(t) = \frac{-v_o}{\sqrt{\frac{K}{M} - \left[\frac{B}{2M}\right]^2}}\, e^{-\frac{B}{2M}t} \sin \sqrt{\frac{K}{M} - \left[\frac{B}{2M}\right]^2}\; t \tag{9.77}$$

Making the substitutions $\omega_n^2 = \frac{K}{M}$, $\sigma^2 = \left[\frac{B}{2M}\right]^2$

$$x(t) = \frac{-v_o}{\sqrt{\omega_n^2 - \sigma^2}}\, e^{-\sigma t} \sin\sqrt{\omega_n^2 - \sigma^2}\; t \tag{9.78}$$

$$\dot{x}(t) = \frac{v_o\sigma}{\sqrt{\omega_n^2 - \sigma^2}}\, e^{-\sigma t} \sin\sqrt{\omega_n^2 - \sigma^2}\; t - v_o e^{-\sigma t} \cos\sqrt{\omega_n^2 - \sigma^2}\; t \tag{9.79}$$

At, $t = \frac{\pi}{2\sqrt{\omega_n^2 - \sigma^2}}$, $x(t) = X_{max}$

$$= \frac{-v_o}{\sqrt{\omega_n^2 - \sigma^2}}\, e^{-\frac{\pi\sigma}{2\sqrt{\omega_n^2 - \sigma^2}}}$$

So $$\Delta P_{max.} = -\frac{K\,v_o}{A\sqrt{\omega_n^2 - \sigma^2}}\, e^{-\frac{\pi\sigma}{2\sqrt{\omega_n^2 - \sigma^2}}} \tag{9.80}$$

9.8. SERVOACTUATOR DESIGN CONSIDERATIONS

No hard and fast rules for designing hydraulic actuators can be stated. Many actuators must be designed for specific operating or spatial conditions. This is particularly true of actuators used in servo systems. However, experience has demonstrated that some approaches to general design problems are superior to others. The following discussion, therefore, falls into the category of "good advice." This good advice should be heeded only so far as it is relevant to the particular problem. It certainly should be modified by the judgment of each designer based on his own experience. In particular, many of the design improvements are costly. They should be used only where the need justifies the expense.

A real attempt should be made to minimize the number of seals in servo-actuators if reliability of operation is to be obtained. Wherever feasible to install, bent or coiled tubing will give more trouble-free service than will flexible hose. For these reasons, the actuator installation of Fig. 9.14 can give more reliable operation than the design of Fig. 9.11.

Seals can be eliminated by drilling passages in the actuator body to carry oil from the valve to the far end. Metal plugs driven into the ends of drilled holes will give better service than will elastomer seals. For the sake of servo stability and the reduction of seals and connectors, the valve should always be mounted directly to the actuating cylinder.

Friction in a servoactuator can cause chattering and stick-slip instability. Seal friction can be minimized by using TFE cap strips over elastomer dynamic seals. Properly installed and seated piston rings minimize friction between piston head and the cylinder barrel. Any leakage across the piston ring will help to stabilize the servo by providing damping. Of course, the leakage must not be so great as to result in an unacceptable power loss.

The best bearings should be used on the actuator body pivot and on the rod end. If space is available, these should be self-aligning roller bearings, at least on one end. Where space is limited, a self-aligning ball bushing may be used. Any sleeve bearings should be amply large. The rod and sleeve should be carefully matched to each other. Control linkage bearings in both the forward and follow-up paths should be pre-loaded ball bearings. Every effort should be made to ensure that no bearing slop occurs anywhere in the control loop.

In the installation of servoactuators, the surrounding structure should be designed so that the load path between the actuator support mount and the load support mount is simple, direct and sturdy. To get satisfactory dynamic response of the servo, the tie between the support mounts must allow for the absolute minimum of deflection. Figure 9.21 shows three typical mounts to minimize deflection.

Structural materials are elastic. A servoactuator will push back on its structural mount exactly as hard as it pushes (or pulls) on its load. The combination of the force on the elastic mount will result in a deflection of the structural support. If the servo command is mechanical, Fig. 9.14, and, if the source of the command is on a place not subject to this deflection, then the structural deflection will be equivalent to an input displacement command. If the porting between the valve and the actuator is known, then the linkage movements which result from the structural deflection can be traced out. If these movements tend to open the valve further, the system will certainly chatter. If the valve has a high flow gain, the system may be unstable. If, however, the deflection tends to close the valve, the system will be stable. Even for the stable condition, the amount of this structural feedback should be minimized through the use of sturdy supporting structure. Structural feedback reduces the effective pressure gain of the valve and reduces the snappiness of the servo action.

One of the most important design goals is to have components, valve, actuator, linkages, etc., which cannot be adjusted from the outside with simple hand tools. The proper performance of a servo system requires that all parts, whether they be hydraulic, electrical or mechanical, work together in proper harmony. This condition is usually obtained only as a result of careful tuning and/or adjusting in the laboratory or in the final assembly process. Once the final adjustments are made they should be sealed or covered so that they cannot easily be changed. An even better solution is to replace the adjustable parts with matching parts which have no adjustments. This caution foils the mechanic or engineer who loves to tinker with equipment.

The servoactuator package itself should be ruggedly designed so that it will not be damaged in handling. Parts and linkages must be mounted on bosses cast or forged into the actuator body. They should never be mounted on light sheet metal clips which can be bent or broken. Electrical feedback devices such as potentiometers or linear variable differential transformers should be buried inside the actuator and all wiring carried through drilled passages or through hydraulic tubing. Never should these be mounted on light clips and operated by a rod mounted parallel to and several inches away from the actuator rod.

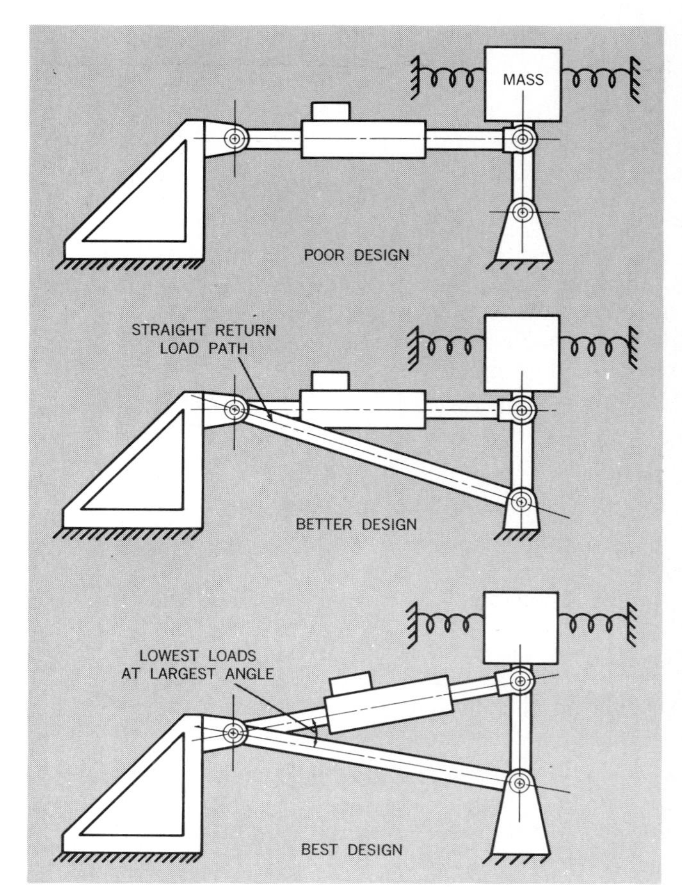

Fig. 9.21. Servoactuator mounting techniques.

9.9. OSCILLATING MOTORS

A vane motor, Fig. 9.22 can be used to provide oscillatory motion of less than one complete rotation. This form of motor is very compact. Because the axial length of the vane is not limited, wide vanes can be used to develop great torques. Because, however, the seal across the vane has theoretically zero width, these actuators are usually used at operating pressures of 1500 psi or less.

Vane motors usually have one vane, Fig. 9.22, or two vanes, Fig. 9.23 but may have three or more vanes. Each additional vane increases the output torque capability but reduces the angle of rotation. Single vane motors can rotate about 280 degrees. Double vane motors are limited to about 100 degrees.

The vane and shaft of oscillating motors are the critically loaded elements. The vane is loaded as a cantilever beam with uniformly distributed load. The shaft must carry the same loads in bending as well as transmitting these loads in torsion. Bearings on the shaft must be large enough to handle these loads without allowing excessive deflection.

The torque developed by oscillating vane motors can be computed from

$$T = \Delta P[(r_v - r_s)\ l]\ \frac{r_v - r_s}{2} \qquad 9.81$$

where: T = torque
r_v = radius of vane
r_s = radius of shaft
l = axial length of vane

$$V = \frac{r_v^2 - r_s^2}{2}\ \theta\ l \qquad 9.82$$

where: V = volume swept
θ = angle swept, radians

It follows that the volume/radian is:

$$V_R = \frac{r_v^2 - r_s^2}{2}\ l \qquad 9.83$$

and

$$T = \Delta P V_R \left[\frac{r_v - r_s}{r_v + r_s}\right] \qquad 9.84$$

9.10. ROTARY MOTORS

Nearly any hydraulic pump can be used as a motor. However, as a practical matter, piston pumps are usually used at high (2000 psi and up) pressures. A

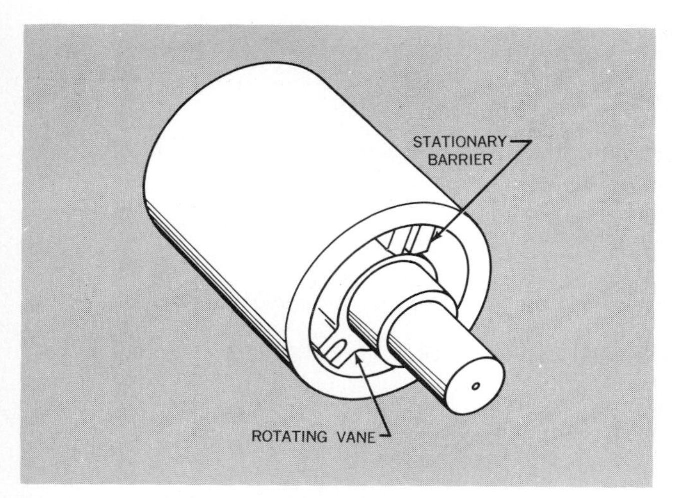

Fig. 9.22. Single vane rotary actuator.

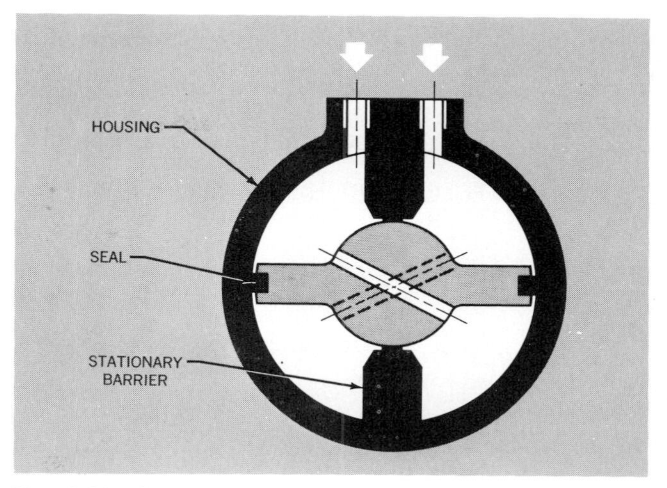

Fig. 9.23. Two vane actuator.

variety of motors is available for lower pressures. A hydraulic motor is specified when the load must be continuously rotated, as in a radar drive, or where stroke length or accuracy requirements preclude the use of linear actuators.

The high rotational speed of a hydraulic motor must be converted to low speed and high torque for most applications. This means that gear speed reducers must be used between the motor and the load. For critical dynamic applications, however, the weight and space loss due to the speed reducer is offset by the great stiffness of the hydraulic motor. In essence, the motor breaks the fluid column into a number of discrete increments. The shorter fluid column provides a stiffer drive. Where precise load positioning or great resistance to impact loads are of primary importance, a motor may be better than a cylindrical actuator.

Hydraulic motors may be used in a number of ways. A fixed displacement motor may be driven from a variable displacement pump. In this case the motor will deliver constant torque but the speed will vary to meet load requirements.

If a variable displacement motor is driven by a fixed delivery pump, the motor output torque can be varied to meet the load requirements while maintaining constant speed. A fixed motor will be slaved to the drive speed of a fixed pump.

A fixed delivery motor can be operated from a constant pressure source using a servovalve to control speed. However, unless changes are made to conventional piston motors, servovalve control will introduce an undesirable stick-slip condition.

When a piston motor is operated directly from a pump or when it is controlled from a shut-off valve, one motor port is held at essentially zero pressure while pressure is raised on the other port to start rotation. A typical motor under these conditions will start to rotate at a differential pressure of about 50 psi. In a 3,000 psi system, this is less than 2% of rated torque.

When the same pump is operated from a servovalve, the port pressures are balanced at 1500 psi for zero rotation. Pressure at one port is lowered and that at the other raised in order to initiate rotation. Under these conditions a differential pressure of about 500 psi will be required before rotation starts. The start will be jerky.

The high differential pressure and the jerky start are caused by high starting friction forces caused by the 1500 psi across the pistons in the null position. The solution lies in eliminating the pressure differential across the pistons by pressurizing the motor housing. The pressure internal to the housing balances the null pressure on the piston and eliminates the heavy friction load on starting.

9.11. LOW SPEED HYDRAULIC MOTORS

In many tasks, the use of a high speed hydraulic motor coupled to the load through a gear speed reducer is unacceptable because of high cost, or bulkiness. Low speed, high torque hydraulic motors have been developed specifically for these tasks. Metzger[3] has given a complete description of available motors.

Low speed motor applications stress the need for high mechanical efficiency. Other basic requirements are (1) smooth, infinitely variable speed changing, (2) full torque output over the whole speed range, (3) minimum torque variation as speed is changed (4) minimum internal leakage, (5) low starting friction, (6) high stall torque and (6) high starting torque.

REFERENCES

1. Baier, W. H. "Calculate Instantaneous Piston Velocity This Easy Way" *Hydraulics and Pneumatics,* June 1963.
2. Gartner, J. R., Harrison, H. L. and Terman, T. V. "Optimum Piston Area for Minimum Time Response" *Hydraulics and Pneumatics,* V20 N6, June 1967.
3. Metzger, J. "Low Speed, High Torque Motors: How They're Used—What's Available" *Hydraulics and Pneumatics,* Sept. 1967.

CHAPTER 9

PROBLEMS

1. What is the lowest spring rate of a balanced actuator with a 3-inch *ID* having a total stroke of 8 inches, when using MIL-H-5606 at 125°F? Rod *OD* is 1 inch.

2. An actuator has an inside diameter of 3 inches. The steel body should not be stressed at more than 30,000 psi at a working pressure of 2500 psi. What should the wall thickness be? How much will the radius increase when the working pressure is applied?

3. A balanced actuator with load has 7000 lbs. weight of moving parts. A_p = 4.0 sq. in., P_s = 3000 psi. Line loss each way is 150 psi at 8 gpm. At that flow maximum hp is delivered to the load. What are the positions and velocities of the actuator 0.01 and 0.05 seconds after valve opening?

4. An unbalanced actuator having a 2:1 area ratio handles an extending load of 3000 lbs. Total valve pressure drop is 600 psi for 20 gpm. Valve supply pressure is 2750 psi. What is the optimum area for traveling 2 inches in minimum time? What is the minimum time?

5. An actuator dashpot is 1-inch in diameter. Radial clearance is 0.003 inches. Piston and load weight = 1000 lbs. Piston diameter is 3.5 inches. Ignore P_1, P_2 and $6W_m$. Initial actuator velocity is 80 ips. If system uses MIL-H-5606 at 80°F, what length of dashpot is required? What is the stopping time?

6. A balanced actuator with a 3-inch *ID* and a 1-inch *OD* has a total weight of 600 lbs for load, rod and piston. The total stroke is 8 inches. If the actuator is at mid-stroke and is traveling at 10 ips when the valve is suddenly closed, what is the maximum transient pressure if the oil is MIL-H-5606 at 150°F? Assume zero damping.

CHAPTER TEN

Seals and Sealing Devices

10.1. INTRODUCTION

Eliminating leakage is probably the most difficult of all hydraulic design tasks. In some military weapons systems, for instance, nearly 40% of all in-service complaints relate to hydraulic system leaks. In industry, leaks are messy and can create hazards to workers. In some process industries, leakage is completely unacceptable because of contamination to the end product.

Dynamic seals, on actuator rods and pump shafts, accomodate relative motion and retain pressurized oil inside the system. Where fixed parts must be removed for maintenance and repair, static seals are needed. Static seals are also used to seal holes that were made for the fabrication or assembly of component parts.

A wide number of materials have been used for both static and dynamic seals. Most modern hydraulic seals are made of elastomers, virgin or loaded PTFE (polytetrafluoroethylene), metal shapes or combinations of these. Seals for rotating parts commonly use carbon as a facing material.

10.2. STATIC SEALS

To contain pressure, the seal and its mating parts must be in contact at a pressure level significantly higher than the pressure being sealed. This high sealing pressure can be developed by the installation method, Fig. 10.1.

Static seals can also be installed with only enough initial sealing pressure to contain low pressure oil. High oil pressure completes the seal. The sealing device must magnify the oil pressure to give a leak tight joint, Fig. 10.2.

O-rings used for static seals are usually compounded to a Shore A durometer hardness of 70. This is harder than customary for dynamic seals. The force required to squeeze the O-ring is given as a function of durometer hardness, Fig. 10.3. There must be space for the elastomer to flow at right angles to the squeeze or a situation akin to hydraulic lock will occur. Because elastomers shrink at low temperatures, the squeeze on static O-ring seals should be great enough to compensate for this shrinkage.

10.3. DYNAMIC SEALS

In addition to containing pressure, dynamic seals must withstand relative motion without dragging, wearing, galling or welding. Some dynamic seals, such as piston rings, even sacrifice sealing efficiency for low friction and good wear characteristics.

For hydraulic use compression rod seal packings can be made of asbestos, asbestos reinforced with metal wire, wool or leather impregnated with an elastomer, PTFE fibers, graphite yarns, lead or copper foils. The packing may be rings or coils having a square or rectangular cross-section. These and other shapes are shown in Fig. 10.4. A screw-type compression member forces the packing into position.

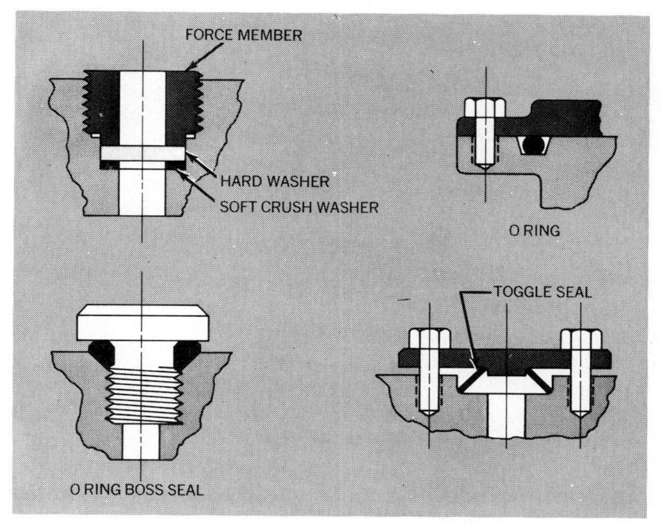

Fig. 10.1. Installation actuated static seals.

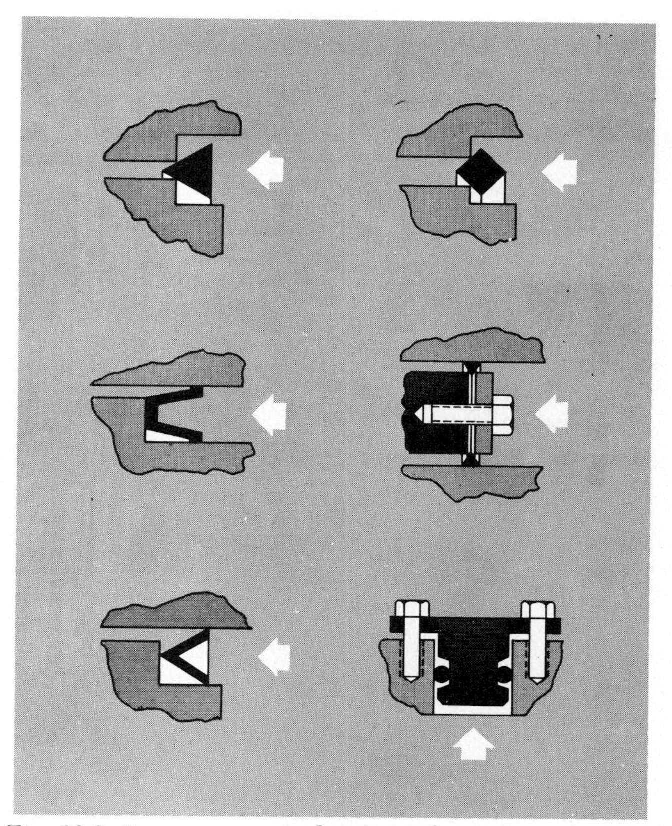

Fig. 10.2. Pressure energized static seals.

Packing can also be molded to shape. Molded packings require adapters or special compression members to allow them to expand properly, Fig. 10.5.

Compression packings, with the possible exception of graphite yarns, need lubrication. In normal installation, the packing is compressed enough to minimize leakage but not enough to eliminate it. The slight leakage lubricates the packing and keeps it

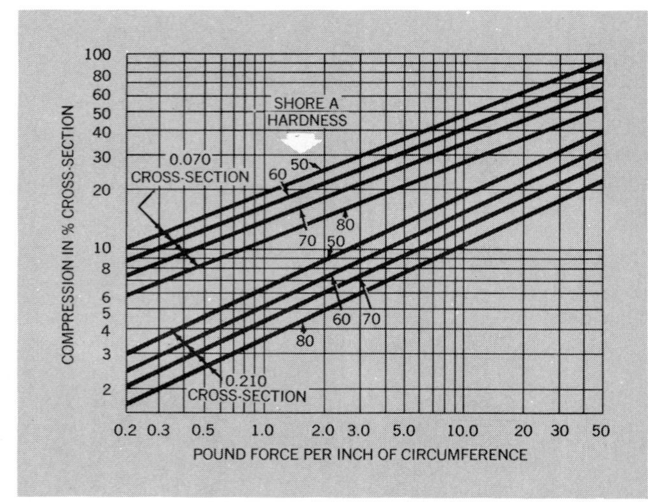

Courtesy Parker Seal Co.

Fig. 10.3. Force required to squeeze an O-ring of known durometer hardness.

cool. Where, because of high operating temperatures or pressures, leakage lubrication is not satisfactory, external lubrication for the packing must be supplied. This lubrication can be supplied from a grease cup or forced oil pressure.

10.4. DYNAMIC O-RING SEALS

Nearly all rod seals, and many piston seals in aircraft use an O-ring. This elastomeric toroid has wide acceptance in industrial hydraulics because of its low cost, small size and ease of handling. Elastomers act as both solid materials and as very high viscosity fluids. Under pressure they flow and deform until the internal stresses equal the external. This flowing action is similar to the flow of highly viscous fluids. However, no true liquid can maintain internal shearing forces under static conditions so that a loaded elastomer also acts as a solid.

The O-ring is installed to give some initial compression when the mating parts are assembled. The initial compression is necessary for sealing under low pressure differential. It also compensates for shrinking of the ring at low temperature. As the pressure difference across the seal is increased, the O-ring will distort, Fig. 10.6. Unless the clearance between the mating parts is made very small, the seal may extrude. This extrusion might cause no problem in a static seal. Under dynamic use, there will be nibbling and erosion. For practical clearances, this extrusion becomes unacceptable at about 1500 psi across the seal.

At pressures above 1500 psi, backup rings, Fig. 10.7, are used to prevent extrusion. These backup rings reduce the clearance gap. At present, most backup rings are made of PTFE plastic. PTFE containing some filler which prevents feathering flow of the plastic is also frequently used.

Friction forces are higher and O-ring wear is greater when the actuator is operating over small amplitudes.

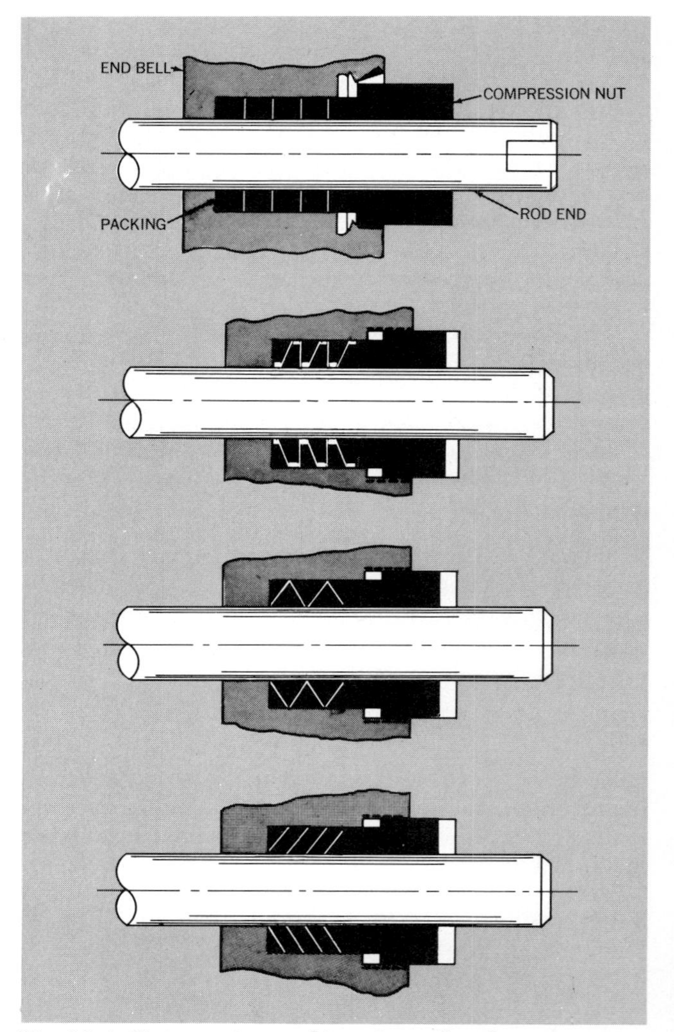

Fig. 10.4. Compression packings for rod seal.

The higher friction forces and more rapid wear can be explained in terms of sliding velocity of the seal. At very small motion amplitudes, the seal probably does not slide at all. The motion is accommodated by flowing of the elastomer. Wear at these very small strokes probably occurs as a snagging effect of the asperities of the metal surfaces on the elastomer.

For slightly greater strokes the O-ring probably slides. However, even at fairly high cycling frequencies, the sliding velocity never becomes great. The pressure between the O-ring and its contacting metal surfaces squeezes oil from between the sealing members. Consequently, the O-ring slides on a dry metal surface. Such sliding creates high friction forces and contributes to rapid wear of the O-ring.

With longer strokes, the sliding velocity increases until a film of oil is dragged between the seal and the metal surface. Now, the friction forces and wear rate drop sharply.

To minimize friction and wear in control system servoactuators, many designers use cap strips of PTFE plastic between the O-ring and the metal surface.

In these designs, Fig. 10.8., the compressed elastomer provides the sealing force but the actual sealing occurs between the cap strip and the metal.

O-ring seals in actuators which only operate over long strokes do not have the short stroke wear and friction problems. However, these cylinders are usually fairly large in diameter. If the surfaces over which the seal passes are not uniformly lubricated with oil, the friction forces can vary considerably around the circumference of the seal. Under such conditions, part of the O-ring may be sliding while another part, less well lubricated, may try to roll. The O-ring can then suffer a shear failure between the section which is sliding and the section which is rolling. This is called a spiral failure.

Because a well lubricated O-ring may never roll, some designers have questioned the desirability of using the toroidal shape for elastomeric seals. Square rings, lobed rings, tee rings, Fig. 10.9, and other configurations have been successfully used. Where friction must be reduced, a square ring may utilize a PTFE insert.

Great care must be taken to avoid the pinching off of parts of the seal. Tapered guides, Fig. 10.10, allow for insertion of elastomers. Avoid designs which cause moving seals to encounter cross ports. If avoidance is impossible, provide clearance, Fig. 10.11, to prevent cutting the O-ring at a sharp intersection edge.

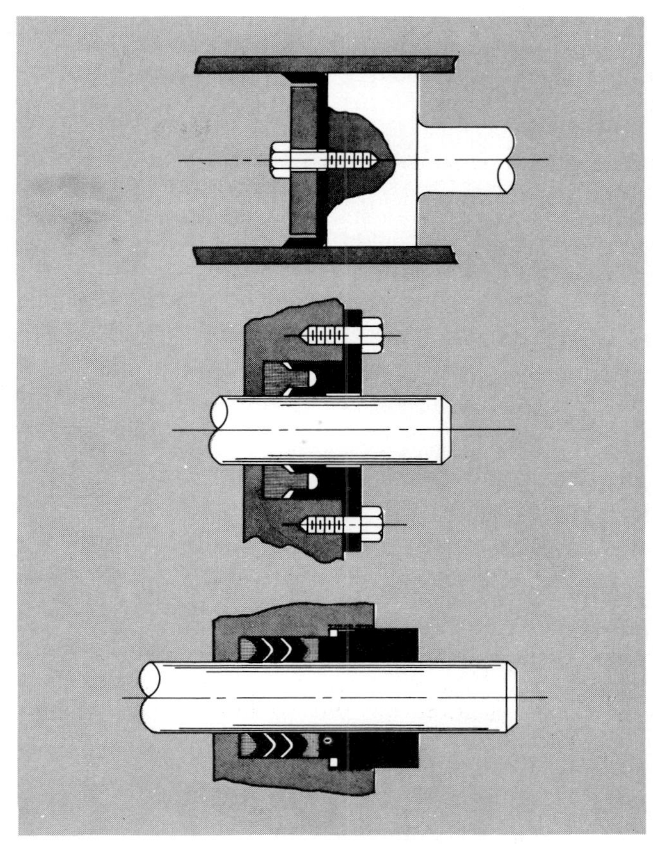

Fig. 10.5. Molded compression packings.

10.5. PISTON RINGS

Rings of PTFE or metal are frequently used on actuator piston heads. In this use, the leakage inherent with non-compressible ring seals is occasionally an advantage. Less frequently, ring seals are used as the high pressure side of a double seal system on piston rods.

Rings of PTFE provide low friction levels. Because PTFE possesses a small amount of elasticity, solid rings can sometimes be expanded or compressed enough to be slipped into retaining grooves. Grooves formed by assembly will also allow the use of solid rings. These rings are usually made with a square cross-section. They tend to have a relatively high ratio of diameter to radial thickness. Occasionally a solid ring is used free-floating in its groove. More typically, an elastomeric ring or metal marcel spring is used to keep the piston ring pressed against its mating surface. Solid rings are usually installed with an interference fit of about 0.2% for small diameters to 0.1% for large diameters. The actual diametrical interference runs from 0.002 inches to 0.008 inches. Rarely is more than one ring used per groove. With multiple rings, one piston ring is used for each 1000 psi pressure differential.

To simplify assembly, rings can be split. PTFE rings can be scarf cut, butt cut or step cut, Fig. 10.12. A flat spring or marcel spring is nearly always used to expand the piston ring. Typically, more

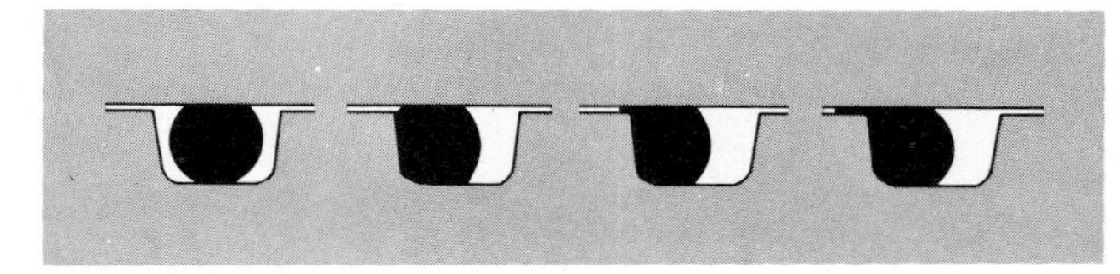

Fig. 10.6. O-ring under increasing pressure.

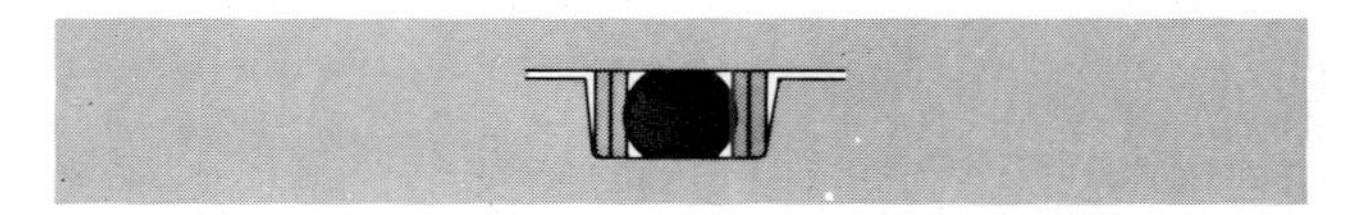

Fig. 10.7. Backup rings prevent extrusion of O-ring.

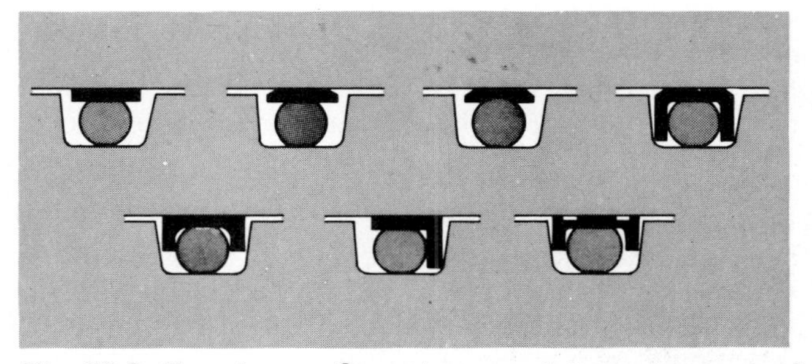

Fig. 10.8. Cap ring configurations to minimize friction and wear.

than one ring is used per groove and the splits are at some sizable angle apart.

PTFE is a very stable plastic. But it does get mechanically weak at elevated temperatures. Filling the plastic with metal particles or with glass or ceramic fibers helps to stabilize dimensions. As pressure goes up permissible operating temperature is reduced.

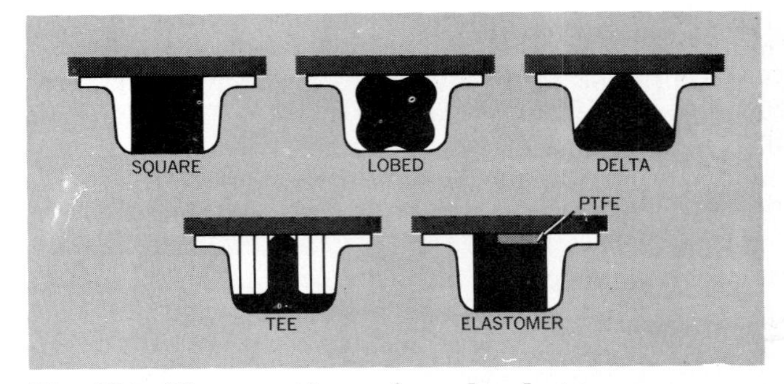

Fig. 10.9. Elastomer shapes for rod and piston seats.

10.6. METAL PISTON RINGS

Metal piston rings are nearly always split. Rings may be butt cut, step cut at an angle or step cut axially, Fig. 10.13. Leakage past a butt cut ring is primarily through the gap. This amount of leakage is usually too great for critical applications so that butt cut rings are only used for lower grade actuators. They are rarely used as rod seals. In order to make the sealing force uniform when the ring is compressed into a cylindrical barrel, the ring is carefully made non-cylindrical. The ring, in the free shape, is defined by the following relationship,

$$R = \frac{D}{2} + \frac{\Delta G}{3\pi}\left[1 + \frac{\theta}{2} \sin \theta\right] \qquad 10.1$$

where R = free ring radial dimension
D = cylinder ID or constraining ring OD
ΔG = change in gap
θ = angle from the origin, radians

The origin of the shape is at a point 180° from the gap. The ring is symmetrical about an axis between the origin and the gap. Equation 10.1 can be written

$$R = \frac{D}{2} + \Delta G x \qquad 10.2$$

$$\text{where } x = \frac{1 + \theta/2 \sin \theta}{3\pi}$$

Fig. 10.14 shows how x varies with θ.

An angle step cut is one means for minimizing leakage. The step section has an overlap from each end of the ring. While the angle of cut with respect to the axis of the ring is the same for each end, the radius used in cutting the inside part of the lap is much smaller than the radius used to cut the outside part. When the parts are brought together, a theoretical line contact occurs. This results in high sealing pressure between the ends. This form of ring is intended to seal in one direction only, Fig. 10.13. When sealing in both directions is needed, two rings, oppositely aligned, each in its own groove, are used.

An axially step cut ring is usually used in combination with an inner ring, Fig. 10.13. The mating between the inner and outer rings must be quite exact. The overlap section at one end of the ring is made flat. The other end is convex in the axial direction. Again a theoretical line contact is achieved. When the sides of these rings are made perfectly flat, sealing in both directions is possible.

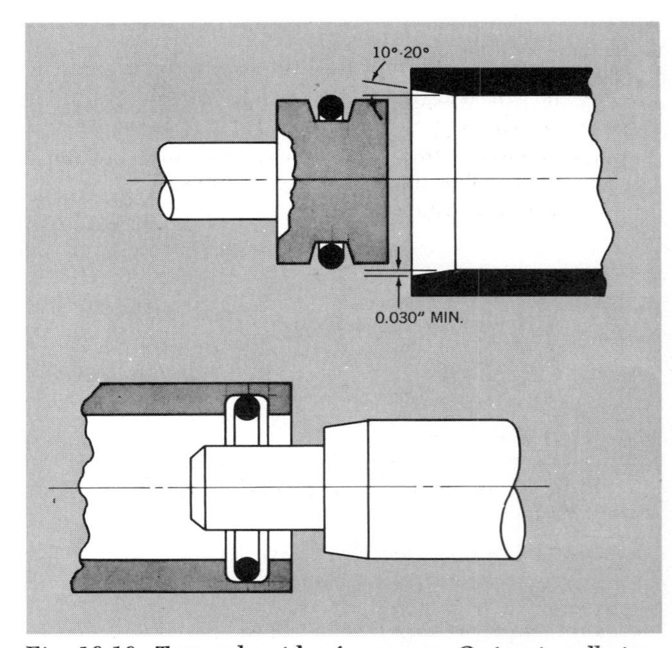

Fig. 10.10. Tapered guides for proper O-ring installation.

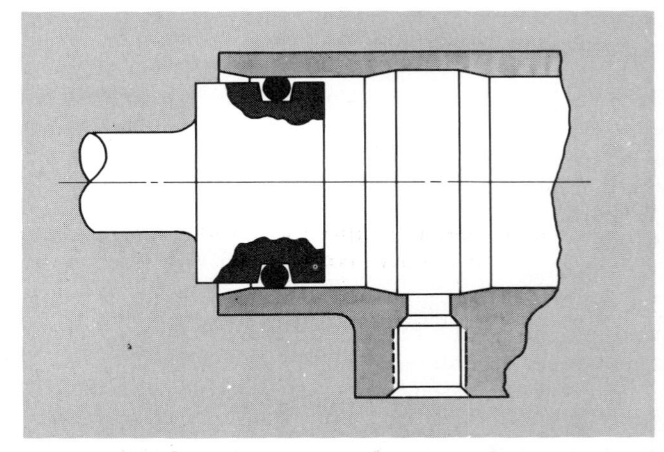

Fig. 10.11. Clearance at port eliminates sharp edge and possible O-ring failure.

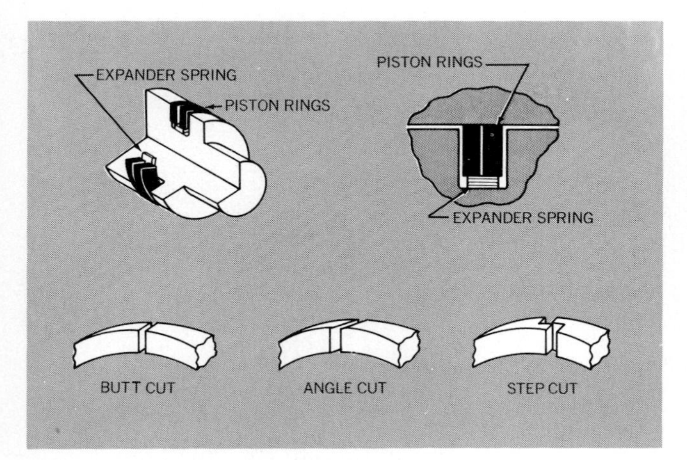

Fig. 10.12. PTFE piston rings.

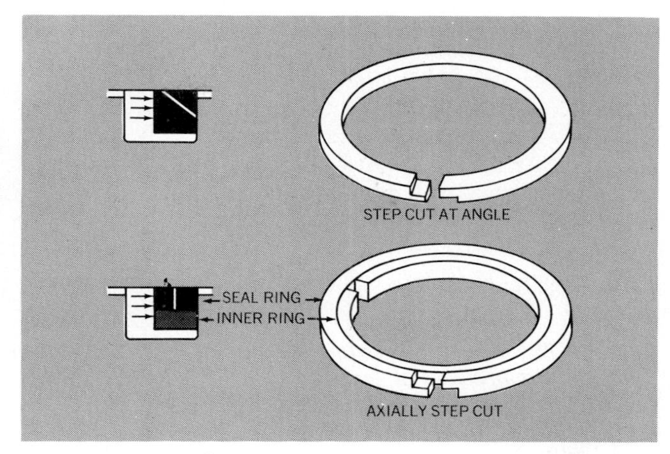

Fig. 10.13. Metal piston rings.

Leakage past piston rings is axially between the outside of the ring and the inside of the cylinder (obviously, the opposite is true for a rod seal) and radially between the side of the seal and the side of the groove. The axial leakage follows the relationship for annular orifices.

$$Q = \frac{\pi \; D \; b^3 \; \Delta P}{12 \; \mu \; l}\left[1 + \frac{3\epsilon^2}{2}\right] \qquad 10.3$$

where ϵ = relative eccentricity
b = radial clearance
μ = absolute viscosity
l = length of leakage path
ΔP = differential pressure

The radial leakage follows a version of the Hagan-Poiseuille law

$$Q = \frac{w \; b^3 \; \Delta P}{12 \; \mu \; l} \qquad 10.4$$

where w = circumference of sealing face

Because a piston ring is a leaking seal, the pressure differential is reduced along the sealing length. This results in the pressure force unbalance, Fig. 10.15. The unbalanced forces are, of course, reacted by the groove side and the sealing cylinder ID. In a dynamic seal the amount of axial force unbalance is reduced by the friction forces (both coulumb and viscous) opposing the motion. Theoretically the radial unbalance is also opposed by friction between the sealing rings and the groove side. Note that the *axial* force is opposed by the coulomb friction effects of the *radial* forces and vice versa.

When used under conditions of force reversal, the piston rings must move axially from one side of the groove to the other in order to effect proper sealing. The forces acting on the ring are

$$F_A - f \; F_R = Ma + Bv \qquad 10.5$$

The speed of response of the ring is directly related to the magnitude of the acceleration. Assuming F_A and F_R to be fixed for any set of design configu-

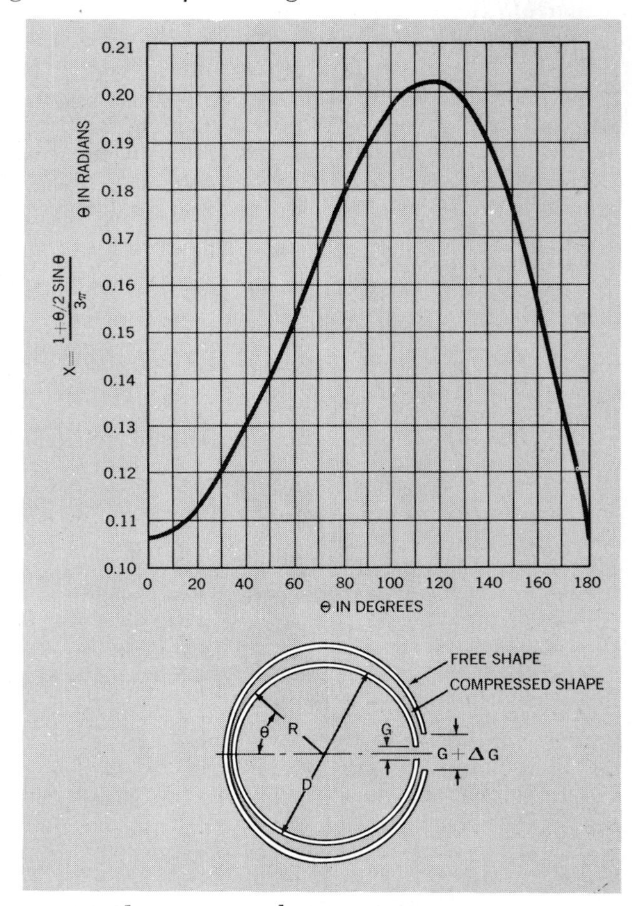

Fig. 10.14. Shaping a metal piston ring.

rations, accelerating can be maximized by (1) choosing materials so that the coefficient of coulomb friction f is as small as possible, (2) having the mass M of the ring as small as possible and (3) using an oil of low vicosity so that the coefficient of viscous friction B is small. Unfortunately, an oil of low viscosity will also leak more freely.

If the radial force unbalance is too great, heavy wear may occur between the sealing faces. This

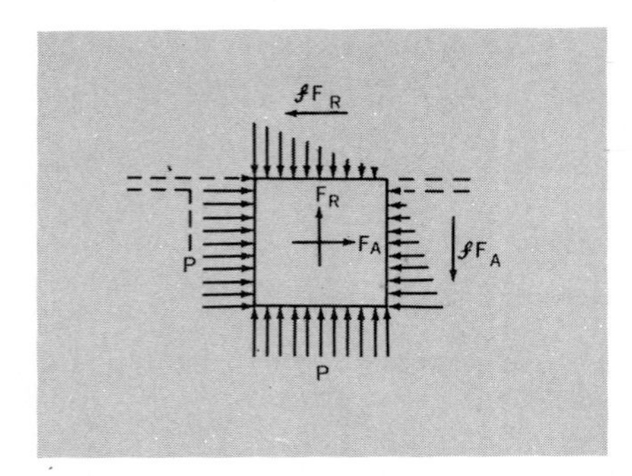

Fig. 10.15. Forces on piston ring are unbalanced because of leakage.

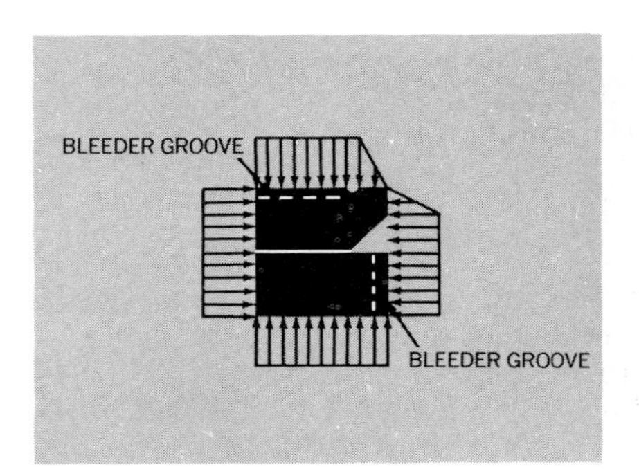

Fig. 10.16. Two piece seal with pressure balancing.

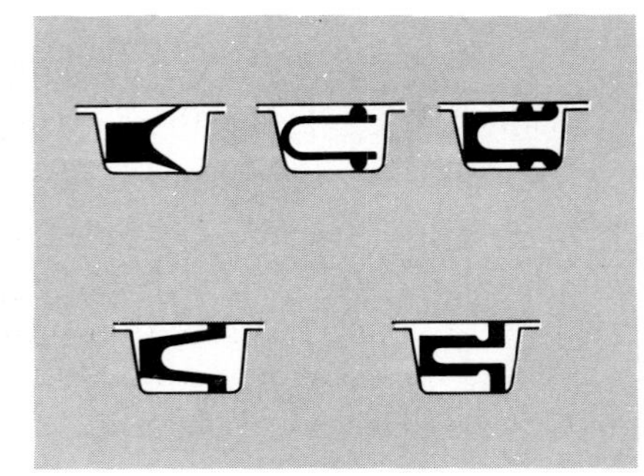

Fig. 10.17. Metal lip seals

wear can be minimized by reducing the force unbalance. If a circumferential groove machined in the ring is connected to high pressure by suitable axial grooves, no net pressure differential can exist over the ring area between the groove and the pressure side, Fig. 10.16. All of the pressure drop occurs over the remaining small area of the sealing surface. The average of this pressure drop times the small area yields a much smaller radial force unbalance. Occasionally side grooving will also be required to avoid an unbalanced force condition.

The important points to remember when designing, installing or analyzing metallic piston rings:

1. The split piston ring must be installed with freedom to move both axially and radially. This is necessary to assure proper seating.
2. Piston rings are pressure actuated. The higher the pressure differential, the better the seal.
3. The split in the ring and the distortion in the ring when installed provide the necessary circumferential conformability.
4. Proper free-state shape is necessary if the installed radial forces are to be uniform around the circumference.
5. The distortion necessary to proper installation must relate to the ring dimensions if the ring is not to be over stressed.

$$S = \frac{0.4815\ E \left(\frac{\Delta G}{d}\right)}{\left(\frac{D}{d} - 1\right)^2} \qquad 10.6$$

where ΔG = Change in gap
D = Ring O.D.
d = Ring radial width
E = Young's modulus
S = Stress

6. Sealing surfaces of the split rings and of the grooves must be flat, smooth and parallel.

10.7. OTHER METAL DYNAMIC SEALS

A number of metal lip seals have been used for both piston and rod seals, Fig. 10.17. In general, these seals are each pressure energized. Wear and sealing effectiveness are closely related to fluid characteristics and metal-on-metal combinations.

Occasionally a labyrinth seal, Fig. 10.17, is used. With non-compressible fluids a labyrinth seal is a series of annular orifices. The pressure drop or leakage

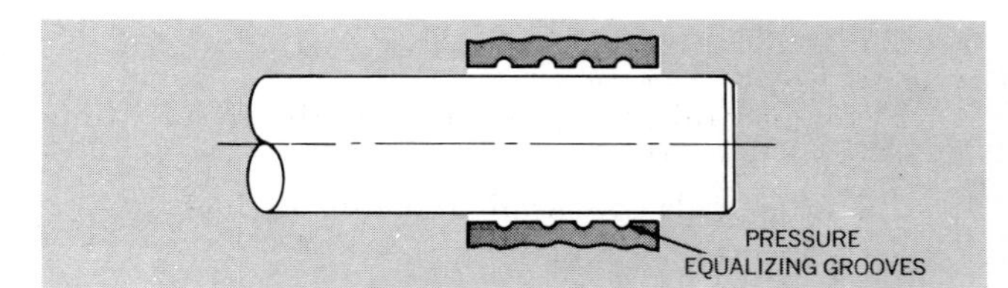

Fig. 10.18 Labyrinth seal is a series of annular orifices.

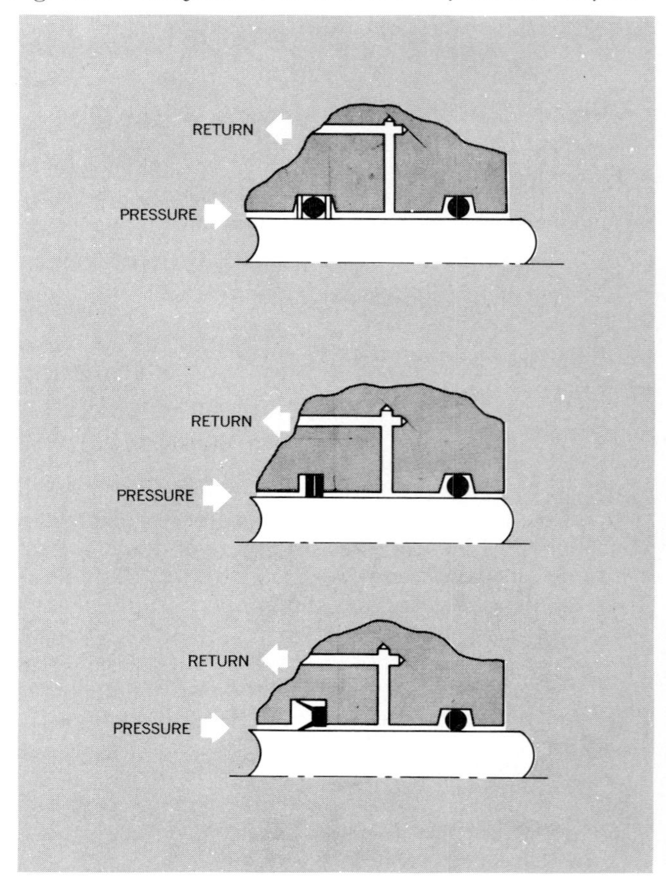

Fig. 10.19. Two stage seals minimize external leakage.

flow is computable by equation 10.1. The circumferential grooves prevent pressure differentials around the seal. Therefore the labyrinth design probably provides a better seal than a single annular orifice of the same length.

10.8. DOUBLE DYNAMIC SEALS

For many design conditions the use of double seals is desirable. When using an angle cut piston ring or a metal lip seal ring with pressure contained in two directions, one seal must be used for each direction.

Usually, leakage out of a rod seal cannot be tolerated. If a piston ring or other leaky seal is desired, it can be backed up with an elastomeric seal, Fig. 10.18. A drain between the seals connects to the system return line. Leakage past the piston ring then returns to the tank or reservoir. The elastomeric seal only contains oil at a pressure slightly above atmospheric or above the reservoir pressurization level.

If double seals are used, the rate of leakage may be of concern. Leakage rates which are too high might drain most of the flow capacity of the pump. One form of magnetically operated tell-tale device, Fig. 10.19, shows when leakage exceeds a set value. When the flow gets too great, the pressure difference across an orifice in the return line can be used to move a piston against a spring. The piston contains a permanent magnet which holds an indicator button in against another spring. When the magnet is moved away from the button, the magnetic flux can no longer overcome the spring force and the button is forced out into view.

10.9. ROTATING SEALS

Where a rotating shaft must pass into or out of an area of pressurized oil, a different form of seal must be used. O-rings quickly break down when used on rotary service and labyrinth or piston seals leak too much. One of the most useful seals for this service is the axial mechanical seal, Fig. 10.20. A combination of pressure and spring force causes two carefully mated parts to bear on each other with relative rotary motion. The O-rings shown are static seals. Effectiveness of the axial mechanical seal depends upon the degree of finish and parallelness of the sealing surfaces and the compatibility of the mating materials. In hydraulic service the sealing elements are usually hard carbon bearing against hard steel or cast iron. In use, these materials lap each other into a very fine fit. This provides a very small axial laminar leakage path. The quality of this fit often reduces the flow to where the oil surface tension can complete the seal and there is no leakage. When needed to reduce wear, the pressure-induced sealing forces can be reduced by balancing.

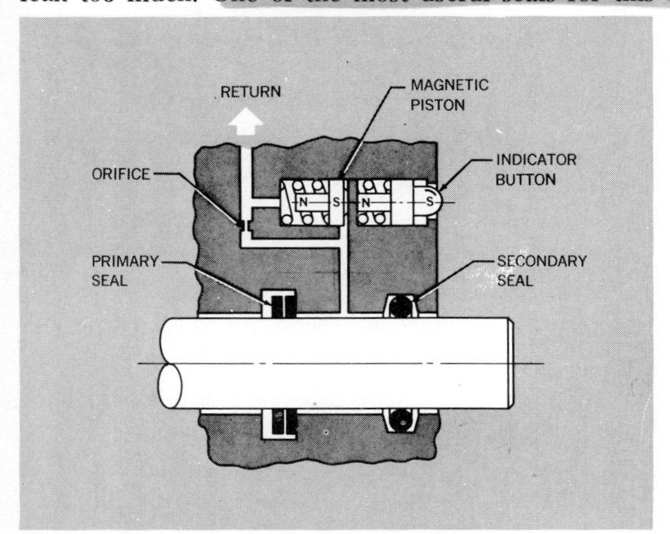

Fig. 10.20. Two stage seal with a leakage indicator.

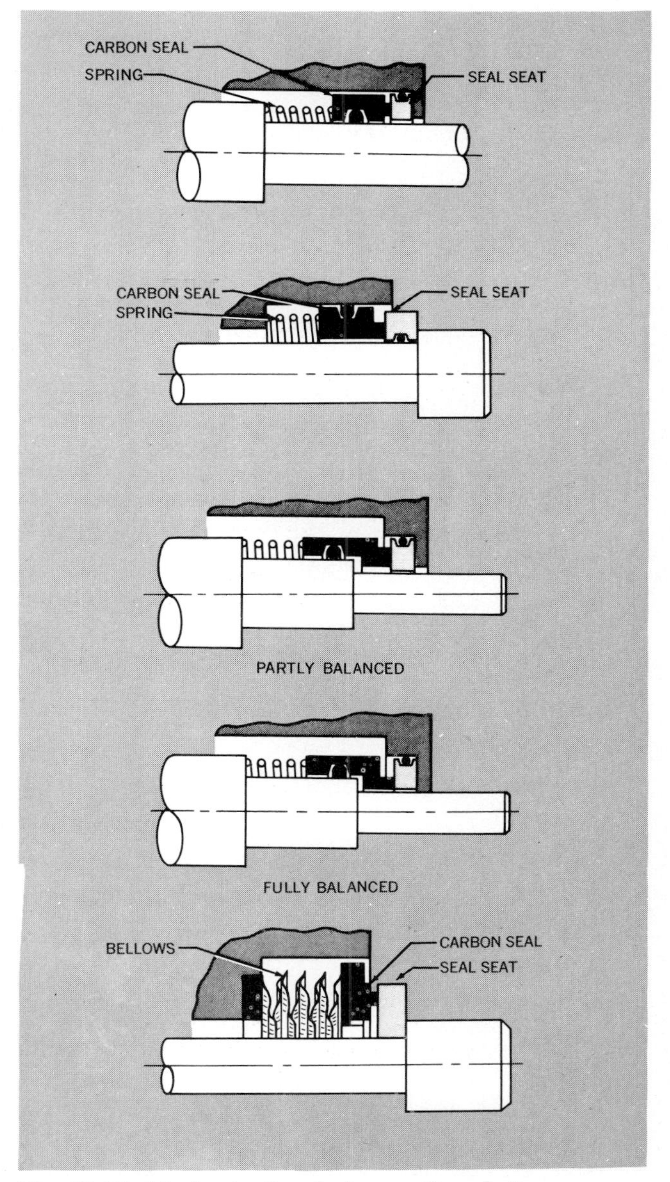

Fig. 10.21. Mechanized seals for rotating shafts.

CHAPTER ELEVEN

Hydraulic Servomechanisms

11.1 Definition

11.2 Other Regulating Systems

11.3 Bang-Bang Servos

11.4 Proportional Servos

11.5 Mechanical Input Servo

11.6 Characteristics of Dynamic Systems

11.7 Spring-Mass System

11.8 Electrical Input Servos

11.9 Frequency Response Techniques

11.1. DEFINITION

Tremendous power amplification in hydraulic systems has led to the development of control systems to apply this power accurately. Most of these systems fall within the general description of servo-mechanisms. These servos are used to control position, velocity, or force. They are usually capable of responding to rapidly changing input conditions.

A servomechanism, as used in these discussions, is a control system in which a controlling input is modified and/or amplified and applied to a load. A measure of the output is fed back and compared to the input signal. The difference between the input and the measure of the output is used as an input to drive the system to reduce this difference to zero. The difference between the input and the measure of the output is called the error signal.

Fig. 11.1 diagrams a generalized hydraulic servo as it might be used to control an inertial load. The output of the servo is the position of the actuator or motor. This position is measured and compared with the input signal to provide an error signal to the servo amplifier. The minus sign at the summing point indicates that the measure of the output is algebraically subtracted from the input. The summing sign indicates the algebraic summing of control system signals, so the use of the minus sign to remind that this is a subtraction point is good practice.

When a system, Fig. 11.1., is held at a fixed input level, the system then acts as a regulator. In this mode, the system acts to minimize the effect of any external disturbance applied to the output. As an example, consider a sudden load change on the actuator or motor. The force developed by the load change will compress the oil in the actuator. This compression will allow the actuator output rod to move. The feedback device will report this displacement to the summing point and an error signal will be generated. The error signal, when amplified, will move the valve in a way which will command the actuator to return to its original position. When this is done the system will again be at rest.

11.2. OTHER REGULATING SYSTEMS

Feedback regulating systems are common in power hydraulic systems. When coupled with hydraulic circuitry, variable delivery pumps, relief valves, and pressure reducing valves each contribute all of the same elements which are found in a servomechanism. A pressure compensated variable delivery pump, Fig. 11.2, provides flow as a function of the difference (error signal) between the system pressure and a nominal pressure setting. The pump flow, acting upon the system dynamics, creates system pressure. Therefore, control of the flow is control of system pressure.

11.3. BANG-BANG SERVOS

The simplest servomechanism in hydraulic systems uses 3-way or 4-way on-off valves and simple switching circuitry, Fig. 11.3. The motor is either moving at full velocity or is stopped. The direction of motion tends to reduce the error signal to zero but the rate of motion bears no relationship to the magnitude of the error signal.

Input command is a displacement of the hand lever. If the actuator rod is initially considered fixed, then the input command, operating through the linkage, will close one or the other of two normally open switches. The switches are mounted on levers which can only swing outward so that only one switch can be activated at a time. The switch which is closed allows current to flow to a solenoid. The tractive solenoid opens a 4-way valve and causes the actuator to move. As the actuator moves it will tend to return the switch and its lever to the neutral position. When the actuator reaches a position in accordance with the amount of input handle displacement, the switch will open and the system will come to rest.

Because a bang-bang servo moves with full velocity right up to the switching point, the momentum of the motor and load will sometimes carry the system to the position where full velocity in the other direction is commanded. In order to avoid this limit cycling condition, the actuator velocity is kept low or a "dead-band" is introduced at the switching null. This dead band is made wide enough so that the system may come to rest under the zero error signal condition.

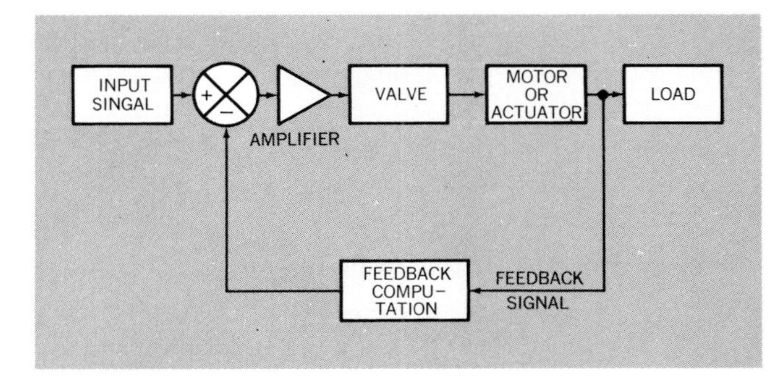

Fig. 11.1 Hydraulic servo block diagram.

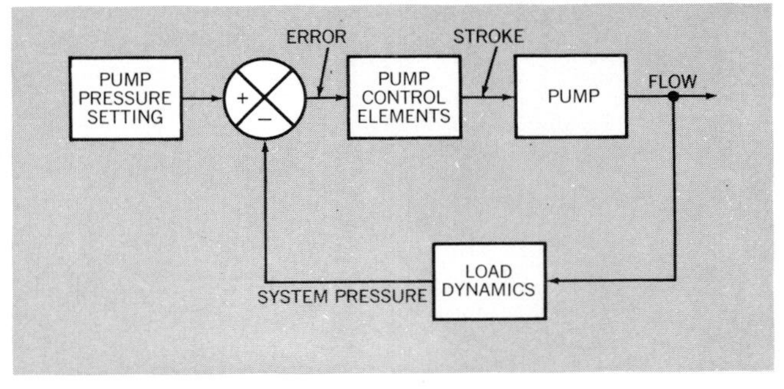

Fig. 11.2. Pressure compensated pump regulator system.

11.4. PROPORTIONAL SERVOS

If the servosystem is designed so that the direction of motion is a function of the sign of the error signal and the motor or actuator output is proportional to the magnitude of the error signal, then the system is a proportional servo.

$$\theta_o = K\,\theta_e \qquad 11.1$$

where: θ_o = output
θ_e = error signal

The proportional servo is different from the bang-bang servo in that the speed of the motor (for a positional servo) is proportional to the distance (error) from the desired position.

A variation of the proportional system is proportional-plus-derivative control.

$$\theta_o = K\left[\theta_e + \frac{d\theta_e}{dt}\right] \qquad 11.2$$

Another form of servo is the proportional-plus-integral control.

$$\theta_o = K(\theta_e + \int_t^o \theta_e\,dt) \qquad 11.3$$

What do these relationships mean? In 11.1 the output is concerned only with the present situation in the servosystem. With relationship 11.2, the output is involved not only with the present, but because the rate with which the present is varying with time is included, with a cautious peek into the future. With 11.3, the output has a chance to dwell on the accumulated occurrences of the past as well as considering the actions of the present.

11.5. MECHANICAL INPUT SERVO

All of the elements of the block diagram of Fig. 11.1 are contained in the power servo shown in Fig. 11.4. There are also a few other elements in this configuration. The input link is the input signal generator. The servovalve doubles as the amplifier. The actuator is the motor. The follow-up linkage is

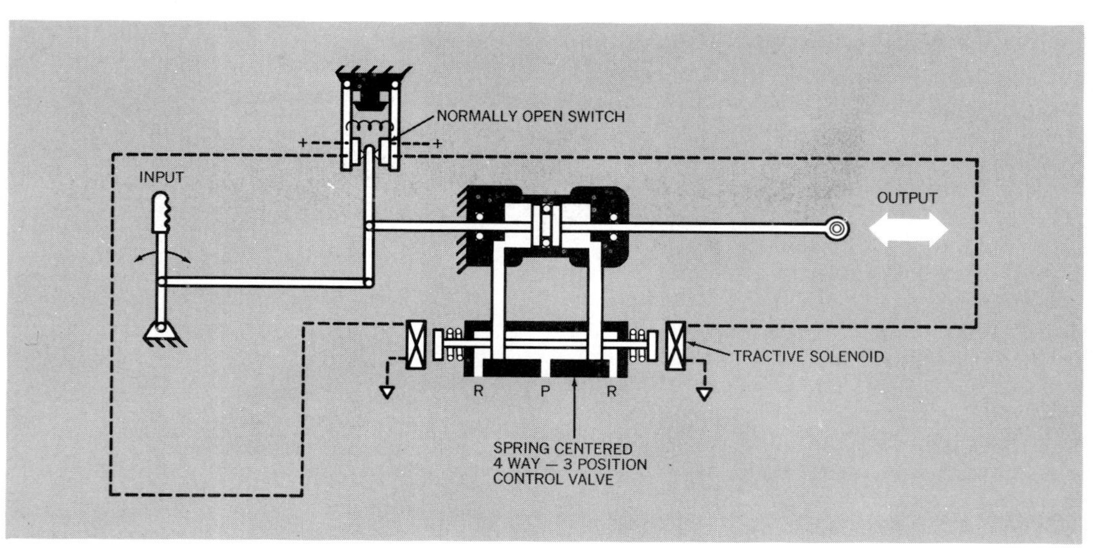

Fig. 11.3. Bang-bang servo.

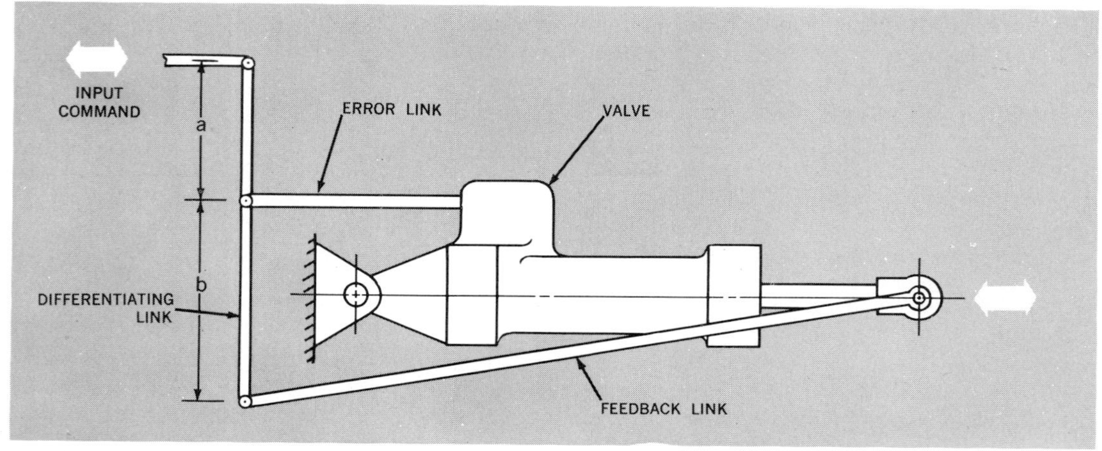

Fig 11.4. Mechanical input servoactuator.

the feedback path and the differentiating link is the summing point.

In operation, the differentiating link is initially considered to be fixed about its lower pivot point by the feedback linkage. For small input displacements, the valve input (or error signal) displacement is $b/(a+b)$ times as large as the input displacement command. The error signal causes a velocity in the actuator. As the actuator displacement increases with time, the feedback motion is transmitted to the differentiating link. This link is now considered to rotate about its connection with the input link. Therefore, actuator output motion causes a displacement of the valve which tends to close the valve. The movement of the valve as a result of actuator displacement is $a/(a+b)$ times the actual output displacement.

Fig. 11.5 is the block diagram for this servo. Before entering the servoloop, the input displacement is amplified by the ratio $b/a+b)$. This amplified displacement is the input signal θ_i. It is mechanically compared to the feedback by use of the differentiating link. The difference between the input and the measure of the output which is supplied by the feedback link is $\theta_i - \theta_o = \theta_e$, a displacement of magnitude X_v, which moves the valve.

As is known, a change in the valve conductance (an opening or closing of the valve) will result in a change in flow rate. Assuming the system initially at rest, the error signal θ_e, which is a displacement X_v, will open the modulating valve and allow flow to go to the actuator. Because $Q = Av$, the flow will result in a velocity of the actuator. The feedback link, however, does not measure the actuator velocity. It only reports on the output displacement. Therefore, the feedback link measures the time integral of the actuator response to the valve command. As mentioned before, the feedback system has an amplification of $a/(a+b)$.

11.6. CHARACTERISTICS OF DYNAMIC SYSTEMS

Some insight concerning the operation of dynamic systems is gained by studying simple arrangements of springs, masses, and dampers. Since both electrical and hydraulic systems are analogous to mechanical systems, any concepts developed on simple mechanical systems are equally applicable to hydraulic or electrohydraulic systems. To avoid any complications due to the effects of gravity, a rotating mechanical system will be used. The three elements considered are J, the polar inertia of a rotating mass; G, the rate of a torsion spring; and B, the viscous damping coefficient. Coulomb, or dry, friction will not be introduced because it has no direct analog in hydraulic or electrical systems.

Consider a simple damped spring, Fig. 11.6, where the spring has just been released from an initial torque. The spring has a rate G and an initial twist of θ_o radians. The damper generates a resisting force B which is proportional to the velocity $\frac{d\theta}{dt}$. Under these conditions, the velocity toward the null position is assumed to be proportional, because of the damping, to the angle of twist remaining. If this relationship is expressed in LaPlace notation where $s = d/dt$, then

$$\theta(s) = \frac{\theta_o}{\frac{G}{B} + s} \qquad 11.4$$

When transformed to the time domain, this becomes

$$\theta(t) = \theta_o e^{-\frac{G}{B}t} \qquad 11.5$$

If $\theta_{(t)}$ is plotted against time, the curve is an exponential having an original amplitude θ_o. With time, $\theta_{(t)}$ goes to zero. A tangent to the curve at any point will intercept the horizontal (time) axis at a time B/G further out. This value B/G is called the "Time Constant." It is defined as the time in which the angle would reach zero if its velocity were held constant. Without going into the proof, the assumption is made that the transient response of any first degree system can be expressed mathematically by an exponential function of time. The change in the variable, here the angle of twist, during any time constant period, is 0.37. By corollary, the value of the variable at the end of any time constant is 0.63 of its value at the beginning of the time constant.

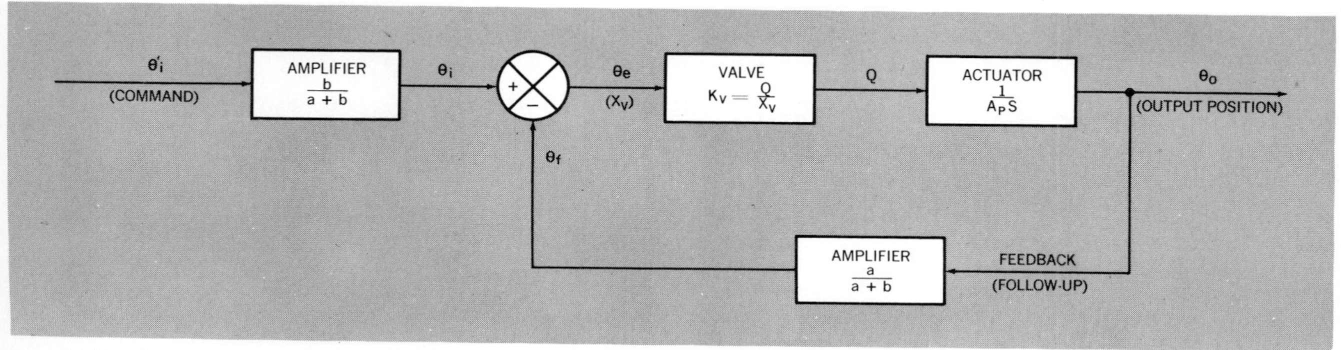

Fig. 11.5. Block diagram of mechanical input servoactuator.

11.7. SPRING-MASS SYSTEM

If a mass is substituted for the damper of Fig. 11.6 the system becomes Fig. 11.7. This system has only energy storages, a mass and a spring, and no energy dissipater. Therefore, any disturbance to the system will be perpetuated indefinitely. The system is an undamped oscillator. The equation is

$$\theta(s) = \theta_o \frac{s}{s^2 + \frac{G}{J}} \qquad 11.6$$

and the response to an initial twist θ_o is a sine wave having an amplitude of $\pm\theta_o$ and a frequency of $\sqrt{\frac{G}{J}}$ radians per second.

The damped spring-mass system, Fig. 11.8, approaches real systems more closely. Here the equation of motion is

$$0 = G\theta + B\frac{d\theta}{dt} + J\frac{d^2\theta}{dt^2} \qquad 11.7$$

If a solution of the form Ae^{st} is assumed, then

$$GAe^{st} + BsAe^{st} + Js^2Ae^{st} = 0 \qquad 11.8$$

Cancelling the common term Ae^{st} leaves

$$s^2 + \frac{B}{J}s + \frac{G}{J} = 0 \qquad 11.9$$

By completing the square and solving for S

$$s = -\frac{B}{2J} \pm j\sqrt{\frac{G}{J} - \frac{B}{2J}^2} \qquad 11.10$$

This is of the form $s = \sigma + j\omega$ where $j = \sqrt{-1}$, σ is the real part of the root and ω is the *damped* natural frequency. For the special case of zero damping, the undamped natural frequency is

$$\omega_o = \sqrt{\frac{G}{J}} \qquad 11.11$$

The damping ratio ζ is

$$\zeta = \frac{\sigma}{\omega_o} = \frac{B}{2J}\sqrt{\frac{J}{G}} \qquad 11.12$$

Using these relationships, equation 11.9 can be written in a frequently used form

$$\frac{s^2}{\omega_o^2} + \frac{2\zeta}{\omega_o} + 1 = 0 \qquad 11.13$$

The decay in amplitude of the sine wave of Fig. 11.8 per oscillation depends entirely upon ζ

$$\text{Decay per oscillation} = e^{\left(\frac{-2\pi}{\sqrt{1/\zeta^2 - 1}}\right)} \qquad 11.14$$

Fig. 11.9 shows how the response of the damped-mass-spring system varies with ζ. The curve for $\zeta = 0$ is an undamped, continuous oscillation. When $\zeta = 1$ the damping is termed "critical." With critical

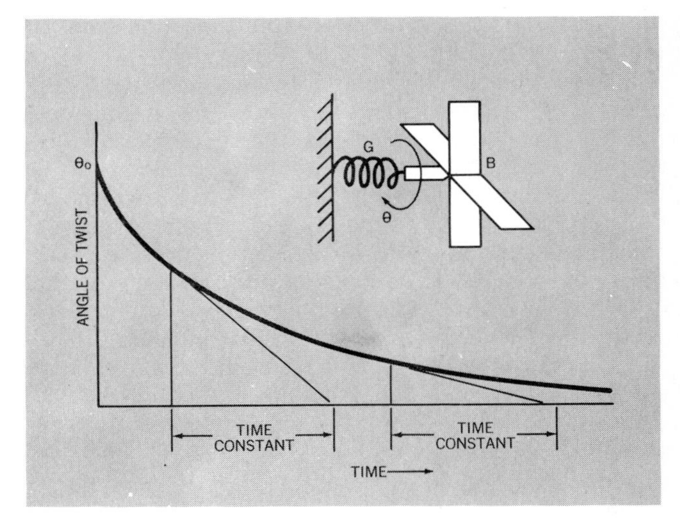

Fig. 11.6. Response of damped spring to initial displacement.

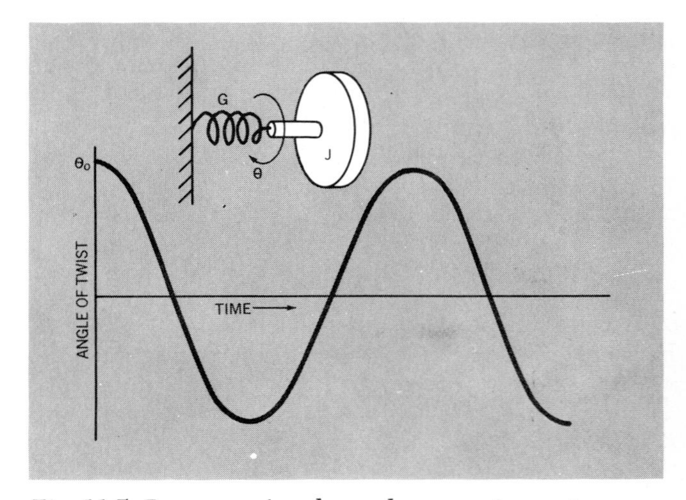

Fig. 11.7. Response of undamped mass-spring system.

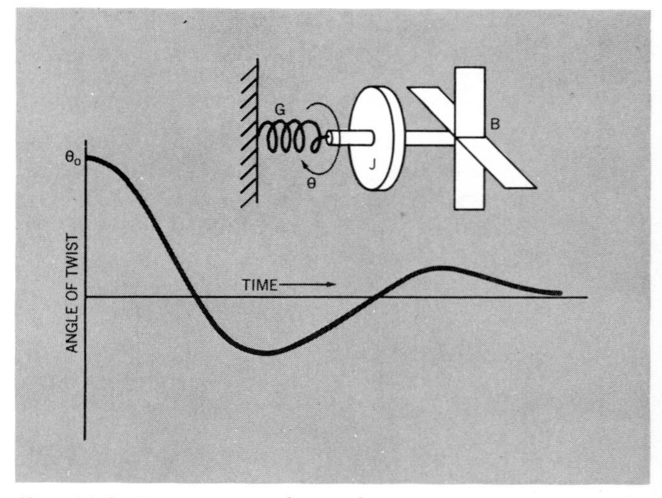

Fig. 11.8. Response of damped mass-spring system.

damping, the response curve just fails to overshoot the zero axis.

The relationship between the input and the output of any dynamic system can be expressed in many different ways. A convenient analytical expression is the *Transfer Function.*

The transfer function of a system is obtained by writing the differential equation relating the output to the input, replacing the operation d/dt with the complex number s and solving for the ratio of output to input.

If, as examples, equations 11.1, 11.2, and 11.3 had been expressed as ratios of θ_o/θ_e, then these equations would fit this definition of a transfer function. Most transfer functions of interest in hydraulic servos are of the form $KG(s)$. This form indicates that the transfer function consists of both a simple gain, such as the ratio of a lever or the area of an actuator; and a portion which is dependent upon s. If s were replaced by $j\omega$ then that portion of the transfer function would be dependent upon the frequency (in radians per second) of the signal being transmitted.

In the power servo, Figs. 11.4 and 11.5, the transfer function from the error signal to the output is the product of the transfer functions of the elements through which the signal passes

$$\frac{\theta_o}{\theta_e} = \frac{Q}{X_v} \cdot \frac{1}{A_p s} = \frac{K_v}{A_p s} \qquad 11.15$$

This is derived by realizing that a flow rate Q results from a valve displacement X_v. The flow from the valve results in a velocity of the actuator rod. However, the transfer function gain v/Q is equal to $1/A_p$. The s in the denominator appears because the desired output is the position of the ram. Position is obtained by integrating the velocity with time. Just as s represents a time differential, $1/s$ is a time integration. This transfer function also assumes that the valve output flow rate will vary linearly with valve displacement.

The transfer function of equation 11.15 is called the "forward transfer function" because it considers only those elements in the forward flow path from the error signal to the output. Because the block diagram has a feedback path to the summing point, a closed loop is formed. The nature of the transfer function for a closed loop is developed in the following manner. Assume that the forward transfer function θ_o/θ_e is denoted by μ and the feedback transfer function θ_f/θ_o is denoted by β. The error signal θ_e is the difference between the input command θ_i and the feedback signal θ_f. Then

$$\theta_e = \theta_i - \theta_f$$

$$\theta_e = \mu\theta_o \qquad \text{and} \qquad \theta_f = \beta\theta_o$$

So

$$\theta_i = \frac{\theta_o}{\mu} + \beta\theta_o \qquad 11.16$$

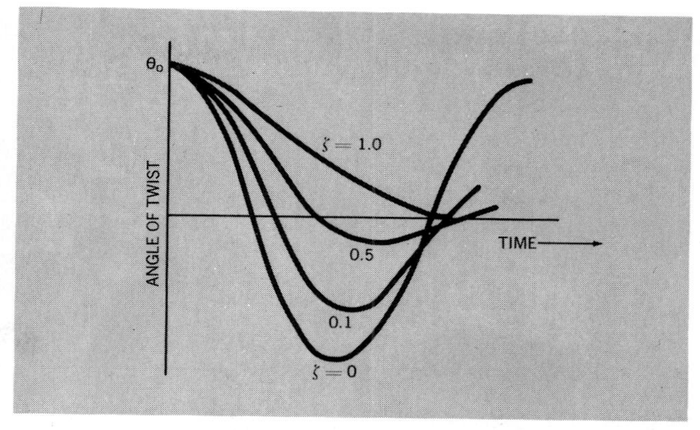

Fig. 11.9. Response of damped mass-spring system as a function of damping ratio.

and

$$\frac{\theta_o}{\theta_i} = \frac{\mu}{1 + \mu\beta} \qquad 11.17$$

Therefore, the "closed loop transfer function" is derived by dividing the forward transfer function by one plus the product of the forward and feedback transfer functions. In Fig. 11.5 the feedback transfer function is a pure gain $a/(a+b)$. Casting the closed loop transfer function in the form of equation 11.17

$$\frac{\theta_i}{\theta_o} = \frac{a+b}{a}\left[\frac{1}{\left(\frac{a+b}{a} \cdot \frac{A_p}{K_v}\right)s + 1}\right] \qquad 11.18$$

In this particular configuration, the initial amplification $b/(a+b)$ is in series with the servo. The overall transfer function is the product of 11.18 and the initial amplification.

$$\frac{\theta_o}{\theta_i} = \frac{b}{a}\left[\frac{1}{\left(\frac{a+b}{a}\frac{A_p}{K_v}\right)s + 1}\right] \qquad 11.19$$

Of course, for large displacement outputs, the values of a and b would have to be multiplied by the cosine of the angle between the actual position of the differentiating link and its original position in order to get the correct ratios. Also, not all valves have a linear relationship between displacement and flow. With such valves, the transfer function is valid only for very small displacements.

Another commonly used configuration for mechanical control is Fig. 11.10. In this system, the actuator displacement and the feedback motion are the same. The servo has a displacement gain of unity because the output displacement has to be the same as the input command. The valve body, being attached to the actuator, closes over the valve spool. In this arrangement, the forward path μ, is $K_v/A_p s$ and β, the feedback, is unity so

$$\frac{\theta_o}{\theta_i} = \frac{1}{\frac{A_p}{K_v}s + 1} \qquad 11.20$$

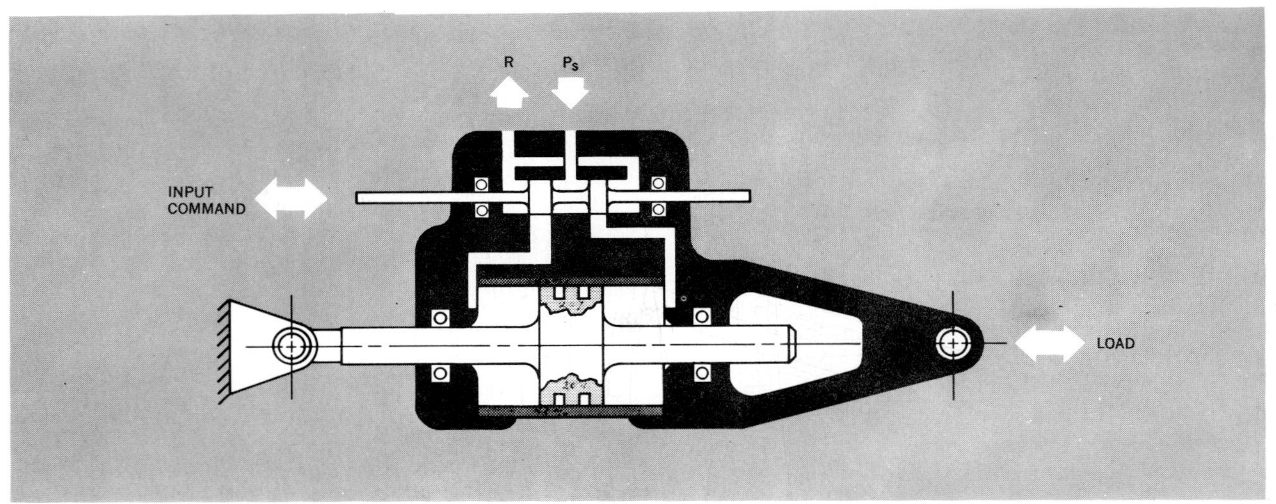

Fig. 11.10. Servoactuator with mechanical feedback.

In these discussions, no mention has been made of a number of factors which can affect the operating characteristics of a servoloop. Depending upon the application, one or more of the following effects must be included in the computation:

Load mass and spring rate
Actuator body mass
Actuator mounting spring rate
Oil column spring rate
Valve dynamics
Valve, actuator and load damping

If these effects are included, the transfer functions become high order functions of s. They can usually not be handled by pencil and paper computations but can only be solved by the use of an electronic computer. However, by judicious choice of the elements to be included in the transfer function, much information can be derived and some real insight into the dynamics of the system can be gained without letting the mathematics become unreasonably burdensome.

To see how this works, consider the body-mounted valve and actuator combination of Fig. 11.11. The flow across one metering port of the 4-way valve is the sum of the change in actuator end volume and of the amount the oil is compressed as the pressure builds up. This can be expressed as

$$Q_1 = \frac{dV_1}{dt} + \frac{V_1}{\beta_e}\frac{dP_1}{dt} \qquad 11.21$$

Similarly, the flow out past the other metering edge of the valve can be written

$$Q_2 = \frac{dV_2}{dt} + \frac{V_2}{\beta_e}\frac{dP_2}{dt} \qquad 11.22$$

The flow which goes to move the load is the algebraic sum of (a) the product of the valve opening and the valve flow gain and (b) the product of the load pressure and the valve flow-pressure coefficient.

$$Q_L = K_Q x_v - K_c P_L \qquad 11.23$$

At the actuator, the load flow can be expressed as the sum of (a) the actuator volume change rate, (b) the load pressure times a leakage coefficient C_L, and (c) a factor to account for the compressible flow due to change in load pressure.

$$Q_L = A_p x_p s + c_L P_L + \frac{V_t}{4\beta_e} P_L s \qquad 11.24$$

Where: $V_t = V_1 + V_2$

The force balance at the actuator is, on one side, the sum of (a) the total mass (load plus actuator rod) of the moving parts (M_t) times its acceleration, (b) the friction coefficient (B_p) (in pounds-per-velocity unit) between the piston and the actuator when multiplied by the actuator velocity, (c) the load spring rate (K_L) times the actuator displacement and (d) the load force. This is equated to the actuator output force

$$A_p P_L = M_t x_p s^2 + B_p x_p s + K_L x_p + F_L \qquad 11.25$$

These combine to form:

SEE BOX I **11.26**

If the damping coefficient B_p is judged to be very small and the only loads are the reactions of the mass and spring, equation (11.26) can be written

SEE BOX I **11.27**

Equation 11.27 is an open loop forward transfer function which describes the servoactuator of Fig. 11.11 in simplified form. The denominator is the product of a simple integration as shown by the *s* and a quadratic in *s*. The quadratic is of exactly the same form as the relationship of the simple damped-mass-spring system of equation 11.13 and Fig. 11.8. The presence of this quadratic introduces to the servo loop the tendency to oscillate which was noted in the discussion of the simple system. Just as in the simple system, the tendency to oscillate is controlled by the size of the damping ratio ζ (zeta). When a quadratic is multiplied with an integration, the resulting system, when put into a closed loop, can become unstable and oscillate violently if the damping is too low or the gain is too high.

For a 3-way valve controlled actuator, Fig. 11.12, the transfer function relating piston displacement to valve command motion is:

$$\frac{x_p}{x_v} = \frac{\dfrac{K_Q \beta_e A_p}{\beta_e A_p^2 + KV_h}}{s\left[\dfrac{s^2}{\omega^2} + \dfrac{2\zeta}{\omega}s + 1\right] + \dfrac{K\beta_e(K_c + c_L)}{\beta_e A_p^2 + KV_h}} \quad 11.30$$

where $\omega = \sqrt{\dfrac{\beta_e A_p^2 + KV_h}{V_h M_t}}$

$$\zeta = \frac{\beta_e M_t (K_c + c_L)}{2\sqrt{V_h M_t(\beta_e A_p^2 + KV_h)}}$$

Or, if the spring is ignored.

$$\frac{x_p}{x_v} = \frac{\dfrac{K_Q}{A_p}}{s\left[\dfrac{s^2}{\omega^2} + \dfrac{2\zeta}{\omega}s + 1\right]} \quad 11.31$$

where $\omega = \sqrt{\dfrac{\beta_e A_p^2}{M_t V_h}}$

$$\zeta = \frac{K_c + c_L}{2 A_p}\sqrt{\frac{\beta_e M_t}{V_h}}$$

Where a valve controls a rotary motor, Fig. 11.13, a similar set of equations can be written

$$Q_i = \frac{dV_1}{dt} + \frac{V_1}{\beta_e}\frac{dP_1}{dt} \quad 11.32$$

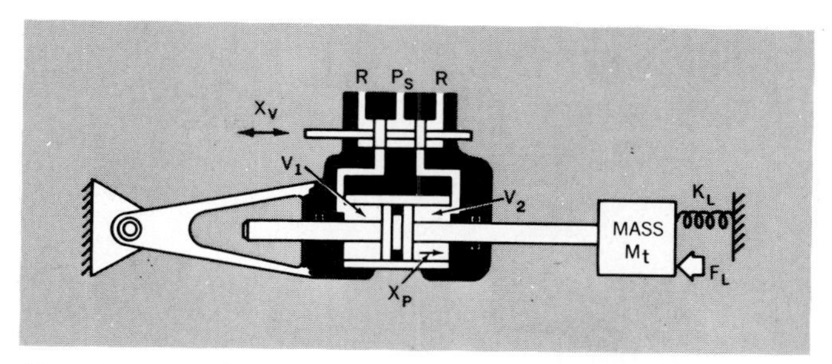

Fig. 11.11. Balanced actuator with 4-way valve.

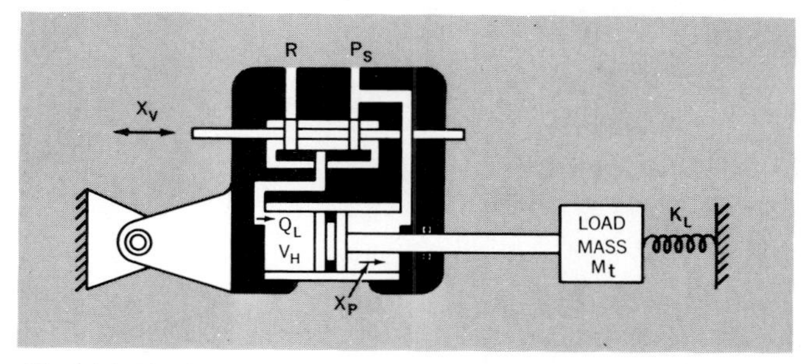

Fig. 11.12. Actuator and 3-way valve.

V_1 = pressure side volume

$$Q_o = \frac{dV_2}{dt} + \frac{V_2}{\beta_e}\frac{dP_2}{dt} \quad 11.33$$

V_2 = return side volume,
$V_t = V_1 + V_2$
$Q_L = \dfrac{Q_i - Q_o}{2}$

These combine to form

$$Q_L = D\theta_s + c_L P_L + \frac{V_t}{4\beta_e} P_L s \quad 11.34$$

D = volumetric displacement/radian
θ = radians
c_L = leakage coeff. in³/sec/psi

BOX I TRANSFER FUNCTIONS

$$x_p = \frac{\left[\dfrac{K_Q}{A_p}\right] x_v - \dfrac{K_c + c_L}{A_p^2}\left[1 + \dfrac{V_t}{4\beta_e(K_c + c_L)}s\right] F_L}{\dfrac{V_t M_t}{4\beta_e A_p^2}s^3 + \left[\dfrac{(K_c + c_L)M_t}{A_p^2} + \dfrac{B_p V_t}{4\beta_e A_p^2}\right]s^2 + \left[1 + \dfrac{B_p(K_c + c_L)}{A_p^2} + \dfrac{KV_t}{4\beta_e A_p^2}\right]s + \dfrac{K(K_c + c_L)}{A_p^2}} \quad 11.26$$

$$\frac{x_p}{x_v} = \frac{\dfrac{K_Q}{A_p}}{\dfrac{V_t M_t}{4\beta_e A_p^2}s^3 + \dfrac{(K_c + C_L)M_t}{A_p^2}s^2 + \left[1 + \dfrac{KV_t}{4\beta_e A_p^2}\right]s + \dfrac{K(K_c + c_L)}{A_p^2}} \quad 11.27$$

$$\theta = \frac{\dfrac{K_Q}{D}x_v - \dfrac{K_c + c_L}{D^2}\left[1 + \dfrac{V_t}{4\beta_e(K_c + c_L)}s\right]T_L}{\dfrac{V_t J}{4\beta_e D^2}s^3 + \left[\dfrac{(K_c + c_L)J}{D^2} + \dfrac{BV_t}{4\beta_e D^2}\right]s^2 + \left[1 + \dfrac{B(K_c + c_L)}{D^2} + \dfrac{G\,V_t}{4\beta_e D^2}\right]s + \dfrac{G(K_c + c_L)}{D^2}} \quad 11.36$$

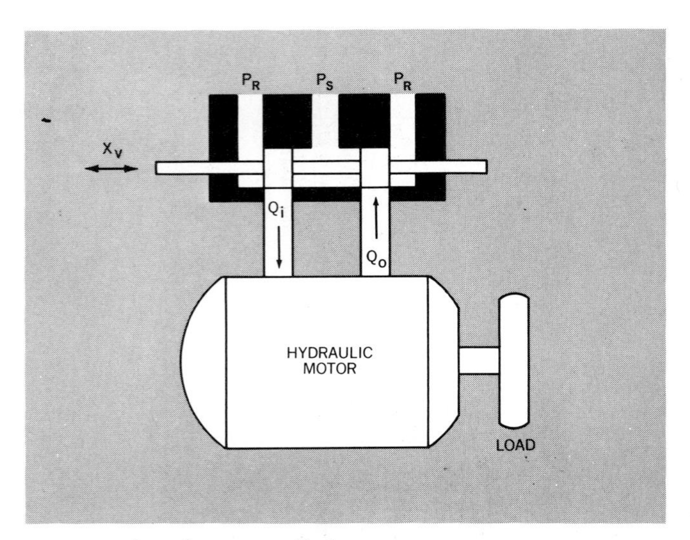

Fig. 11.13. Valve controlled motor.

The torque summation is

$$T = J\theta s^2 + B\theta s + G\theta + T_L \qquad 11.35$$

J = motor and load inertia
B = damping coefficient
G = torsional spring rate
T_L = load torque

An equation similar to 11.26 can be written

SEE BOX I 11.36

If B and G are considered negligibly small

$$\theta = \frac{\dfrac{K_Q}{D} x_v - \dfrac{K_c + c_L}{D^2}\left[1 + \dfrac{V_t s}{4\beta_e(K_c + c_L)}\right] T_L}{s\left(\dfrac{s^2}{\omega^2} + \dfrac{2\zeta}{\omega} s + 1\right)} \qquad 11.37$$

$$\omega = \sqrt{\frac{4\beta_e D^2}{V_t J}}$$

$$\zeta = \frac{K_c + c_L}{D}\sqrt{\frac{\beta_e A}{V_t}}$$

Two transfer functions can now be written

$$\frac{\theta_M}{x_v} = \frac{K_Q/D}{s\left(\dfrac{s^2}{\omega^2} + \dfrac{2\zeta}{\omega} s + 1\right)} \qquad 11.38$$

$$T = 0$$

$$\frac{\theta_M}{T_L} = \frac{-\dfrac{K_c + c_L}{D^2}\left[1 + \dfrac{V_t}{4\beta_e(K_c + c_L)} s\right]}{s\left(\dfrac{s^2}{\omega^2} + \dfrac{2\zeta}{\omega} s + 1\right)}$$

$$x_v = 0 \qquad 11.39$$

Equation 11.38 is the usual form of transfer function used. The transfer function gain will vary with the valve flow gain. If the gain is nonlinear, then the gain at valve null is used. Both the natural frequency and the damping ratio vary with the square root of the effective bulk modulus. The damping ratio also varies with the valve flow-pressure coefficient and with the amount of leakage internal to the motor.

11.8. ELECTRICAL INPUT SERVOS

Automatic control systems are used among other places for airplane and ship control and in the processing industries. Most of these systems use gyroscopes, accelerometers, pressure sensors and other measuring devices. The information from these devices is usually an electrical signal. Because of these control requirements which involve electrical signals, a whole study of electrical input power hydraulic servosystems has been developed.

The basic control elements of the electrohydraulic system are similar to the elements of the mechanical input servosystem. The control input, instead of a displacement, is usually a voltage. The output of the servo is measured by devices which express displacement, velocity and acceleration as voltages. The output voltages are subtracted algebraically from the input voltage to generate an error voltage. The error voltage (or a current which is proportional to it) is used to drive the electrohydraulic (transfer) servovalve.

The amplifier, Fig. 11.14, produces a current (differential current) which is proportional to the error signal. The valve has a transfer function as given by equation (8.43) or (8.44) depending upon the frequency range of interest. The actuator and electrical feedback convert flow to velocity and integrate the velocity to a displacement.

The forward transfer function $\dfrac{\theta_o}{\theta_e}$ is

$$\frac{\theta_o}{\theta_e} = \frac{i}{e}\,\frac{K_v}{1 + Ts}\,\frac{1}{A_p s} = \frac{\dfrac{K_A K_v}{A_p}}{s(1 + Ts)} \qquad 11.40$$

The feedback transfer function is unity so the closed loop transfer function is

$$\frac{\theta_o}{\theta_i} = \frac{\dfrac{K_A K_v}{A_p}}{s^2 + \dfrac{1}{T} s + \dfrac{K_A K_v}{A_p T}} \qquad 11.41$$

As with the mechanical input servo, the inclusion of the load dynamics, the structural mounting spring and the internal dynamics would raise the transfer function to a cubic or higher function.

A modification of the electrical input servo is the electrical input-mechanical feedback servo shown in Fig. 11.15. Diagram Fig. 11.16 of this servo is shown for the unloaded condition. The transfer function is

$$\frac{X_o}{X_i} = \frac{K_{TM}}{K_{fs}\,K_{fi}}\left[\frac{1}{\dfrac{s^2}{\omega^2} + \dfrac{2\zeta s}{\omega} s + 1}\right] \qquad 11.42$$

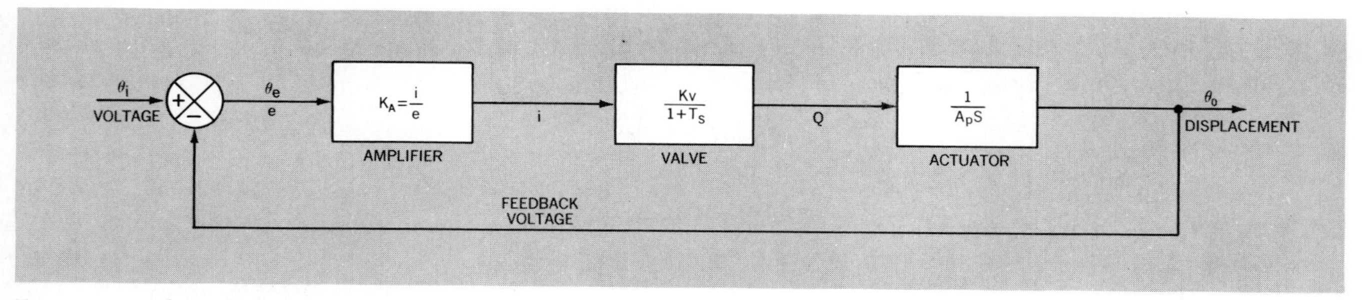

Fig. 11.14. Electrohydraulic servo block diagram.

$$\omega = \sqrt{\frac{K_p}{T_{HA}}}$$

$$\zeta = \frac{1}{2\sqrt{K_p T_{HA}}}$$

$$K_p = \frac{T_{HA}\ K_1\ K_{fs}\ K_{fl}}{A_p}$$

11.9. FREQUENCY RESPONSE TECHNIQUES

The preceding discussions have all related to what is known as transient responses. The response was to an initial displacement which dies out as a function of time—except, of course, for the undamped oscillatory system.

Another method of studying these dynamic systems is called the frequency response method. With this technique, sinusoidal inputs of a constant amplitude are fed into the system. The frequency of the input is varied over the range of interest. Assuming that the output waveform is reasonably sinusoidal, the comparison between the input and output will usually look like Fig. 11.17. As the input frequency is increased, the system will be unable to follow the command. The output amplitude will fall off and the output will tend to lag behind the input. The elements of interest are (a) the ratio of the output amplitude to the input amplitude as a function of frequency and (b) the amount the output lags behind the input as a function of frequency.

At low frequencies, the output will follow the input without any loss of amplitude. The ratio of the amplitudes at these low frequencies is arbitrarily designated as unity. The amplitude ratio at higher frequencies is then stated as an amplification or attenuation of the basic ratio. These are usually given in decibels where

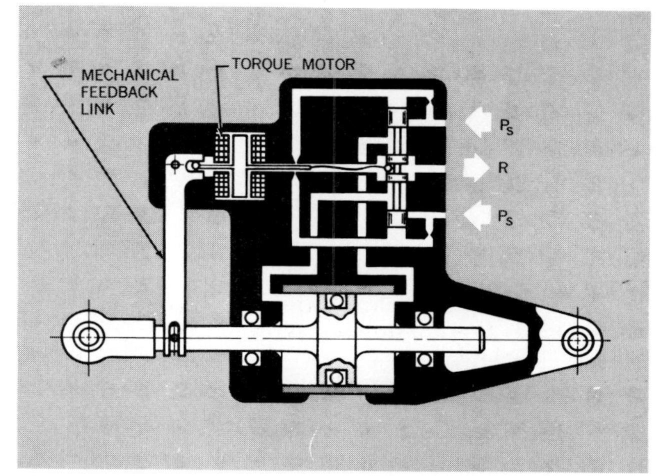

Fig. 11.15. Electrical input-mechanical feedback servo.

$$db = 20\ log\ \frac{R_f}{R_o}$$

At the base frequency, the ratio of the ratios is unity so the logarithm is zero. Therefore, if amplitude ratio in decibels is plotted against frequency, the basic ratio will be on the zero point of the decibel scale.

One way of measuring the lag of the output behind the input is to define one complete sinusoidal cycle as 360°. The lag can then be given as a number of degrees at a specific frequency. Obviously, when the out-

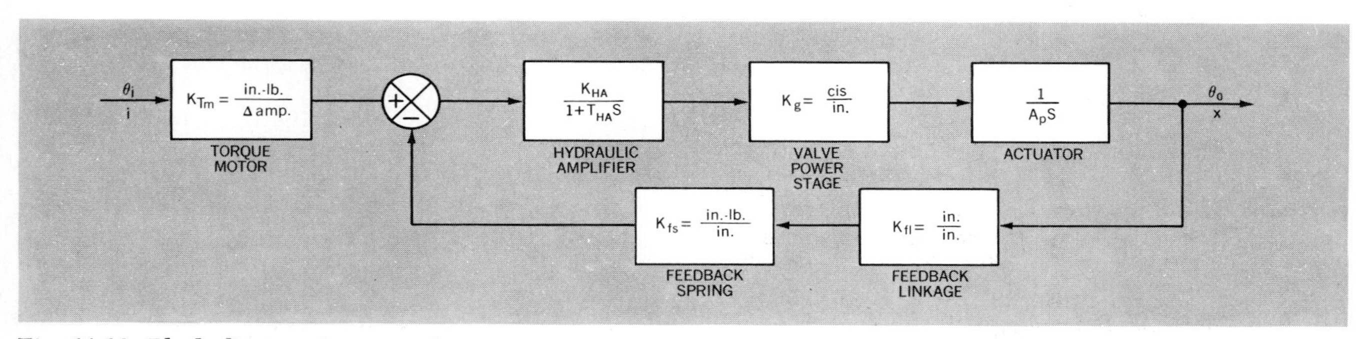

Fig. 11.16. Block diagram for servo shown in 11.15.

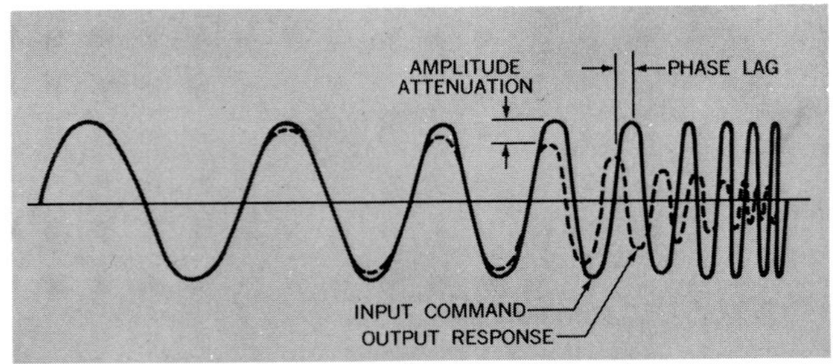

Fig. 11.17. Comparison between input and output waveforms.

put lags the input by 180°, the output motion directly opposes the input motion.

In equation 11.15 the open loop transfer function of a simple valve and actuator with feedback is given as

$$\frac{\theta_o}{\theta_e} = \frac{K_v}{A_p s}$$

Here, K_v/A_p can be called the "gain" of the set. The gain is constant for any set of components. Where the gain of the system is unity, the curve of the transfer function on a db-frequency plot would cross the zero db line at a frequency of one radian per second. To show this, substitute $j\omega$ ($\sqrt{-1}$ radians) for s, then the gain must be unity when $j\omega = 1$. When the ratio *is* unity, db = 0. For K_v/A_p not equal to unity, the crossover point is at a frequency equal to K_v/A_p. The slope of the curve is $1/\omega$. On the db frequency plot, this slope is 6 db per octave or 20 db per decade. The open loop phase lag is 90° throughout the frequency band. Fig. 11.18 shows these relationships.

The transfer function of an electrohydraulic valve has been shown to be

$$\frac{\theta_o}{\theta_e} = \frac{Q/i}{1 + Ts}$$

The plot of this transfer function is a little more complicated. The gain is Q/i. The denominator can be approximated by two curves which assume that s is replaced by $j\omega$ and that

$$1 + j\omega T = 1 \qquad \text{when} \qquad \omega T < 1$$

$$= j\omega T \qquad \text{when} \qquad \omega T > 1$$

These curves produce straight lines on the logarithmic plot, Fig. 11.19. The actual plot is 3 db below these approximations when $\omega = 1/T$ and 1 db below when $\omega = 2/T$. The phase plot is approximated by a straight line starting at 0° at $\omega\ T = 0.1$ and ending at 90° when $\omega T = 10$. The true plot actually crosses this line three times but is never greatly in error.

If the electrohydraulic servovalve is in series with an actuator, the transfer function of this set would be

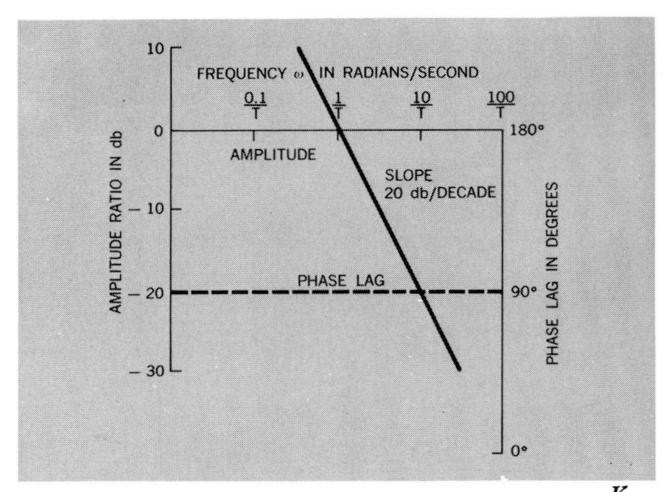

Fig. 11.18. Bode plot of valve actuator combination $\frac{K_v}{A_p s}$

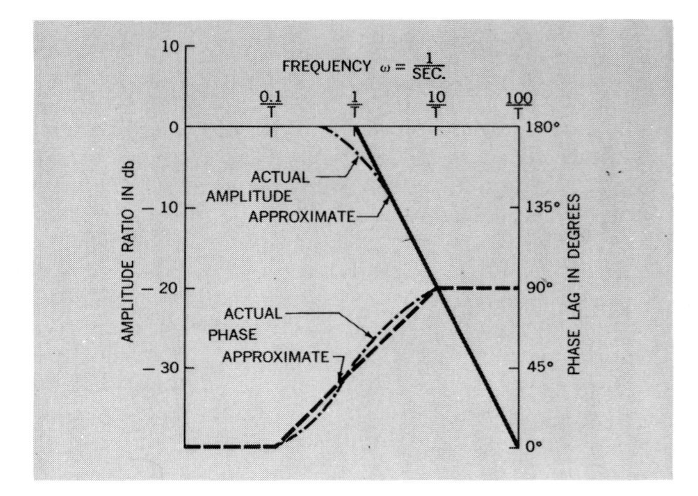

Fig. 11.19. Bode plot of $1/(1 + Ts)$ *transfer function.*

$$\frac{\theta_o}{\theta_e} = \frac{1}{A_p s}\,\frac{K_v}{1 + Ts} = \frac{\dfrac{K_v}{A_p}}{s(1 + Ts)} \qquad 11.42$$

Again, assuming that the gain is unity, the s in the denominator contributes a slope $1/\omega$ which intercepts the 0 db line at $\omega = 1/T$. At this point the $1 + Ts$ in the denominator contributes another $1/\omega$ to the slope and the approximate slope is $1/\omega^2$ or 40 db per decade. At $\omega = 1/T$ the actual curve, as shown in Fig. 11.20, is 3 db below the approximated curves. In this system, when the gain is not unity, the 0 db point is established by the need for the amplitude ratio to be unity at 0 db. Therefore, for a gain less than unity, the crossover point will be at a frequency equal to the gain. When the gain is greater than unity, the crossover occurs at a frequency equal to the square root of the gain.

The phase plot starts at 90°. At $\omega = 0.1/T$ the phase lag starts to increase. At $\omega = 10\ T$ the phase lag is 180° and stays at that angle.

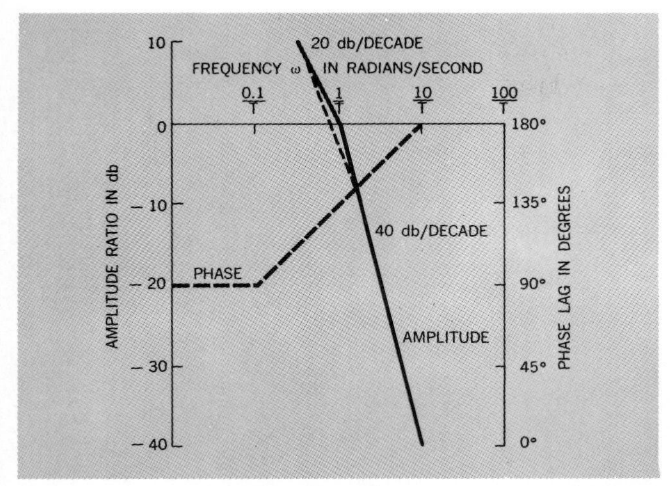

Fig. 11.20. Bode plot of valve actuator combination $1/s(1 + Ts)$

More complex functions can be plotted by an extension of the same technique. Note that an s or a $(1 + Ts)$ in the numerator will generate an amplification and a phase lead. These plots of amplitude and phase angle as a function of frequency are known as Bode (pronounced Bo-Dee) plots.

Bode plots can be made for either open loop (forward or feedback transfer function only) or closed loop systems in which the feedback signal modifies the input command.

All of the discussion of Bode plots so far has been of open loop systems only. The study of open loop systems is valuable because, while such systems can oscillate, they cannot become wildly unstable.

Closed loop systems having transfer functions with second order denominators can become oscillatory. Third order and higher denominators can, under some circumstances, cause the closed loop system to become unstable. Under these conditions the output oscillations become larger and larger until something breaks or corrective action is taken. This instability occurs when the amplitude ratio in db is positive (greater than zero) at the time that the phase lag is 180°. Under these conditions in a closed loop system, the output will reinforce the input and instability quickly occurs. This condition can be investigated experimentally by testing a servo with the feedback path open. A Bode plot is made from the test data. If the open loop Bode plot has an amplitude ratio of 0 db at the frequency at which the phase lag is 180°, the closed loop system would oscillate. If the open loop Bode plot is greater than 0 db at 180° phase lag, the closed loop system would be unstable. For a system which is apparently stable, the amplitude attentuation when the phase lag is 180° is called the "amplitude margin." Wherever possible, the amplitude margin should be made 6 db or more. Another way of describing the amount of system stability is to note the difference between the actual phase lag and 180° when the amplitude ratio is unity (0 db). This angle

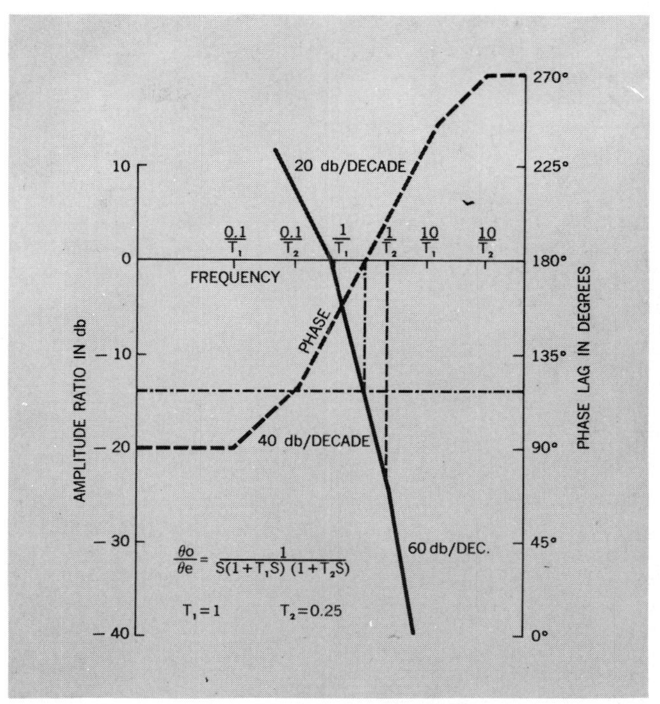

Fig. 11.21. Open loop Bode plot of cubic system.

is called the phase margin. An open loop phase margin of 40° is considered a good design goal.

Similarly, if the open loop Bode plot for any gain showed a safe amplitude margin at the frequency when the phase lag was 180°, the gain at which oscillations would occur can be judged by lowering the 0 db line until the amplitude plot is on the 0 db line at a phase lag of 180°. Fig. 11.21 shows a plot of

$$\frac{\theta_o}{\theta_e} = \frac{1}{s(1 + T_1 s)\ (1 + T_2 s)}$$

$$T_1 = 1$$

$$T_2 = 0.25$$

As can be seen, the amplitude plot is at —14 db when the phase plot is at 180°. For a slope of $1/\omega$, a gain change of 2 can be achieved by lowering the 0 db line 6 db. For a slope of $1/\omega^2 = 12$ db per octave, the gain change is 2 for a change of 12 db. The 180° phase point corresponds to 14 db so that a gain of a little more than 2 would cause the closed loop system to be unstable. A phase margin of 40° corresponds to a gain which provides a maximum closed loop amplification of 1.4. Such a system has minimum overshoot and fast response.

All of the Bode plots so far have been open loop plots. When the loop is closed, the plots assume a different appearance. If no computation or amplification is done in the feedback loop ($\beta = 1$), the relationship of equation (17) becomes

$$\frac{\theta_o}{\theta_i} = \frac{\mu}{1 + \mu} \tag{11.44}$$

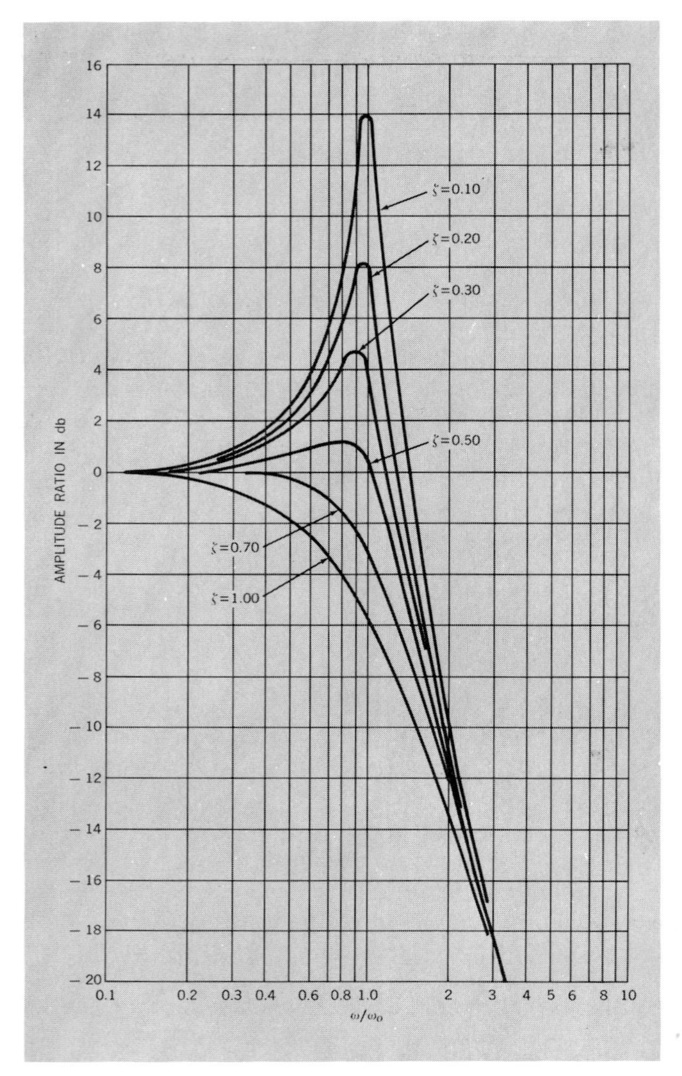

Fig. 11.22. Amplitude response of quadratic system as function of damping ratio.

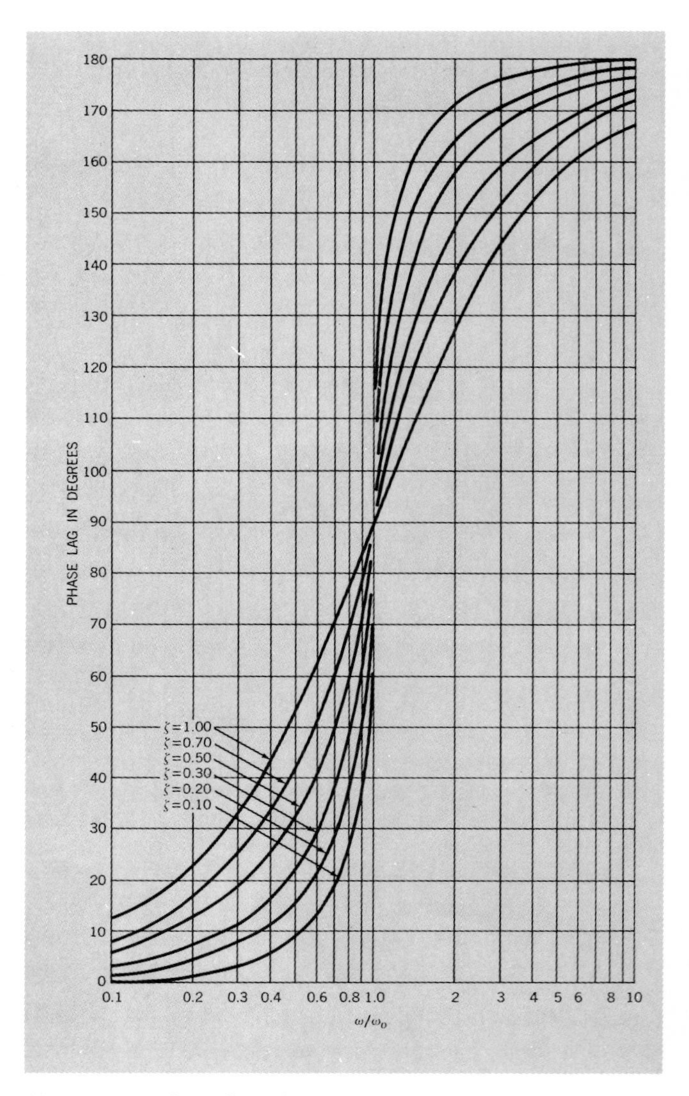

Fig. 11.23. Phase lag characteristics of a quadratic system as a function of damping ratio.

The open loop plot of Fig. 11.18 is really a plot of μ versus frequency. As can be seen, when $\omega < 1/T$, μ is much larger than unity. Therefore, in this region

$$\frac{\theta_o}{\theta_i} = \frac{\mu}{1 + \mu} \approx 1 = 0 \; db \qquad \textbf{11.45}$$

When $\omega > 1/T$, μ rapidly becomes much less than unity. For larger frequencies, then

$$\frac{\theta_o}{\theta_i} = \frac{\mu}{1 + \mu} \approx \mu \qquad \textbf{11.46}$$

In the amplitude plots, the response is approximated by a 0 db line until the first break point at $\omega = 1/T$, is reached. The balance of the plot is similar to the open loop plot. Because, as was seen in Fig. 11.9, the response of second order and higher systems vary with damping ratio, the variation from the approximation can be quite large. Fig. 11.22 shows how second order systems vary in amplitude response with ζ. The phase lag of second order systems is similarly affected by damping ratio. Fig. 11.23 shows this variation.

CHAPTER 11

PROBLEMS

1. What is the decay per oscillation of sine waves in which $\zeta = 0.1, 0.5, 0.7, 0.9$?

2. Write the transfer functions for the following block diagrams:

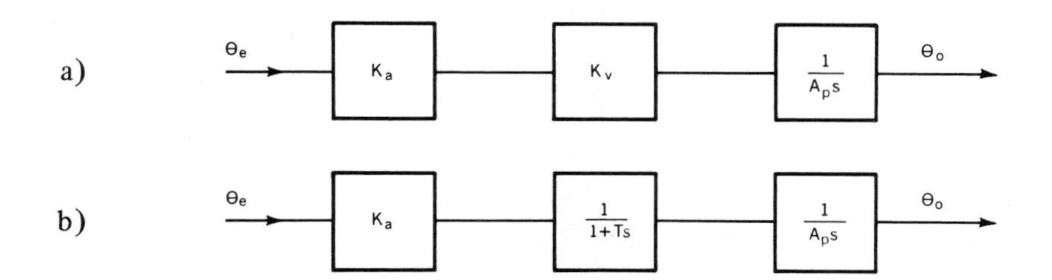

3. Write the transfer functions for the following block diagrams:

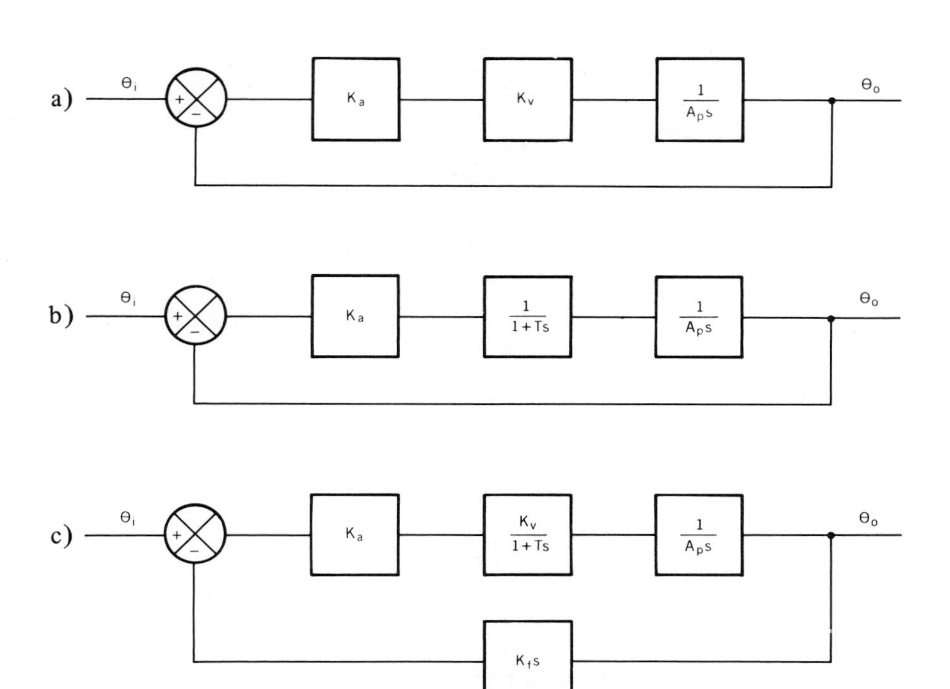

4. Make Bode plots of the open loop transfer functions of Problem 2.

$K_a = 10, K_v = 500, A_p = 2, T = 0.02.$

5. Make Bode plots of the closed loop transfer functions of Problem 3. Use same values for constants and gains as in No. 4 plus $K_f = 2$.

CHAPTER TWELVE

System Concepts

12.1 INTRODUCTION

For maximum efficiency and lowest cost, each component in a system must be carefully matched to the others. Because the system must be matched to its task, an airplane control hydraulic system becomes greatly different from a hydraulic system on a die casting machine. The hydraulics which power a mobile scoop shovel do not use the same components as the hydraulics on a numerically controlled machine tool.

Regardless of application, most fall into one or more of a few general categories. Depending upon the power source, the systems are either constant flow CQ or constant pressure CP. The form of control devices determine whether the system is *pump-stroke controlled* or *valve controlled*. Either type may have an output which is primarily a function of flow such as displacement, velocity or acceleration or the output may be a force or a torque which is tied more directly to pressure. The key elements may be either analog or digital in operation. Hydraulic systems have been built to operate in a manner which is comparable to alternating current electrical systems.

12.2 CONSTANT FLOW SYSTEMS

A CQ arrangement which has wide application is shown in Fig. 12.1. Here a fixed delivery pump supplies power to the load through a 4-way valve. When the load does not require part or all of the flow, the balance is returned to the reservoir through a pressure relief valve. In this arrangement, operation is of constant flow when the load absorbs all the pump output. When the load does not absorb all the flow, system pressure will rise until the relief valve setting is reached. The system will then operate as a constant pressure system until full flow is again required.

In practical applications, some version of the system, Fig. 12.2 is used rather than the elemental configuration of Figure 12.1. A check valve is placed at the pump outlet to prevent an over-riding load from driving the pump as a motor. A screen is often used in the pump suction line to protect the pump or a filter, as shown here, placed in the supply line to protect the valve and actuator. A sequence valve, which is similar in operation to a relief valve, establishes priority between motors. Actuator *A* must extend fully or must pick up a full load pressure in the extend direction before the sequence valve will open and allow extension of actuator *B*. When the control valve is moved to the alternate position, the load forces retraction of actuator *B*. The oil in *B* flows freely past the check valve.

The flow restrictor controls both the extend and retract speed of actuator *A* and the retract speed of actuator *B*. With the flow control valve in the position shown, the arrangement is termed "metering-out." In the meter-out design, all of the pump flow is available for actuation of the cylinders, but the pressure drop during actuation is shared between

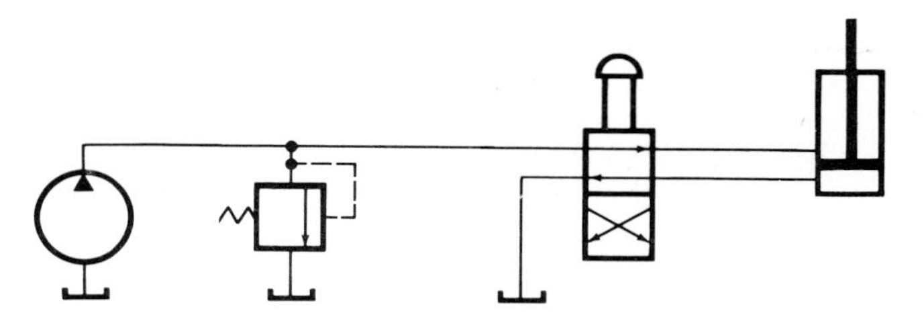

Fig. 12.1. Simple constant flow circuit.

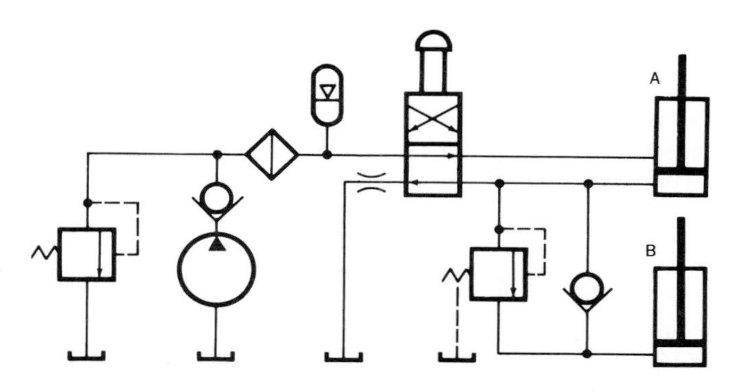

Fig. 12.2. Sequence circuit with meter-out control.

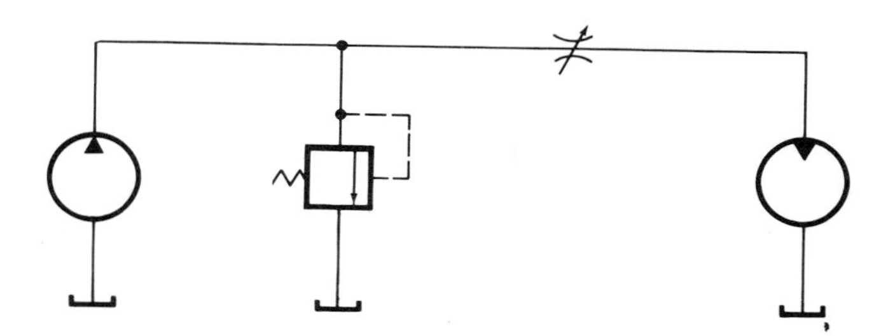

Fig. 12.3. Meter-in flow control.

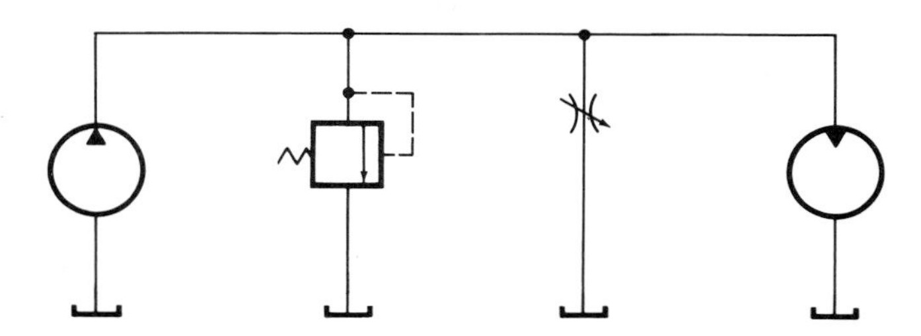

Fig. 12.4. Bleed-off flow control.

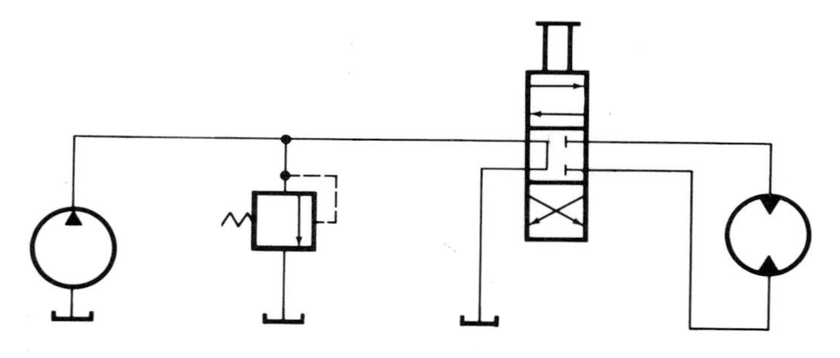

Fig. 12.5. Open-center valve control.

the cylinder and the flow restriction. Fig. 12.3 and 12.4 show simple CQ circuits having "meter-in" and "bleed-off" flow controls.

The meter-in circuit is similar to the meter-out circuit except that the motor outlet is at tank pressure. In the meter-out system, the flow control produces a back pressure on the actuator outlet. The bleed-off circuit allows the full pressure differential to be used in the motor but the motor and flow control valve share the pump flow.

One major disadvantage of the system in Fig. 12.2 is that flow which is not being used represents waste power. In addition to requiring an unnecessary amount of input power, this waste flow generates heat internally. Sometimes this heat raises the oil temperature to a point where there could be damage if there is no heat exchanger or cooling radiator.

Two basic methods of minimizing this power loss with CQ-CP systems are in general use. The first, Fig. 12.5, uses a valve which connects the pressure line to the return line when no load flow is needed. Pump output flow then goes directly back to the reservoir without going to the load or being discharged through the relief valve. The only pressure which is developed comes from the losses in the lines. The power loss during quiescent periods is thus held to a minimum.

The other technique, Fig. 12.6, operates similarly except that a de-pressurizing or unloading valve is installed between the pump suction and pressure ports. This valve can be operated manually or automatically by pressure or electrical signal. The use of an unloading valve eliminates the power losses in the lines. The price paid for this is the additional complexity of the unloading valve. Fig. 12.7 is a modification of the basic unloading valve circuit. Here the unloading valve is connected across a check valve. When the pump is started, the unloading valve is closed and system pressure builds up to the design value. At this point the unloading valve opens and the pump flow is directed to the suction line. The accumulator maintains system pressure until a load demand drops the pressure to the point where the unloading valve closes and allows flow from the pump to the load. With this version, a high working pressure is instantaneously available to the load.

Where the loads are cyclic, the designer can often minimize initial cost and operating expenses by use of an accumulator. Consider the stroking circuit, Fig. 12.2. Assume a single 5 second stroke every minute and that an average flow of 10 cubic inches per second will satisfy the rate requirements. With no accumulator in the system, the pump and drive motor must be sized to the 10 cis requirement. Assuming a 1000 psi working pressure requirement at the load, normal efficiencies in pump and motor require the use of a pump capable of about 3.2 gpm and a 2 hp motor. 91.5% of the motor energy would be wasted in heating the oil unless an unloading circuit were used.

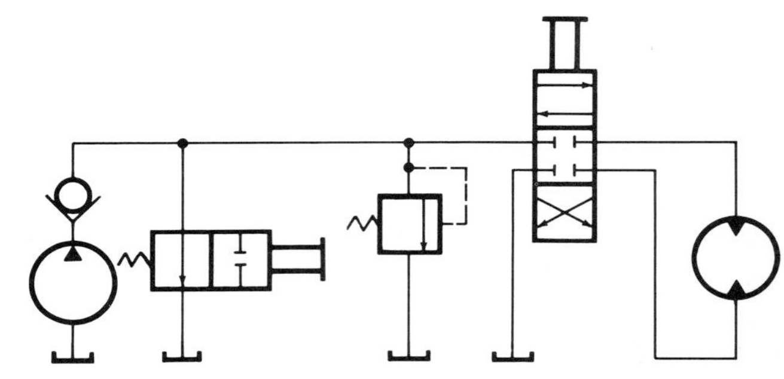

Fig. 12.6. Unloading valve circuit.

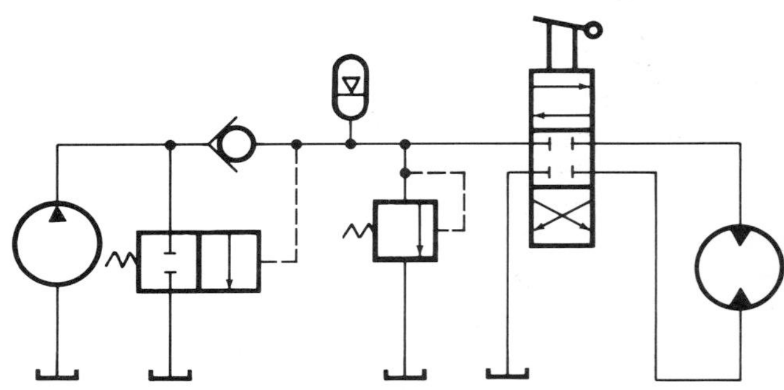

Fig. 12.7. Accumulator circuit.

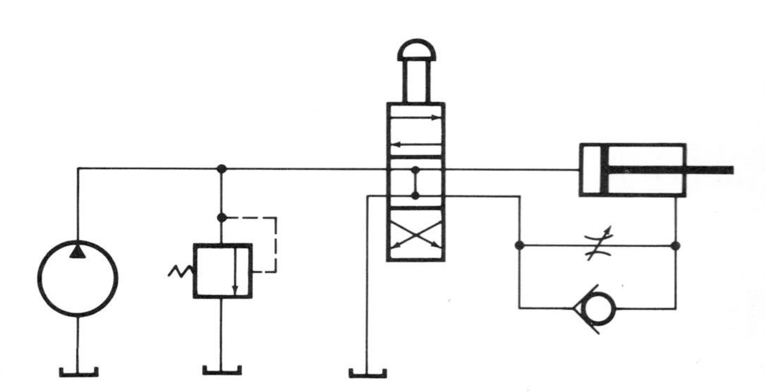

Fig. 12.8. Slow advance, rapid return circuit.

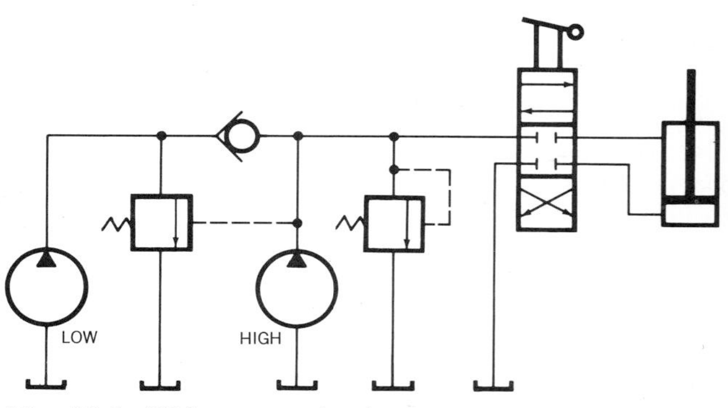

Fig. 12.9. Hi-lo pump circuit.

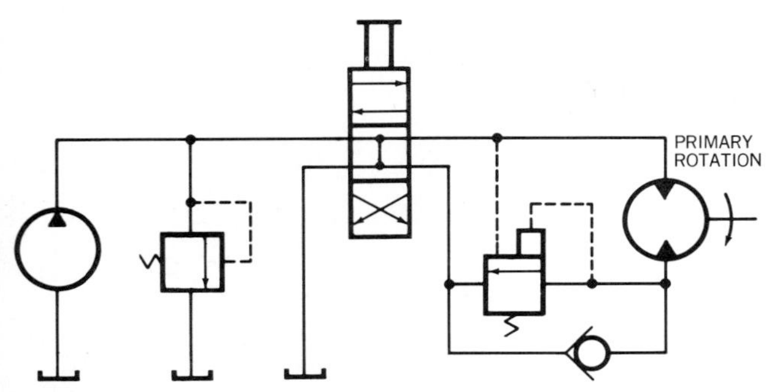

Fig. 12.10. Brake circuit.

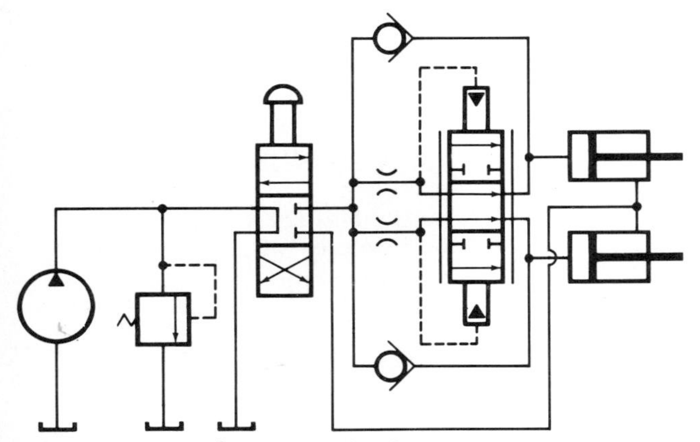

Fig. 12.11. Synchronizing circuit.

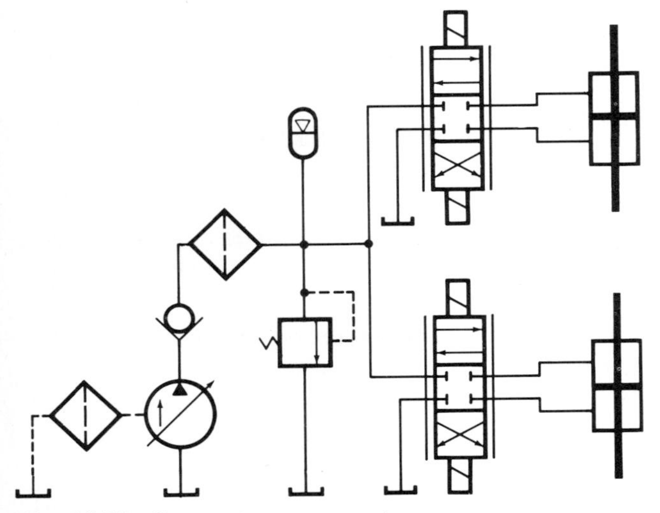

Fig. 12.12. Constant pressure system.

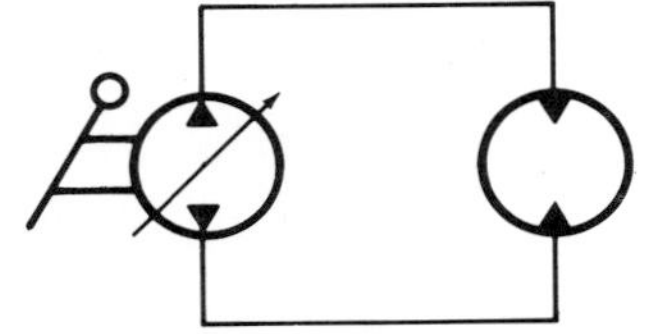

Fig. 12.13. Simple hydrostatic transmission.

If a gallon accumulator charged to 1000 psi is installed and the relief valve set at 1500 psi, the pump and motor sizes can be sharply reduced. When system pressure reaches 1500 psi, the accumulator can supply 77 cubic inches of oil with an isothermal expansion before the pressure drops to 1000 psi. The accumulator alone can handle the total requirement so that the pump and motor can be sized to supply the 50 cubic inches of oil in the 55 seconds between strokes. A pump which supplies only 0.3 gpm can handle this need. A half horsepower electrical motor is adequate to the task. In this example the accumulator allows the use of a pump one tenth the original size and a motor a quarter the size of the one originally installed.

The number of variations of configurations which can be built with the simple building blocks of the *CQ* system is endless. A few of them will be shown here to illustrate some methods of performing specific tasks. Fig. 12.8 shows a variable flow control valve coupled with a check valve to provide an actuator which extends slowly and retracts quickly. Although the variable flow control valve is between the control valve and the actuator, this is a modified meter-out circuit.

Occasionally a system must provide a rapid advance with low force to be followed by a slow final movement at high force. This requirement is typical of many presses and pelleting machines. One way of providing this is with a two pump system, Fig. 12.9. A large volume, low pressure pump is in parallel with a low volume, high pressure pump. The low pressure pump provides most of the stroking flow. When the load pressure has built up to the pump pressure limit, the low pressure relief valve connects the low pressure pump outlet to the reservoir. The high pressure is retained by the check valve. The balance of the load flow is provided by the high pressure pump with the load pressure limited by the high pressure relief valve.

Fig. 12.10 shows a simple circuit with a brake valve. The brake valve operates as a relief valve with two sensing inputs. One sensing input from the motor primary inlet line is connected to a large piston which is balanced by the brake valve spring. The other sensing input from the motor primary outlet line is connected to a small piston which is in parallel with the big piston. Because of the differences in area, the small piston requires a higher sensing pressure to open the brake valve against the spring setting than does the large piston. A check valve is provided in parallel with the brake valve to allow secondary operation of the motor in the reverse direction.

In the condition shown, the load pressure must build up to the large piston relief setting before the brake valve will open and flow can go through the motor. If the load should attempt to drive the motor, the inlet pressure will drop and the brake valve will close and the driving load will be snubbed. When the control valve is put in the mid position, any driving of the system by the load inertia will raise the primary outlet pressure until the small piston relief pressure is reached. The load can then be brought to a halt using the brake valve back pressure to slow the rotation of the motor. Operation is without braking in the reverse direction.

Synchronizing the strokes of hydraulic cylinders operating in parallel calls for the use of special circuits. Usually, the differences in friction from cylinder to cylinder will cause erratic and uneven operation unless special care is taken. Of course, synchronized operation within the limits imposed by the compressibility of the oil is possible with series operation of balanced actuators. However, because the available pressure must be shared among the actuators in series, the output force per actuator is usually too low.

Fig. 12.11 shows one way of achieving parallel synchronization. The two cylinders must extend in stroke synchronization. Flow from the pump divides into two parallel paths, one for each actuator. Each flow path goes through a flow control valve and then through an infinitely variable, pressure operated synchronizing valve. The synchronizing valve is operated by pressures downstream of the flow restrictors. If too much flow passes one of the flow restrictors, its downstream pressure will drop and the synchronizer valve will move so as to add restriction in that line and to minimize flow.

12.3 CONSTANT PRESSURE SYSTEMS

Constant pressure systems may use a fixed delivery pump which produces more flow than the load ever demands. A more typical configuration uses a pressure compensated variable delivery pump. The allowable leakage flow and open center flow in the system must be added to the load flow to establish minimum pump flow capability. As shown in Fig. 12.12, these systems are often used to supply multiple servo systems. The control systems of submarines, boats or airplanes are typical users of CP systems. While the use of the pump pressure compensator would seem to eliminate the need for a relief valve, most designers include one to handle the case of a driving load becoming the dominant input source. A relief valve can also prevent damage due to thermal expansion of the fluid.

12.4 HYDROSTATIC TRANSMISSIONS

A hydraulic pump driving a hydraulic motor provides a simple technique of achieving a continuously variable, stepless gearing ratio between a power source and a load. Fig. 12.13 shows the simplest form of this transmission, a variable delivery pump driving a fixed displacement motor. Occasionally a fixed delivery pump is used with a variable displacement motor and, less frequently, both motor and pump may be variable displacement.

By methods similar to those used previously in this book, transfer functions can be developed relating the output motor displacement θ_o to the variable pump stroking displacement x. For the transmission (variable delivery pump and fixed displacement motor) without any load

$$\frac{\theta_o}{x} = \frac{K}{s} \frac{1}{1 + Z_L \left[L + \frac{V_t}{\beta_e} s\right]} \qquad 12.1$$

where:
$K = \frac{d_p \, \omega_d}{d_m}$
d_p = pump displacement rate at full stroke $in.^3$/radian
d_m = motor displacement rate $in.^3$/radian
ω_d = pump rate—radians/sec.
$Z_L = \frac{P_L}{Q_m}$
P_L = load induced pressure
Q_m = load flow
L = system leakage factor
V_t = total volume of oil under compression

The motor rate transfer function is the derivative of equation 12.1.

$$\frac{\omega_o}{x} = \frac{K}{1 + Z_L \left[L + \frac{V_t}{\beta_e} s\right]} \qquad 12.2$$

If an inertial load J were added to the motor output

$$\frac{\theta_o}{x} = \frac{K}{s} \frac{1}{\frac{J\,V_t}{U\,d_m\,\beta_e} s^2 + \frac{J\,L}{U\,d_m} s + 1} \qquad 12.3$$

where U = motor torque per unit pressure ($in.^3$)

Were the load to be a damped mass-spring combination

$$\frac{\theta_o}{x} = \frac{K}{s} \frac{Js^2 + Bs + K}{\left[\left[J + \frac{J\,K\,V_t}{U\,d_m\,\beta_e}\right] s^2 + \left[B + \frac{J\,K\,L}{U\,d_m}\right. \right.} \\ \overline{\left. + \left[\frac{B\,K\,V_t}{U\,d_m\,\beta_e}\right] s + \left[K + \frac{B\,K\,L}{U\,d_m}\right]\right]} \qquad 12.4$$

In the absence of damping, this becomes

$$\frac{\theta_o}{x} = \frac{K}{s} \frac{J/K\, s^2 + 1}{\left[\left[\frac{J}{K} + \frac{V_t\,J}{U\,d_m\,\beta_e}\right] s^2 + \frac{L\,J}{U\,d_m} s + 1\right]} \qquad 12.5$$

Of course a fixed displacement pump and motor may be used together if the motor is controlled by directional and speed control valving. This system is not, however, considered a hydrostatic transmission. The variable pump-fixed motor transmission is usually called a "constant torque" drive. Actually, the torque is always proportional to the system pressure. The pressure is, of course, determined by the load characteristics. The constant torque drive is reversible if the pump can be stroked over center. The drive itself has only a relatively narrow torque range. It is often used with a 2 or 3 speed gear box.

The fixed pump-variable motor transmission is called a constant horsepower drive. In this configuration the power is proportional to the load pressure. The variable-variable combination provides the characteristics of both the constant torque and constant power drives but poses a real problem in control.

12.5 PUMP CONTROLLED SERVOMECHANISMS

The control of a hydraulic motor or actuator directly from a pump can be extended to cover a conventional position servo, Fig. 12.14. Leakage losses in the main servo and flow demands in the small pump stroking servo are supplied by an auxiliary pump. Compared to a valve controlled servo, the pump controlled servo is usually larger, more complex and, perhaps, more costly. On the other hand, the pump controlled servo does not pay the price of the valve pressure drop in order to achieve control of the output motor.

Automatic control devices can be added to the pump controlled servo to provide automatic stabilization and automatic piloting functions or to allow for direct electrical control of the servo in addition to mechanical command. In Fig. 12.15 an electrical input control servo has been added to the basic system. The output of this servo is algebraically position summed with the mechanical command signal. In this fashion the electrical command is said to be in series with the mechanical command. This arrangement is technically identical to the modulating piston form of command already shown in Fig. 9.15. As far as the electrical command is concerned, the positional servo loop is closed at the summing lever and not at the differential lever as in the basic system. Because the loop is closed at the summing lever, the input command point on the differential lever can be considered fixed. Therefore, no evidence of motion commanded by the electrical servo will be fed into the mechanical command system. In this way, the series input system acts like an active, dynamic shock absorber. Assume, for instance that the electric command were generated by the time rate change of the load position. Because the *rate* of change of position becomes large much quicker than the change itself, the electric command system can act more quickly than can a system based only on manual control of position. Therefore, when the position of a heavy fluctuating load must be controlled, the rate-plus-position control can greatly ease the operators task.

If, in addition to the rate stabilizing scheme, an automatic steering or automatic controlling input is required, the system can look like Fig. 12.16. Here, the automatic controlling input is added to the stabilizer input electrically. In addition, however, the mechanical input levers are locked between the differential point and the summing point. Because no mechanical feedback is possible, an electrical feedback device must be added and the feedback summation made electrically in the amplifier. In this arrangement the automatic control input is usually a position command and any rate input is additive to it. Because the mechanical linkage is locked as shown, any output motion of the actuator or motor will be transmitted back to the manual command point. Such an arrangement is usually called a "parallel" command to distinguish it from the series system previously described. The parallel command system is preferred where a load must be capable of being

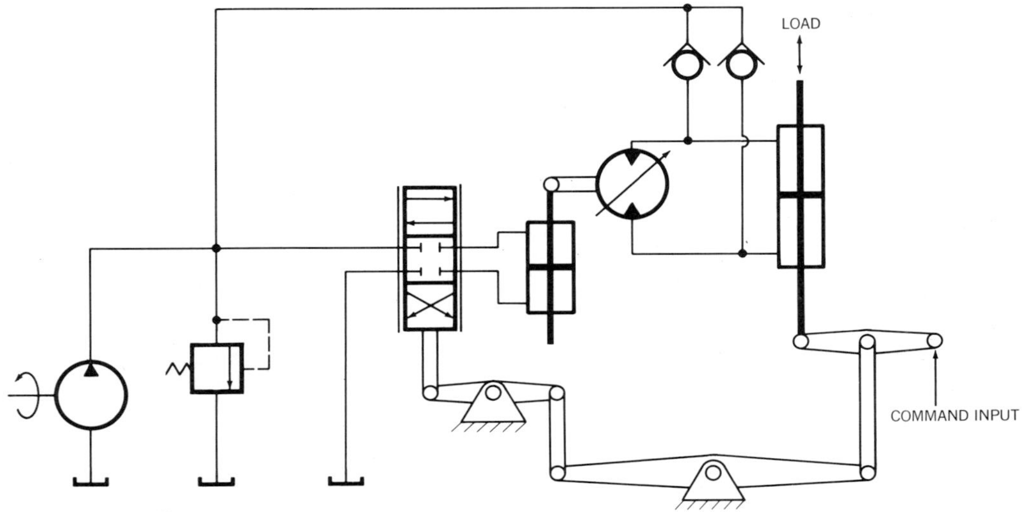

Fig. 12.14. Pump controlled position servo.

controlled either manually or automatically because the position of the manual command linkage is synchronized with the position of the output motor. This would not be true with a series arrangement.

12.6 RUN-DOWN SYSTEMS

Sometimes, to save weight and to reduce complexity, hydraulic systems are designed without pumps. Compressed gas, expelling oil from an accumulator, supplies the power to the control system. Typical of such systems is the fin control system on Little Joe described by Oberst[1]. Little Joe is a solid propellant launch vehicle. Flight attitude is controlled by aerodynamic surfaces on each of four fins. Each control surface is powered by a separate hydraulic system. For the sake of simplicity, low cost and ease of maintenance, an accumulator powered system using easily available components was chosen. During the 2 minute flight, oil flowed from the accumulator to the servo. After use, it was dumped overboard.

The size and working capacity of the system was determined from computer-programmed flight plans which determined flight altitude, surface dynamic pressure and angular displacement as functions of flight time. These data were converted to system

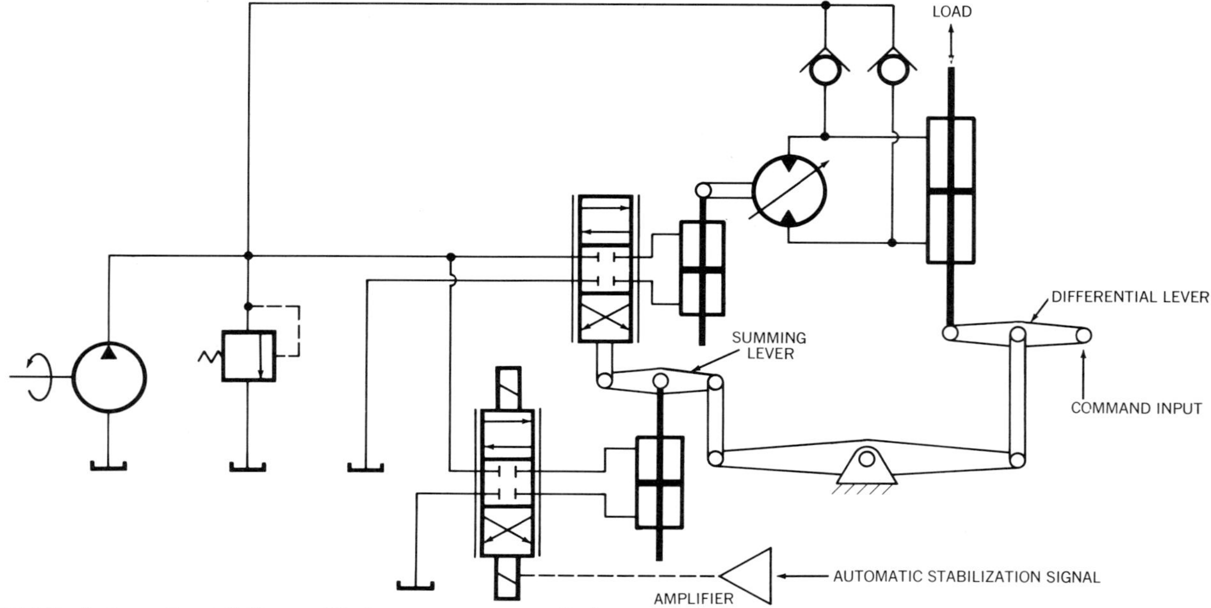

Fig. 12.15. Automatic stabilizer added to pump controlled servo.

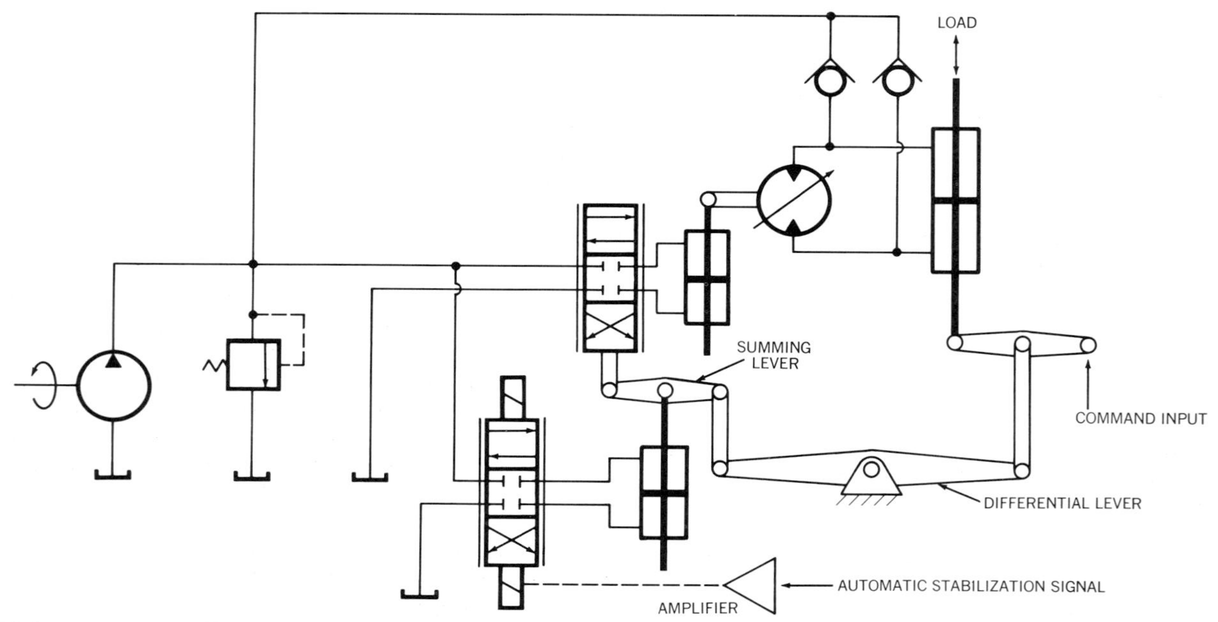

Fig. 12.16 Pump controlled servo with automatic pilot input.

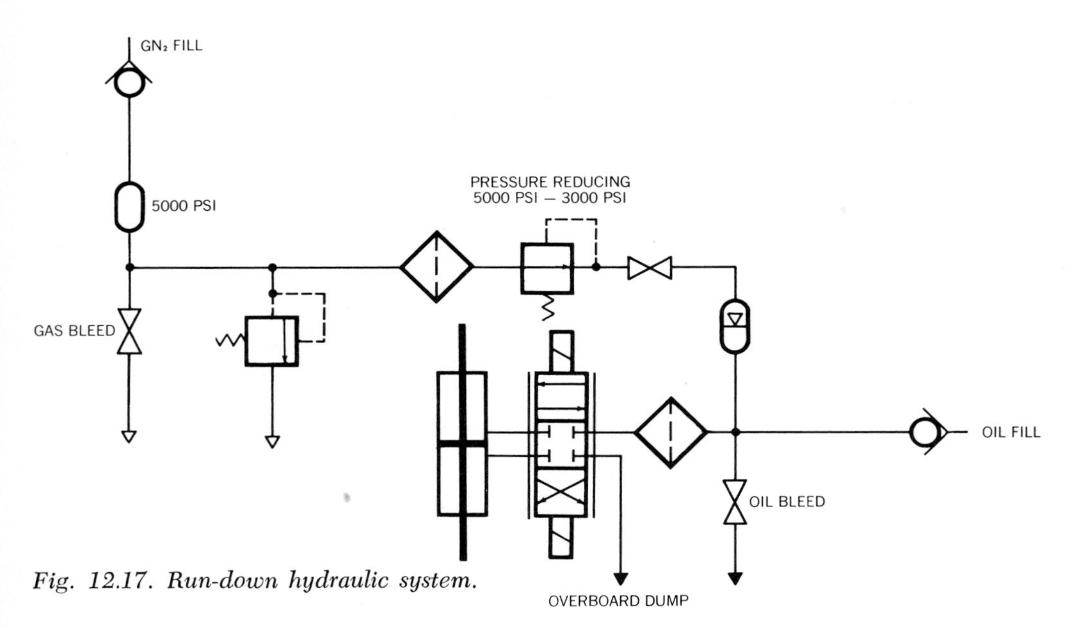

Fig. 12.17. Run-down hydraulic system.

pressure-flow requirements. Fig. 12.17 shows the hydraulic-pneumatic system which was designed to meet these requirements. The basic system energy source is 400 in^3 of gaseous nitrogen (GN_2) stored at 5000 psi. When the system is in use, the GN_2 is regulated down to 3000 psi and introduced to one side of an accumulator. Oil is drawn from the other side, filtered, and then used in a balanced servo actuator controlled by a two stage electro-hydraulic servovalve. A linear potentiometer is used as the position feedback element.

Approaches other than those used on Little Joe can also be considered. One technique would be to store the oil and gas in one accumulator rather than using an accumulator and a gas storage bottle. This would simplify the system but require a larger accumulator for the same total energy availability.

If available energy were very critical, the use of a bang-bang servo operating from a 3 position—4 way on-off valve should be considered. The quiescent flow in a jet pipe or flapper-nozzle servovalve could be eliminated with a valve which shut off completely under zero command. The accuracy of control with a bang-bang servo probably would not be as great as with a proportional servo.

12.7 SELF STARTING SYSTEMS

In the discussion on hydraulic motors, it was stated that most pumps can be operated as motors if the energy inputs were reversed. Hydraulic self starting systems take advantage of this. Instead of using an electrical battery to store energy for starting as does an automobile engine, the hydraulic system can store energy in an accumulator. In its simplest form, Fig. 12.18, an accumulator can be charged initially by a hand pump and subsequently by a starter motor-pump. The high pressure oil in the accumulator can be used to start the engine directly by being ported

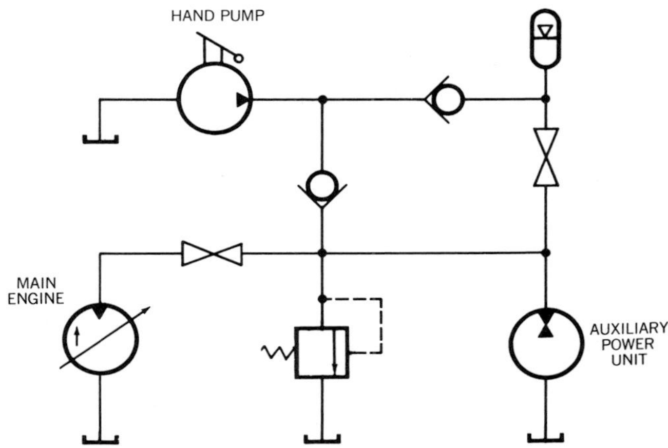

Fig. 12.18. Basic engine starting system.

to the engine starter motor. If the engine is large, the accumulator oil is often used to start an auxiliary power unit (APU). The starter motor-pump on the APU is then operated as a pump to start the main engine and to recharge the accumulator.

This sort of starting system is particularly adaptable to large marine and aircraft jet engines. Such engines often have, in addition to an APU, an auxiliary gear box. Power from the engine shaft, operating through a clutch, then drives hydraulic pumps, electrical generators, etc. The accumulator supplies hydraulic power to start the APU by means of a pump/motor acting as a motor. The APU then drives the pump/motor as a pump to supply energy enough to the main engine motor to start the engine. System hydraulic and electrical power is then provided by pumps, pump motors, and generators mounted on an accessory gear box. When the main engine is not running, the APU pump can supply oil directly or can drive other accessories through the pump/motor mounted on the gear box.

12.8 PULSATING HYDRAULIC SYSTEMS

Pollard[2] has described studies made at Republic Aviation on hydraulic systems which are, in part, analogous to AC electrical systems. Because the analogy is not universal, Pollard refers to these systems as pulsating rather than alternating systems.

The most elemental form of a pulsating system is shown in Fig. 12.19. A piston, driven by a scotch yoke, supplies oil to the system. Because of the form of drive, the oil flow is sinusoidal and alternating in direction as shown in Fig. 12.20. The rate of drive determines the frequency of pulsation. The piston stroke determines the pulsation amplitude. With two transmission lines, the phase difference between the lines is 180°.

In order to use this pulsating flow in conventional output actuators, the flow must be converted to continuous (DC) flow. This conversion is done in a ring demodulator which consists of four check valves. Downstream of the demodulator the flow consists of a continuous series of half sinusoids, Fig. 12.21. A smoothing filter which consists of the fluid inertia and the capacitance of an accumulator changes the half waves to continuous rippled flow.

Another, more complex form of a pulsating system is shown in Fig. 12.22. Here the system energy is supplied by a pressure compensated variable delivery pump. With it is a conventional array of components: filter, check valve, reservoir and relief valve. The conversion from continuous flow to pulsating flow occurs in a rotary alternator valve. This valve ports flow into the lines which carry the pulsating energy. A two line system (single phase) can be used as in Fig. 12.19. However, the use of an alternating valve allows for the easy splitting of the flow into many lines. As shown here, the system has three lines. The transmission line is then analogous to a 4 wire, wye-connected, three phase electrical system.

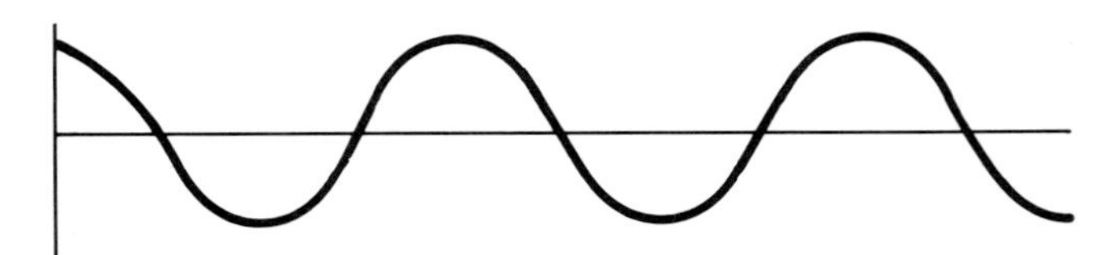

Fig. 12.20. Sinusoidal output from pulsating source.

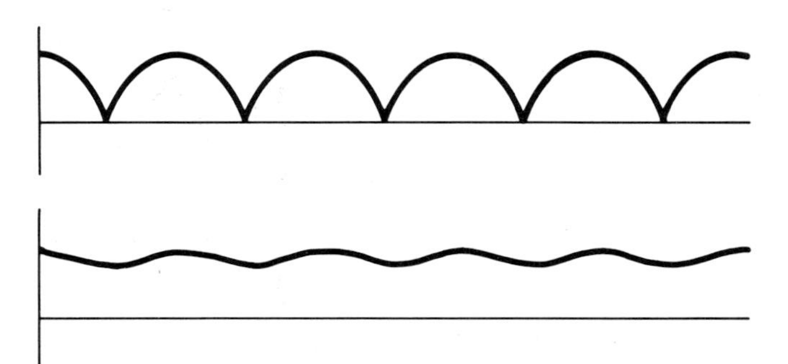

Fig. 12.21. Rectified and smoothed flow from pulsating circuit.

Fig. 12.22 is another component peculiar to pulsating systems, a pressure transformer. The transformer, Fig. 12.23, is similar to a free piston pressure intensifier having (usually) unequal piston areas. The transformer can, as a function of area ratio, increase or decrease system pressure. As distinct from a pressure reducing valve, the transformer converts

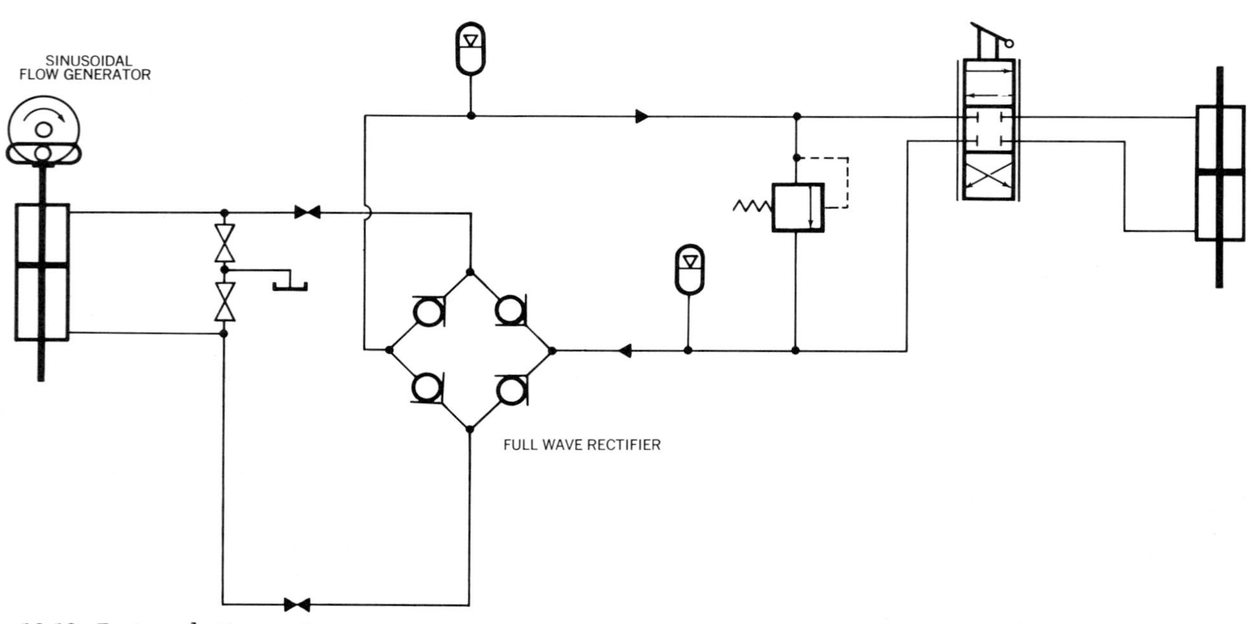

Fig. 12.19. Basic pulsating system.

pressure without losing power. The transformer also isolates one part of the system from the other. Therefore, a transformer-coupled system could operate with two entirely different hydraulic fluids. This attribute could be a real advantage when controlling high temperature functions, in cryogenic operations or in nuclear systems.

Because the fluid downstream of the transformer is isolated from the reservoir, a device to compensate for changes in fluid volume is necessary. This fluid volume compensator is essentially identical to a bootstrap reservoir.

The flow output of the three phase alternator valve would appear as in Fig. 12.24. After passing through the demodulator, flow and pressure would be the average of the three inputs. The ripple frequency would be six times the rotation rate of the alternator valve. This relatively high ripple frequency obviates the need for accumulators to smooth the flow.

A pulsating system provides some interesting advantages over a continuous flow system. The possibility of using two or more fluids could be useful in controlling systems which must operate in very hot or very cold environments. A fluid which is readily pumped and handled could be used in the normal portion of the system. Another fluid, more adapted to extreme temperatures could be used in the working areas.

Also, by the installation of transformers with their isolation capability, nuclear radiation effects could be retained in a shielded area. The same technique could be used to limit the effects of combat damage to restricted parts of the control system.

System contamination can be controlled more easily than in a continuous flow system. In the simple system of Fig. 12.19, no flow occurs; the fluid merely pulses back and forth; therefore, particulate contamination will not be transported around the system. In the configuration of Fig. 12.22, the rate of flow will be ⅓ that of a conventional system. Contamination transport will be slow. Contamination can also be isolated to specific areas through the use of transformers.

12.9 DIGITAL SYSTEMS

The widespread use of digital computers has generated an interest in hydraulic power amplification systems which are directly compatible with digital

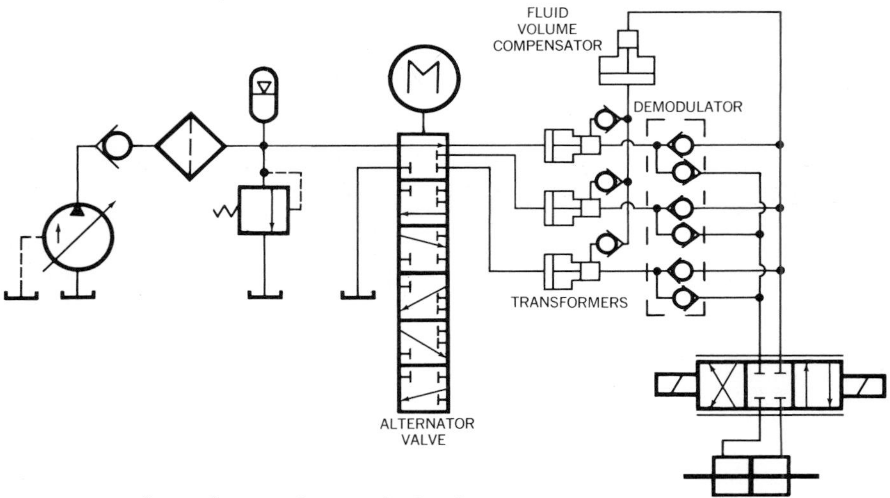

Fig. 12.22. Three-phase pulsating hydraulic system.

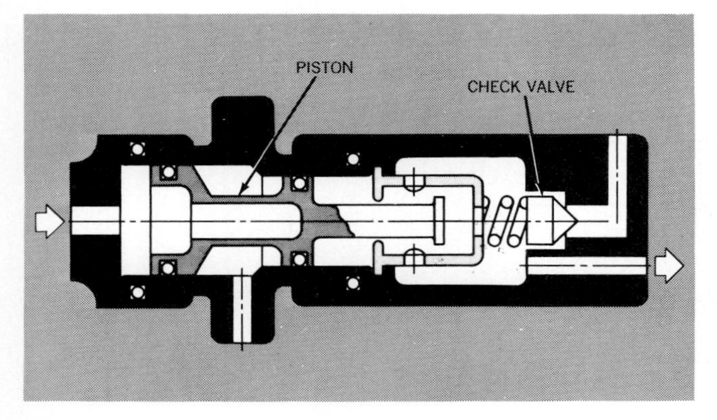

Fig. 12.23. Hydraulic transformer.

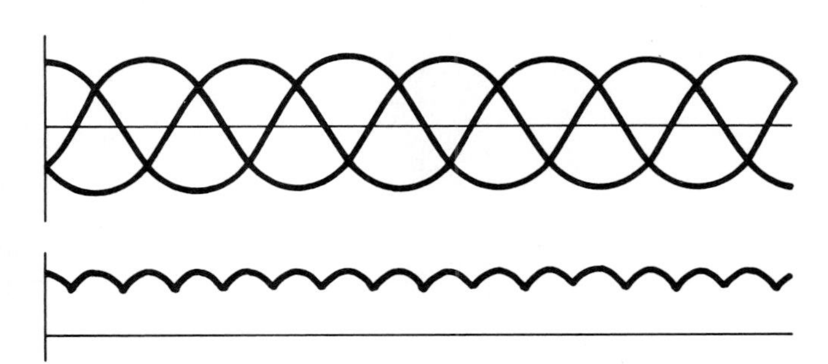

Fig. 12.24. Three-phase sinusoidal and demodulated flow.

computers. With the single exception of the digital actuator, Fig. 9.12, all the hydraulic systems and devices which have been discussed are *analog* arrangements. In an analog device changes are made smoothly and an infinite number of positions are available between any two conditions. In a modulating servo valve, for example, an infinite range of flow variations is available between zero and maximum flow. A linear actuator can assume an infinite number of positions or a rotary motor an infinite number of speeds. Digital devices, by comparison, make changes in discrete steps and only a finite number of steps are available. A simple comparison between analog and digital approaches might be between a ramp and a flight of stairs. Each provides a pathway between differences in elevation. One pathway is smooth and infinitely variable; the other uses a finite number of steps and each step represents a discrete change.

Digital-to-analog and analog-to-digital converters are available to provide the interface between the computer and the hydraulic servo. They are, in general, fairly reliable devices. Yet, the total reliability of a system is usually increased and the cost decreased if such conversion devices are not used. Also, where the absolute range of the output must be great (long stroke actuator or wide variation in output speed of a motor) the digital system is usually more accurate than a comparable analog system. This is because the maximum error in an analog system is normally a fixed percentage of the total output range. In a digital system, the maximum output error is a fixed value no matter what the range of the output variable.

The output of a digital computer can be in the form of a train of pulses or it can be a parallel group of pulses known as a "word".

Fig. 12.25*a* is one form of control servo which accepts a pulse train. Here, the pulse train operates an electro-mechanical stepper motor. This motor rotates through a fixed angle for each pulse. Thus the length of the pulse train controls the total angle of rotation and the pulse rate controls the motor speed. The stepper motor drives a suitable control valve through a differential gear. A measure of the motor output is fed back to the differential to close the output loop.

In the above system, the valve and motor comprise a partial digital-to-analog conversion and the

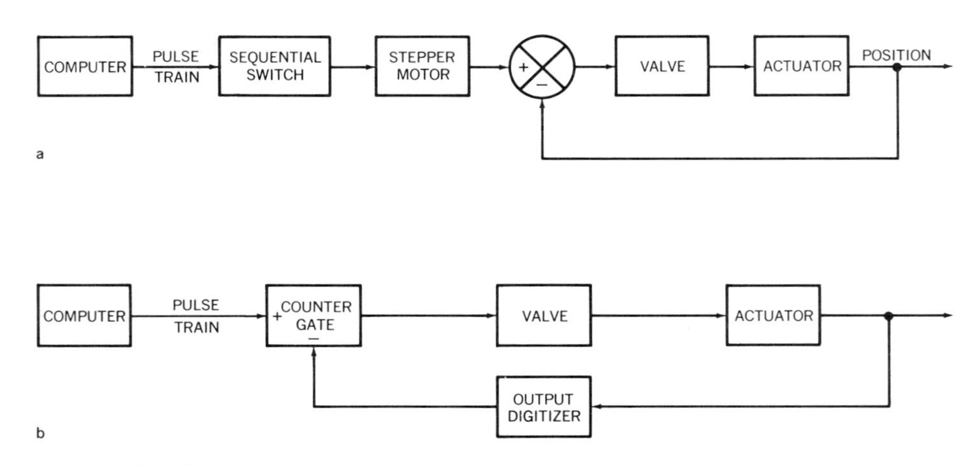

Fig. 12.25. Pulse train digital servo.

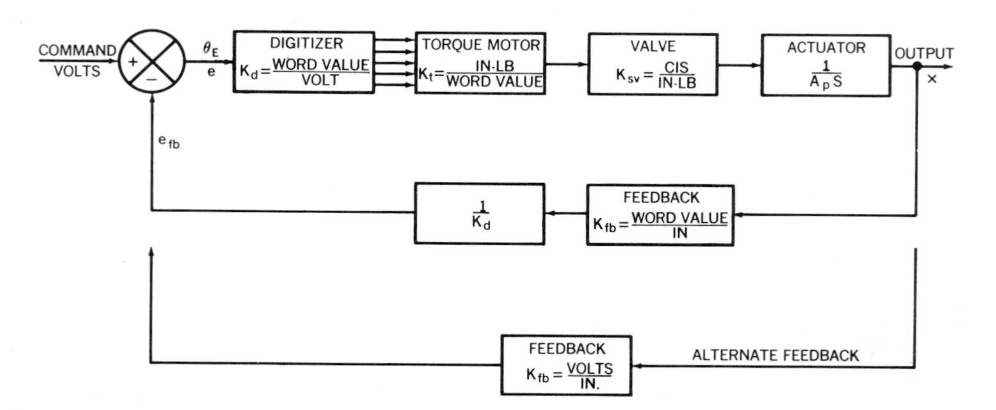

Fig. 12.26. Control loop using digital servo.

output approximates the digital input. A slightly closer approximation can be obtained by the scheme of Fig. 12.25*b* in which each pulse operates a valve momentarily. Therefore each pulse corresponds to a discrete quantity of oil passing through the valve to the actuator. The feedback device must be a number counting system such as a light shining through a calibrated grid onto a photo-diode. If the grid is mounted on the actuator, the system can require that the motor come to rest only when the number of pulses from the photo-diode equals the number of pulses in the input command.

Most modern digital computers operate on parallel binary words. Such a word is simply a number in binary arithmatic. Just as in the decimal system, each digit in a number must be multiplied by that power of ten which corresponds to its position in the number, so each "bit" (*Binary Digit*) in a binary word must be multiplied by its proper power of two in order to evaluate the number. However, the *only* digits allowed in the binary system are zero and one (0,1). Therefore, each bit really just tells whether its corresponding power of two is incuded or not. This is a much simpler arrangement than the decimal system in which each power of ten must be multiplied by some digit which can vary between zero and nine. The following table shows the development and decimal equivalent of the digital number 10110111.

Number (word)	1	0	1	1	0	1	1	1
Power of Two	2^7	2^6	2^5	2^4	2^3	2^2	2^1	2^0
Decimal Value	128	0	32	16	0	4	2	1
								2
								4
								16
								32
								128
Decimal Equivalent								183

One of the real virtues of the binary system is that each bit can be developed by a switching function: black or white, on or off, open or closed. These relationships are easily mechanized. The actuator of Fig. 9.12 can be controlled by 4, 3-way on-off control valves. This particular actuator can achieve positions equivalent to decimal 0 through 15. Any feedback device used with the binary digital actuator would have to develop a binary word as a function of actuator position. Servo null would be achieved when the feedback word numerically equalled the input word.

The U. S. Navy[3] has sponsored the development of a digital-to-analog servo valve which accepts a ±7 bit word. The valve generates a torque in the amplifier stage which is proportional to the numerical value of the input binary word. Fig. 12.26 is the block diagram of a servo utilizing such a valve. In this form the error signal is a voltage and the word is generated inside the servo loop. The forward path amplifier K_d and the feedback amplification l/K_d could be left out and a binary gate used to compare the command and feedback binary words. Versions of this valve so far tested have shown the following characteristics:

Flow Gain = 0.173 cis/word value unit
Frequency Response: 3 db below 0 with 90° phase shift at 40 Hz

REFERENCES

1. Oberst, R. S., "Nitrogen Powers Little Joe's Open Hydraulic System", Hydraulic and Pneumatics, July 1966.
2. Pollard, F. H. "Pulsating Flow Hydraulic Concepts," Proceedings SAE Aerospace Fluid Power Conference, May 1965, pp 413-419, 425.
3. NOTS TP4298 "Development of a ±7 Bit Parallel Binary Digital Servovalve," U. S. Naval Ordnance Test Station. China Lake, California, March 1967.

Appendices

SYMBOLS USED IN HYDRAULIC SYSTEM ANALYSIS

SYMBOL	DEFINITION	UNITS
a	Acceleration	in/sec^2
	Amplitude	in
A	Area	in^2
	Constant	-
	Weight rate	lb/sec
b	Height	in
	Radial clearance	in
B	Constant	-
	Viscous damping coefficient	lb-sec/in
c	Orifice coefficient	-
c_d	Coefficient of discharge	-
c_f	Fixity factor	-
c_L	Valve leakage coefficient	-
C_d	Viscous drag coefficient	-
C_f	Dry friction coefficient	-
C_H	Hydraulic capacitance	in^5/lb
C_L	Pump leakage coefficient	-
C_p	Specific heat at constant pres.	BTU/lb/^{o}F
C_s	Slip coefficient	
C_v	Specific heat at constant volume	BTU/lb/^{o}F
ς	Celerity of sound	in/sec (ips)
d, D	Diameter	in
	Volumetric displacement	in^3/rev
e	Eccentricity	in
e, E	Efficiency	-
E	Energy	lb-in
	Modulus of elasticity	lb/in^2
	Voltage	volts
f, F	Force	lb
f	Frequency	Hz (cps)
	Friction factor	-
g	Acceleration of gravity	in/sec^2
G	Gap length	in
	Torsional spring rate	lb/degree
G_c	Compensator valve gain	-
G_p	Pump gain	-
G_H	Conductance to ground	-
h	Height	in
H	Heat	BTU
	Power	in-lb/sec
	Pressure head	in
i, I	Current	amps
i	Transit time ratio	-

SYMBOL	DEFINITION	UNITS
I	Internal energy	BTU
	Moment of inertia	in^4
j	Vector operator $\sqrt{-1}$	-
J	Mechanical Equivalent of Heat	in-lb/BTU
	Polar moment of inertia	in^4
k	Constant	-
	Porosity	Darcy
	Radius of gyration	in
	Ratio of areas	-
	Ratio of specific heats	-
K	Constant	-
	Gain	-
	Spring rate	lb/in
K_c	Flow/pressure coefficient	-
K_p	Pressure gain	-
K_q	Flow gain	-
l, L	Length	in
L	Leakage coefficient	-
L_H	Hydraulic inductance	lb-sec^2/in^5
m	Constant	-
M	Mass	lb-sec^2/in
	A number	-
n	Exponent	-
N	A number	-
	Rotation rate	rpm
N_c	Cavitation number	-
N_R	Reynolds Number	-
p	Pressure ratio	-
P	Pressure	psi
q	Heat	BTU
Q	Flow	gpm, cips
r, R	Radius	in
R	Gas constant	in-lb-^{o}F/lb
	Velocity slope	-
	Volume ratio	-
R_H	Hydraulic resistance	lb-sec/in^5
s	La Place operator - $d/_{dt}$	-
S	Constant	-
	Stress	lb/in^2
t, T	Thickness	in
	Time	sec
T	Time constant	sec
	Temperature	degrees
	Torque	in-lb

SYMBOL	DEFINITION	UNITS
T	Transit time	sec
T_R	Rankine temperature	°R
v	Velocity	in/sec (ips)
V	Volume	in^3
w	Weight	lb
	Weight density	lb/in^3
	Width	in
W	Circumference	in
	Contamination	-
	Work	lb-in
x	Displacement	in
X	La Place displacement	-
y	Displacement	in
	Eccentric clearance	-
	Valve opening ratio	-
Y	Constant	-
	Valve conductance	-
	Valve opening	in
Z	Section modulus	in^3
Z_o	Characteristic tubing impedance	-
Z_R	Load impedance	-
Z_s	Source impedance	-

SYMBOL	DEFINITION	UNITS
α	An Angle	degrees
β	An angle	degrees
	Bulk Modulus of Elasticity	lb/in^2
	Feedback transfer function	-
γ	Surface tension	lb/in
δ	Displacement	in
	Disturbance force	lb
ϵ	Relative eccentricity	-
ζ	Damping ratio	-
θ	An angle	degrees
λ	Wavelength	in
μ	Absolute viscosity	$lb\text{-}sec/in^2$
	Forward transfer function	-
	Poisson's Ratio	-
ν	Kinematic viscosity	in^2/sec
ρ	Mass density	$lb\text{-}sec^2/in^4$
σ	Damping exponent	-
τ	Tortuosity factor	-
	Valve conductance ratio	-
ϕ	An angle	degrees
	Porosity ratio	-
ω	Frequency	rad/sec

CHAPTER 1

1. Air pressure = 14.7 psi   Water pressure = $\rho gh = Wh$ = 27.7 psi   Total = 41.4 psi

2. Air pressure = 14.7 psi   Water pressure = $\rho gh = Wh$ = 17.3 psi   Total = 32.0 psi

3. $\Delta P = \rho gh = Wh$ = 3.6 psi   Total = 3003.6 psi

4. Rupturing force = Pd Pd = (0.500 - 0.40) 3000 = 1380 lb

   Tension per unit length, per side = S $S = 1380/2t$ = 34,500 psi

5. 130 in^3/min = 2.17 cis $A = 4.9 \times 10^{-4}$ in^2 $\Delta P = \rho QZ/2c^2A^2$ = 1530 psi

6. $V = 1.04 \times 10^4$ in^3 $h = 1 + H = 37$ A_o = 0.7854 in^2 $A_t = V/h = 2.82 \times 10^2$ in^2

   $$T = \frac{2A_tH}{A_og} = 155.2 \text{ sec} = 2.59 \text{ min}$$

7. $\rho Q_i = \rho Q_o + \rho \dfrac{dV}{dt}$ $Q_i = Q_o + A \dfrac{dh}{dt}$

   $$\frac{dh}{dt} = 0.0129 \text{ ips} = 0.77 \text{ ipm}$$

8. ID = 0.680 in A_o = 0.363 in^2 Q = 15 gpm = 58 cis $v = Q/A_o$ = 160 ips

   1-cosβ = 0.293 $F_x = \rho Qv\,(1-\cos\beta)$ = 0.244 lb $F_y = \rho Qv \sin\beta$ = 0.59 lb

9. ω = 6Hz = 37.7 rad/sec V_{max} = 6 in^3 ID = 0.652 in A = 0.335 in^2

   $S = V/A$ = 17.95 in $\omega t = 1$, $a = S\omega^2$ = 25,500 in/sec^2

   $\rho = W/g = 8.4 \times 10^{-5}$ lb-sec^2/in^4 $M = \rho LA = 6.76 \times 10^{-3}$ $F = Ma$ = 172 lb

CHAPTER 2

1)

cs	x .001552 =	NEWTS
5		0.00776
35		0.05432
125		0.1940
440		0.6829
1300		2.0176

2)

SSU	ν = 0.00035t	–	0.303/t	NEWTS	
50	0.0175	–	0.00606	@	.01144
75	0.0263	–	0.00404	=	.02226
	ν = 0.000392t	–	0.21/t		
125	0.049	–	0.00125	=	0.0478
750	0.294	–	0.0003	=	0.2937
2500	0.98	–	0.00008	=	0.98

3)

Temp	ν (from Table 2–4)	ρ(from Table 2–7)	$\mu = \nu\rho$
–20°F	0.27 Newt	8.22 x 10^{-5}	2.219 x 10^{-5} Reyn
20	.09	8.0	0.72
75	.033	7.72	0.255
150	.0125	7.33	0.092
200	.0082	7.08	0.058

4)

FLUID	ASTM SLOPE
MIL-H-5606	0.51
MIL-H-8446	0.455
SAE 10	0.765
SAE 60	0.79

5) 10 gal = 2310 in^3 $\Delta T = 200 + 40 = 240°F$ Coef = 4.6 x 10^{-4} in/in/°F

a) At –40°F, $\rho = 8.26 \times 10^{-5}$ lb – sec^2/in^4

At 200°F, $\rho = 7.44 \times 10^{-5}$

$$V_i = 2310 \left(\frac{8.26}{7.44}\right) = 2560$$

$$\Delta V = 2560 - 2310 = 260 \text{ in}^3$$

b) No direct solution is possible because β varies both with temperature and pressure. Assume $\beta_{ave.} \approx 2.4 \times 10^5$ psi

$$\beta = \frac{\Delta P}{\Delta V / V_o} \qquad \Delta P \approx \frac{\beta \Delta V}{V_o} \approx 2.65 \times 10^4 \text{ psi}$$

CHAPTER 3

1. $Q = 3.26$ gpm = 12.55 cins $\quad A_L = 0.05$

$\rho = 7.7 \times 10^{-5} \quad G_C = 0.01 \quad G_P = 1.25$

$\beta = 2 \times 10^5 \quad V = 45 \quad N = 377$ rad/sec $\quad C_V = 129$

Zero flow

$$\zeta_o = \tfrac{1}{2} G_C \sqrt{\frac{\beta}{VNG_CG_P}} = 0.153$$

$$\omega = \sqrt{\frac{\beta N G_C G_P}{V}} = 145 \text{ rad/sec} = 23 \text{ Hz}$$

Full flow

$$\zeta_{ff} = \tfrac{1}{2}\left(\frac{C_V A_L}{2\sqrt{P_o}} + C_L + G_C\right)\sqrt{\frac{\beta}{VN\,G_CG_P}} = 1.055$$

2. $Q = 4.55$ gpm $\quad$ Required flow = 3.50 gpm $\quad$ Excess flow = 1.05 gpm

$HP = \triangle P/Q = 0.918$

3. $R_{GN_2} = 661 \quad T_o = 70^o = 530^o$ R $\quad P_o = 1000 \quad V_o = 250$ in^3

$$V_1{}^{1.4} = \frac{P_o}{P_1} V_o{}^{1.4} = 114 \text{ in}^3$$

$$P_o V_o = w R T_o$$

$$w = \frac{P_o V_o}{R T_o} = 0.71 \text{ lbs}$$

$$T_1 = \frac{P_1 V_1}{w R} = 726^o \text{ R} = 266^o \text{ F} \qquad T_2 = 150^o \text{ F} = 610^o \text{ R}$$

$$V_2 = \frac{T_2}{T_1} V_1 = 96 \text{ in}^3$$

$V_3 = 176$ in$^3 \qquad P_2 = P_1$

4. $V_o = 300 \quad P_o = 1000 \quad T_o = 100^o\ F = 560^o\ R$

$P_1 = 3000 \quad V_1 = 100 \quad V_2 = 145$

$$P_2 = P_1 \left(\frac{V_1}{V_2}\right)^{1.4} = 1785 \text{ psi}$$

$$w = \frac{P_o V_o}{R\, T_o} = 0.81 \text{ lbs}$$

$$T_2 = \frac{P_2\, V_2}{w R} = 482^o\ R = 22^o\ F$$

$$W = \frac{P_2\, V_2 - P_1\, V_1}{n-1} = 1.025 \times 10^5 \text{ in-lb}$$

5. $P_o = 300 \quad V_o = 150 \quad T_o = 75^o\ F = 535\ R$

$P_1 = 1100 \quad V_1 = 150 \quad T_1 = T_o$

$$w_o = \frac{P_o\, V_o}{R\, T_o} = 0.127 \text{ lbs}$$

$$w_1 = \frac{P_1\, V_1}{R\, T_1} = 0.467 \text{ lbs}$$

$\triangle w = 0.340$ lbs

6. $D = 4$ in. $\quad A_p = 12.566 \text{ in}^2 \quad P_o = 1200$

$V_o = 251.32 \text{ in}^3 \quad T_o = 70^o\ F = 530^o\ R \quad P_1 = 3000$

$V_1 = 100 \quad l = 7.95$ in. $\quad \beta = 3000$ psi

$$K = \frac{A\, \beta}{l} = 4750 \text{ lb/in} \qquad M = 6.48 \times 10^{-3}$$

$$\omega = \sqrt{\frac{K}{M}} = 856 \text{ rad/sec} = 137 \text{ Hz}$$

$$P_3 = P_2 \left(\frac{V_2}{V_3}\right)^{1.1} = 1545 \text{ psi}$$

$$W = \frac{P\, V^{1.1}}{0.1} \left(\frac{V_2^{\,n-1} - V_1^{\,n-1}}{V_1^{\,n-1} - V_2^{\,n-1}}\right) = 1.7 \times 10^5 \text{ in-lb}$$

CHAPTER 4

1. ID = 0.555 in $A = 0.242$ $Q = 1.5$ gpm = 5.78 cips

$v = 23.9$ ips $\nu = 0.25$ Newts

$$N_R = \frac{vd}{\nu} = 53$$

Flow is laminar, so $f = \dfrac{64}{N_R} = 1.2$

2. $Q = 30$ gpm = 115.5 cips ID = 0.902 $A = 0.6372$

$\nu = 0.022$ $N_R = 7450$

Flow is turbulent, so $f = \dfrac{0.316}{N_R^{\ 0.25}} = 0.034$

$$\triangle P = \frac{64 \quad l \rho v^2}{2 N_R \, d} = 28.3 \text{ psi}$$

3. $Q = 20$ gpm = 77 cips $\nu = 0.26$ $\rho = 1.025 \times 10^{-4}$

$\triangle P/l = 0.167$ psi/in

$$T = \frac{1}{\nu}\left(\frac{\triangle P \, Q^3}{l \rho}\right)^{1/5} = 2.287 \times 10^3$$

From chart, $f = 0.036 = \dfrac{\pi^2 \triangle P \, d^5}{8(l \rho Q^2)}$

$d = 0.64$ in.

4. $d = 0.79$ Area = 0.490 $\nu = 0.07$ $\rho = 8.9 \times 10^{-5}$

$\triangle P/l = 0.416$

$$S = \frac{1}{\nu}\left(\frac{\triangle P \, d^3}{l \rho}\right)^{1/3} = 688$$

From chart, $f = 0.038$

$$N_R = \frac{S}{\sqrt{f/2}} = 4800 \qquad v = \frac{\nu N_R}{d} = 420 \text{ ips}$$

$Q = 206$ cips = 53.5 gpm

5. $c = 0.59$ $K_B = 0.09$ $r = 2$ $d = 0.901$

$r/d = 2.22$ $f = 0.012$ $l = \frac{2\pi r}{8} = 1.57$ in

$$\text{Ratio} = \frac{f(^l/d) + cK_B}{f(^l/d)} = 3.53$$

6. $Q = 3.85$ cips $\nu = 0.032$ $\rho = 7.75 \times 10^{-5}$

$d_o = 0.25$ $A_o = 0.049$ $d_u = 0.430$ $A_u = 0.145$

$d_o/d_u = 0.58$ $v_u = 26.6$ ips $c = 0.8$

$$N_R = \frac{v_u\, d_u}{\nu} = 358$$

$$Q = c\,A_o \sqrt{\frac{2\,\triangle P}{\rho}}$$, therefore $\triangle P = 380$ psi

7. $Q = 5.78$ cips $\nu = 0.023$ $\rho = 1.02 \times 10^{-4}$

$d_u = 0.555$ $A_u = 0.243$ $d_o = 0.50$ $A_o = 1.96 \times 10^{-3}$

$d_o/d_u \approx 0.1$ $v_u = 23.8$ ips $c = 0.68$

$$N_R = \frac{v_u\, d_u}{\nu} = 575$$

$$Q = c\,A_o \sqrt{\frac{2\,\triangle p}{\rho}}$$, therefore $\triangle P = 960$ psi

8. a) Sudden contraction

$d_o/d_i = 0.725$ $Q = 25$ gpm $= 96.2$ cips

$\rho = 8.95 \times 10^{-5}$ $A_o = 0.338$ $v_o = 250$ ips

$$\frac{\rho\, v_o^2}{2} = 2.8 \qquad \triangle P \Big/ \left(\frac{\rho\, v_o^2}{2}\right) = 0.24, \text{ from chart}$$

$\triangle P = 0.673$ psi

b) Run

$N_R = 1370$ $f = 0.05$

$$\triangle P = \frac{f\, l\, \rho\, v_o^2}{2\, d_o} = 0.61 \text{ psi}$$

c) Sudden expansion

$$d_i/d_o = 0.875 \qquad \triangle P \Big/ \left(\frac{\rho\, v_o^2}{2}\right) = 0.08$$

$\triangle P = 0.224$ psi

Total $\triangle P = 1.507$ psi

9. $d = 0.281$ $D = 0.283$ $b = 0.001$ $e = 0.001$

$\triangle P = 2950$ $l = .03$ $\epsilon = \frac{e}{b} = 1$

$\nu = 0.018$ $\rho = 7.5 \times 10^{-5}$ $\mu = \nu\rho = 1.35 \times 10^{-6}$

$$Q = \frac{\pi d\, b^3 \triangle P}{12\, \mu\, l}\left(1 + \frac{3\epsilon^2}{2}\right) = 1.33 \text{ cips}$$

CHAPTER 5

1. $E = 29 \times 10^6$ $\mu = 0.3$

$$\beta_t = \frac{t\,E}{D\,(1 - \frac{\mu}{2})} = 1.365 \times 10^6 \text{ psi}$$

$$\beta_e = \frac{\beta\,\beta_t}{\beta + \beta_t} = 1.9 \times 10^5 \text{ psi}$$

2. $$\beta_t = \frac{tE}{D(1-\mu/2)} = 1.67 \times 10^6 \text{ psi}$$

$\beta_o = 2.1 \times 10^5$ psi

$$\beta_e = \frac{\beta\,\beta_t}{\beta + \beta_t} = 1.865 \times 10^5 \text{ psi}$$

$\beta_g = 250$ psi

$$\frac{1}{\beta_e} = \frac{V_g}{V_e}\left(\frac{1}{\beta_g}\right) + \frac{V_o}{V_e}\left(\frac{1}{\beta_o}\right) + \frac{1}{\beta_t}$$

$\beta_e = 2.22 \times 10^4$ psi

3. $\beta_H = 4.5 \times 10^4$ $\beta_o = 3 \times 10^5$ $\beta_g = 125$

$\beta_t = 1.6 \times 10^6$ $V_H = 11.05 \text{ in}^3$ $V_t = 43.5 \text{ in}^3$

$$\frac{1}{\beta_e} = \frac{1}{\beta_t} + \frac{1}{\beta_H} + \frac{V_g}{V_e}\left(\frac{1}{\beta_g}\right) + \frac{V_o}{V_e}\left(\frac{1}{\beta_o}\right)$$

$\beta_e = 1.39 \times 10^4$ psi

4.

	$\beta \times 10^5$	$\rho \times 10^{-5}$	$\varsigma = \sqrt{\beta/\rho} \times 10^4$
MIL-H-5606, at 80°, 0 psig	2.20	7.70	5.35
Industrial phosphate ester at 100°, 0 psig	3.40	10.40	5.70
MIL-H-8446, at 170°, 3000 psig	2.40	8.26	5.28
Aircraft phosphate ester at 220°, 0 psig	2.25	9.35	4.90

5. $\omega = 5090$ rad/sec $\beta = 2.99 \times 10^5$ $\rho = 9.7 \times 10^{-5}$

$\varsigma = \sqrt{\beta/\rho} = 5.55 \times 10^9$

$\varsigma/a = 10.95$

$l = 17.2$ in; 51.5 in; 85.7 in. . . . .

6. $\omega = 4140$ rad/sec $\beta = 1.9 \times 10^5$ $\rho = 8.15 \times 10^{-5}$

$\varsigma = \sqrt{\beta/\rho} = 4.84 \times 10^4$

$l = 18.6$ in., 55.7., 93 in. . . . .

7. $d = 0.430$ $r = 0.145$ $l = 240$ in. $\beta = 2.1 \times 10^5$

$\rho = 7.6 \times 10^{-5}$ $V_t = 34.8$ in^3 $V_c = 8$ in^3 $\varsigma = 5.25 \times 10^4$

$f_t = \frac{\varsigma}{4l} = 54.8$ rad/sec $\frac{V_t}{V_c} = 4.35$ $R = 0.8, 2.5, 4.35 \ldots$

$f_{tc} = Rf_t = 43.8$ rad/sec, 137 rad/sec, 238 rad/sec . . .

8. $\beta_g = P_{go} \frac{C_p}{C_v} = 4200$ psi W at 125° = 132 in^3/lb

$\omega = 5330$ rad/sec $\rho = 1.97 \times 10^{-5}$

$\varsigma = \sqrt{\beta/\rho} = 1.464 \times 10^4$

$l = 4.31$ in., 12.95 in., 21.6 in. . . .

9. ID = 0.842 in. Area = 0.556 $Q = 12$ gpm = 46.3 cips

$v_o = Q/A = 83$ ips $\rho = 7.55 \times 10^{-5}$ $\nu = 0.02$

$\beta = 2.6 \times 10^5$ $\varsigma = 5.86 \times 10^4$

$$\Delta P = \frac{4\rho\varsigma v_o}{\pi} = 467 \text{ psi}$$

$$f = \frac{\varsigma}{4l} = 24.5 \text{ Hs}$$

$$P_d = e^{\frac{-l\nu}{r^2 \varsigma}} P_u = e^{-0.0046} P_u = \frac{P_u}{1.0052}$$

CHAPTER 7

1. 25 microns = 98×10^{-4} in. $\Delta P = 0.78$ psi/cips/in^2

$\rho = 7.6 \times 10^{-5}$ $\nu = 0.022$ $\mu = \nu\rho = 1.67$ microreyns

$$K = \frac{2.4\, Q\, \mu\, dl}{A\, \Delta P} = 0.257 \text{ darcy}$$

2. $\beta = 2.75 \times 10^5$ $\rho = 10^{-4}$ $\nu = 62.84$ cips

$$K = \frac{A\,\beta}{l} = 4.33 \times 10^4 \text{ lb/in} \qquad M = \nu\rho = 6.284 \times 10^{-3}$$

$$f_n = \frac{1}{2\pi}\sqrt{\frac{K}{M}} = 416 \text{ Hz}$$

3. $\beta_o = 2.52 \times 10^5$ $\beta_g = 4200$ $\rho = 7.5 \times 10^{-5}$

$$\frac{1}{\beta_e} = \frac{V_o}{V_e}\left(\frac{1}{\beta_o}\right) + \frac{V_g}{V_e}\left(\frac{1}{\beta_g}\right)$$

a) $\beta_e = 1.6 \times 10^5$ $K = \dfrac{A\,\beta}{l} = 2.52 \times 10^4$ $f = \dfrac{1}{2\pi}\sqrt{\dfrac{K}{M}} = 368$ Hz

b) $\beta_e = 1.21 \times 10^5$ $K = 1.9 \times 10^4$ $f = 319$ Hz

c) $\beta_e = 6.38 \times 10^4$ $K = 10^4$ $f = 232$ Hz

4. $F_b = \frac{4}{3} \pi r^2 \rho g = 0.00933$ lb

$v = 1.38 \sqrt{rg} = 3.84$ ips

$F_d = C_d \left(\frac{\rho}{2}\right) \pi r^2 v^2 = 1.125 \times 10^{-6}$ lb

5. 15 microns = 5.9×10^{-4} in.

$P = \frac{4\gamma}{d} = 1$ psid

6. $v = 1.6 \times 10^{-4}$ $P = 0.5$ psi $d = 1.28 \times 10^{-3}$ $K = 0.005$

$l = 0.030$ $Q = 10$ gpm = 38.5 cips $\rho = 7.75 \times 10^{-5}$

$v = 0.035$ $\mu = 2.72$ microreyns

$A = 2.4 \left(\frac{Q \mu}{K}\right) \left(\frac{d\, l}{dP}\right) = 3.01 \times 10^3$ in^2

CHAPTER 8

1. a) 4.667 hp

   b) 7.66 hp

   c) 16.45 hp

2. 5.45 hp

3. $\frac{0.385\, PQ}{1714} = 4.58$ hp

4. P after losses = 2860 psi $P_V = \frac{P}{A} = 953$ psi

$P_L = \frac{2}{3} P = 1907$ psi $Q = 3$ gpm = 11.55 cips

$K = Q / \sqrt{\triangle P} = 0.374$

a) Load = $AP_L = 5721$ lb

b) 6 ips (3) = 18 cips $\sqrt{P_V} = Q/K$ $P_V = 2310$ psi

Line loss = (18) (140) / 11.55 = 218 psi

$P_L = 472$ psi

Load = $AP_L = 1416$ lb.

5. P after losses = 2650 psi $\quad P_L = \frac{2}{3}P = 1770$ psi $\quad P_V = \frac{P_L}{A} = 885$ psi

$QP_L/1714 = 8.25$ hp

8 gpm = 30.8 cips $\quad K = Q/\sqrt{P_V} = 1.03$

30.8 cips / 3 = 10.27 ips

At 8 ips, Q = 24 cips $\quad$ and $P_V = Q/K^2 = 542$ psi $\quad$ therefore,

$P_{Line} = 234$ psi and $P_L = 2174$ psi

Load = $AP_L = 6522$ lb.

6. Force per land = $2cw\,x\,\Delta P\,\cos\theta$

$w = \pi D/2 = 0.59$

Force = 6.34 lb.

Total force = 12.68 lb.

CHAPTER 9

1. $A = 6.28 \quad l = 8 \quad \beta = 2.1 \times 10^5$ psi

$$K = \frac{A\,\beta}{(0.5 - 0.25)\,l} = 6.6 \times 10^5 \text{ lb/in}$$

2. ID = 3 in $\quad A = 7.069 \quad r_1 = 1.5$ in $\quad S = 3 \times 10^4 \quad P = 2500$ psi

$$S = P\left(\frac{r_2^2 + r_1^2}{r_2^2 - r_1^2}\right), \text{ therefore } r_2 = 1.64 \text{ in}$$

$$\triangle r_1 = P\frac{r_1}{E}\,[\frac{r_2^2 + r_1^2}{r_2^2 - r_1^2}] + \mu = 0.0016 \text{ in.}$$

3. $M = 18.1\ \text{lb-sec}^2/\text{in}$ $P_S = 2700$ psi $A_p = 4$ $Q = 8$ gpm = 30.8 cips

P_L at 8 gpm = 1800 psi $P_V = 900$ psi $\triangle P_1 = 450$ psi $k = 1$

$K = \triangle P_1 / Q^2 = 0.474$

$$c_1 = \left(\frac{K A^3{}_p}{M}\right)\left(1 + k^3\right) = 3.35$$

$$c_2 = \frac{P_s A_p}{M} = 597$$

$$c_3 = \frac{1}{c_1} = 0.298$$

$$c_4 = \sqrt{c_1 c_2} = 44.7$$

At $t = 0.1$ sec:

$$T = c_4 t = 0.447 \qquad \frac{dY}{dT} = 0.43 \qquad Y = 0.125$$

$$\chi = c_3 Y = 0.037$$

$$x = \frac{1}{A_1}\sqrt{\frac{P_s}{K(1+k^3)}}\ \frac{dY}{dT} = 6.2 \text{ cips} \quad \text{ips}$$

At $t = 0.5$ sec.:

$$T = 2.11 \qquad \frac{dY}{dT} = 0.97 \qquad Y = 1.5$$

$x = 0.505$ x = 13.7 ips

4. $k = 2$ $Q_1 = 20$ gpm = 77 cips $Q_2 = 38.5$ cips $P_L = 2150$ psi

$\triangle P_2 = 600 - \triangle P_1$

$K = \triangle P_1 / Q^2 = \triangle P_2 / Q^2 = 600 - \triangle P_1 / 1480$, therefore

$\triangle P = Q_1{}^2\ (600 - \triangle P_1 / 1480) = 460$ psi

$K = 0.078$

$x_f = 2''$ $M = 7.3$

$$A = \left[\frac{1.145\ M}{K\,x_f(1+k^3)}\right]^{1/3} = 1.82 \text{ sq. in.}$$

$$t_f = \frac{1.182\ M}{A^2\sqrt{K P_s(1+k^3)}} = 0.0595 \text{ sec.}$$

5. $$\frac{Mv_o^2}{2} = W_f + \frac{3\pi \mu D_3^2 l^2 v_o}{16 b^3}$$

$b = 0.003''$ $d = 1''$ $l = 2''$ $v_o = 80$ ips

$\rho = 7.7 \times 10^{-4}$ $\mu = 2.31 \times 10^{-5}$ Reyn

Assume $W_f = 30$ lb

$l = 1.43$

$$t = \frac{2l}{V_o} = 0.036 \text{ sec.}$$

6. $D = 3$ in. $d = 1$ in. $A_p = 6.284$ in. $V_o = 10$ ips $l = 8$ in.

Assume 1000 psi in actuator $\beta = 2 \times 10^5$ psi

$$K = \frac{4\,\Delta\beta}{l} = 6.26 \times 10^5 \text{ lb/in}$$

$M = 1.55$

$$\Delta P = \frac{V_o}{A}\sqrt{KM} = 1570 \text{ psi}$$

CHAPTER 11

1. $$\text{Decay} = e^{\frac{-2\pi}{\sqrt{(1/\zeta^2) - 1}}} = e^{-\frac{6.28}{\sqrt{(1/\zeta^2) - 1}}}$$

$\zeta = 0.1$ $=$ $e^{-0.631}$ $= 0.532$

$\zeta = 0.5$ $=$ $e^{-3.62}$ $= 0.027$

$\zeta = 0.7$ $=$ $e^{-6.15}$ $= 0.00212$

$\zeta = 0.9$ $=$ $e^{-27.4}$ ≈ 0.0000

2. a) $\dfrac{\theta_o}{\theta_e} = \dfrac{K_a K_v}{A_p s}$ b) $\dfrac{\theta_o}{\theta_e} = \dfrac{K_a/A_p}{S(1 + Ts)}$

3. a) $$\frac{\theta_o}{\theta_i} = \frac{\dfrac{K_aK_v}{A_p s}}{1 + \dfrac{K_aK_v}{A_p s}} = \frac{1}{1 + \dfrac{A_p}{K_aK_v}\, s}$$

b) $$\frac{\theta_o}{\theta_i} = \frac{\dfrac{K_a / A_p}{s(1+Ts)}}{1 + \dfrac{K_a / A_p}{s(1+Ts)}} = \frac{K_a / A_p}{s^2 + \dfrac{1}{T}s + \dfrac{K_a}{A_pT}} = \frac{1}{\dfrac{A_p T s^2}{K_a} + \dfrac{A_p}{K_a}s + 1}$$

c) $$\frac{\theta_o}{\theta_i} = \frac{\dfrac{K_aK_v/A_p}{s(1+Ts)}}{1 + \dfrac{(K_aK_vK_f/A_p)s}{s(1+Ts)}} = \frac{\dfrac{K_aK_v}{A_p}}{s(1+Ts) + \dfrac{K_aK_vK_f}{A_p}\, s}$$

$$\frac{\theta_o}{\theta_i} = \frac{\dfrac{K_aK_v}{K_aK_vK_f + A_p}}{s\ 1 + \dfrac{A_pT}{K_aK_vK_f + A_p}\, s)}$$

4. a) $$\frac{\theta_o}{\theta_e} = \frac{(10)(500)}{2\ s} = \frac{2500}{s}$$ Crossover at 2500/sec

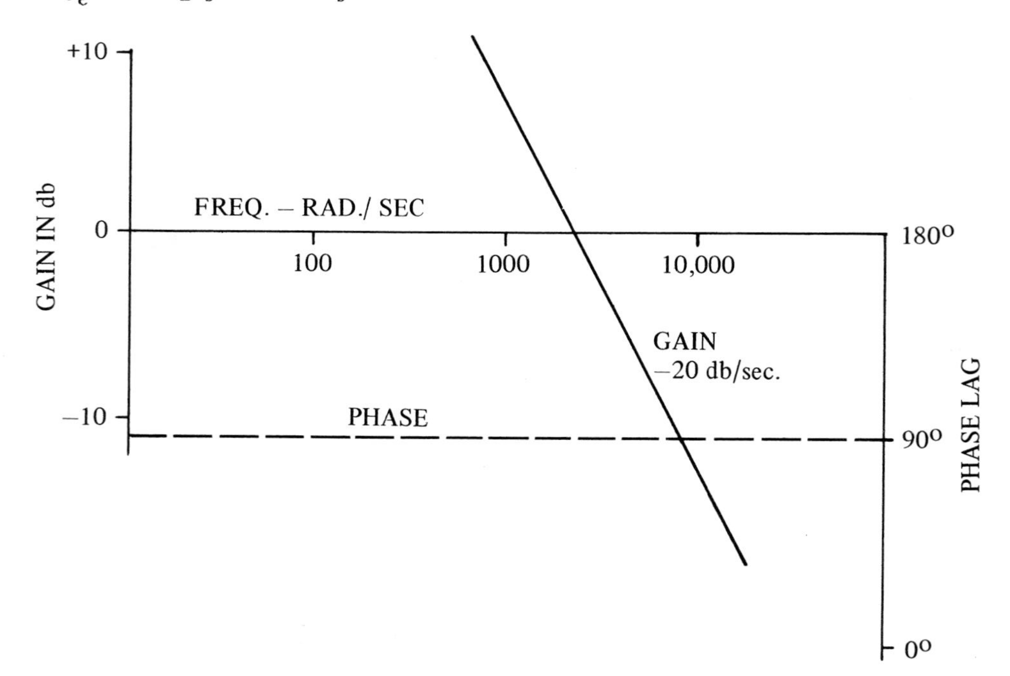

b) $\dfrac{\theta_o}{\theta_e} = \dfrac{5}{s\,(1+ .02s)}$ $\qquad \dfrac{\theta_o}{\theta_e} = 1$ when s = 4.58 $\qquad$ crossover at 4.58/sec.

$\dfrac{1}{T} = 50/\text{sec.}$

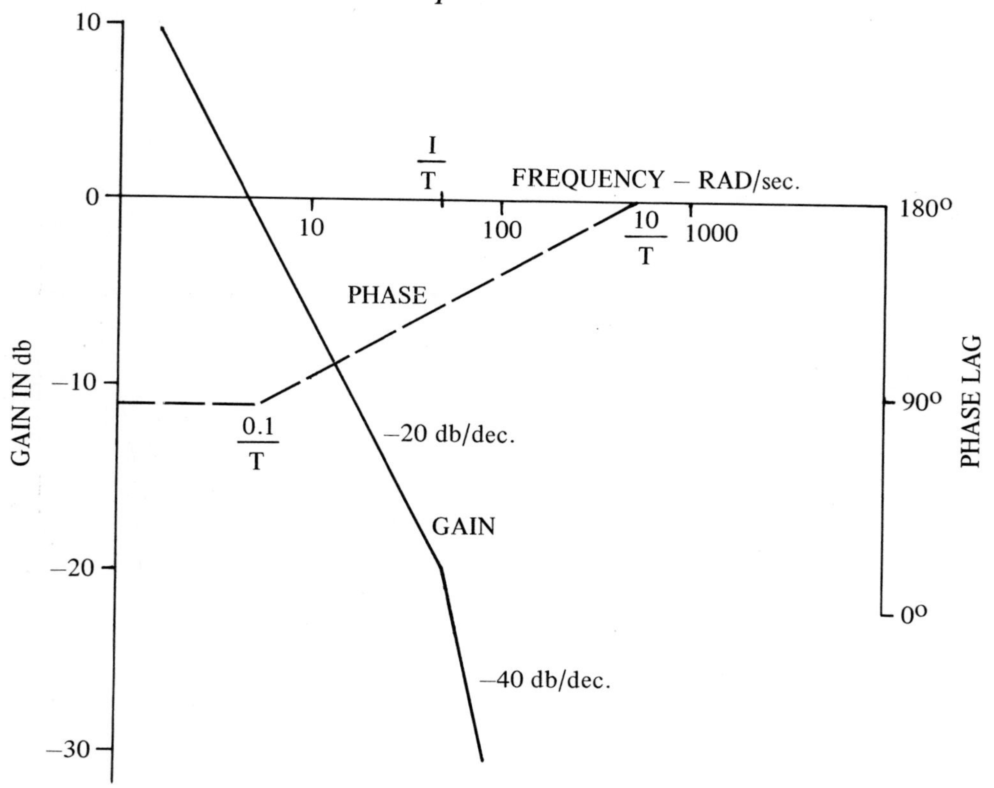

5. a) 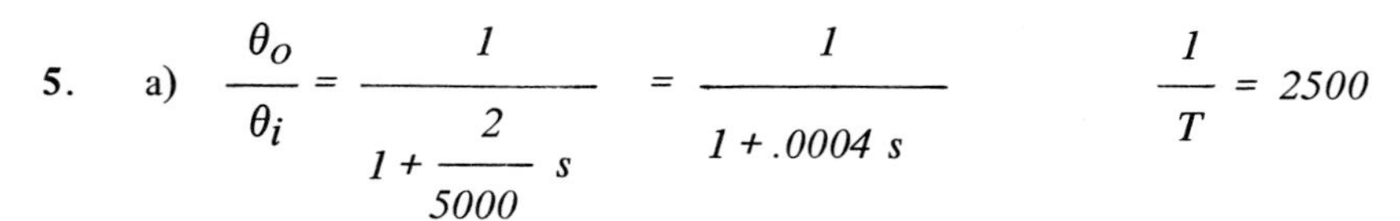

$$\frac{\theta_o}{\theta_i} = \frac{1}{1 + \dfrac{2}{5000}\,s} = \frac{1}{1 + .0004\,s} \qquad \frac{1}{T} = 2500$$

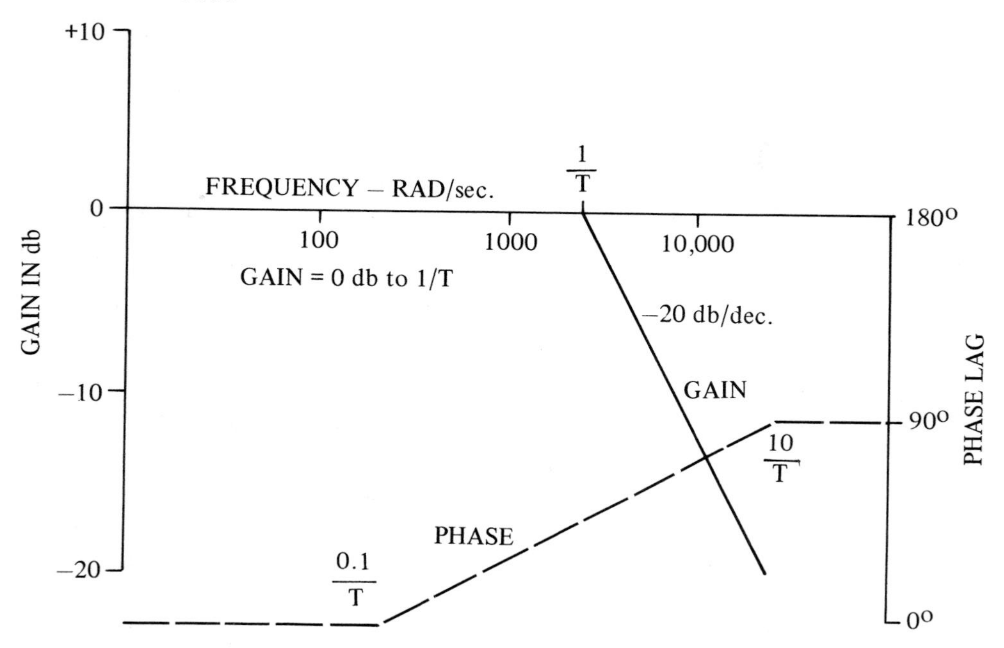

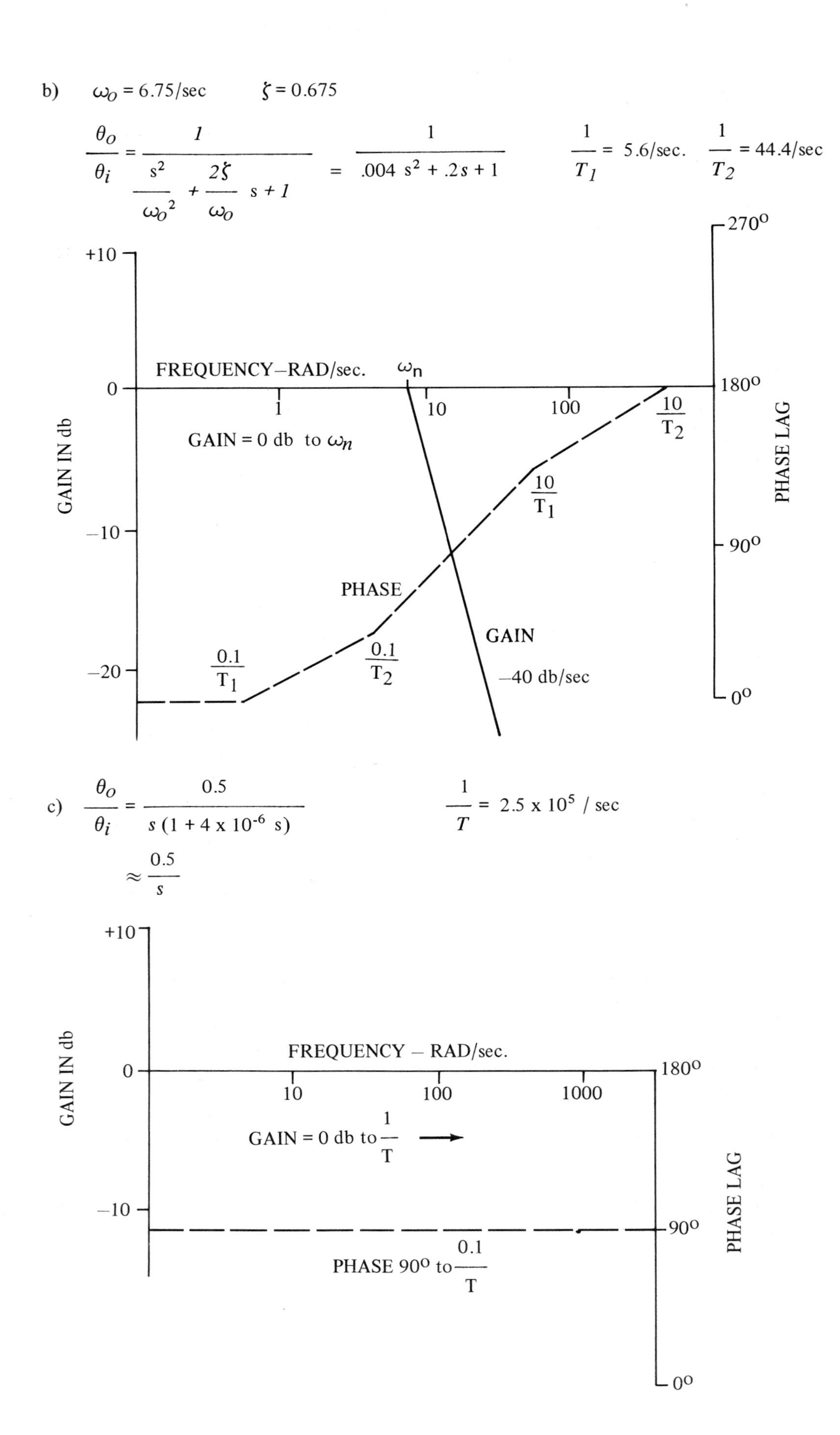

b) ωo = 6.75/sec ζ = 0.675
θo/θi = 1 / (s²/ωo² + 2ζ/ωo s + 1) = 1 / (.004 s² + .2s + 1)
1/T1 = 5.6/sec.
1/T2 = 44.4/sec
+10
0
−10
−20
GAIN IN db
FREQUENCY–RAD/sec.
ωn
1
10
100
GAIN = 0 db to ωn
10/T2
10/T1
PHASE
GAIN
0.1/T1
0.1/T2
−40 db/sec
270°
180°
90°
0°
PHASE LAG
c) θo/θi = 0.5 / s (1 + 4 x 10⁻⁶ s)
≈ 0.5/s
1/T = 2.5 x 10⁵ / sec
+10
0
−10
GAIN IN db
FREQUENCY – RAD/sec.
10
100
1000
GAIN = 0 db to 1/T
180°
90°
0°
PHASE LAG
PHASE 90° to 0.1/T

A

B

C

D

L

M

N

O

P

R

S

T

U

V

W